HOLOCENE HUMAN ECOLOGY in Northeastern North America

INTERDISCIPLINARY CONTRIBUTIONS TO ARCHAEOLOGY

ECOLOGY AND HUMAN ORGANIZATION ON THE GREAT PLAINS
Douglas B. Bamforth

HOLOCENE HUMAN ECOLOGY IN NORTHEASTERN NORTH AMERICA
Edited by George P. Nicholas

THE PLEISTOCENE OLD WORLD
Regional Perspectives
Edited by Olga Soffer

HOLOCENE HUMAN ECOLOGY in Northeastern North America

Edited by

GEORGE P. NICHOLAS

University of Massachusetts
Amherst, Massachusetts and
American Indian Archaeological Institute
Washington, Connecticut

PLENUM PRESS • NEW YORK AND LONDON

Library of Congress Cataloging in Publication Data

Holocene human ecology in northeastern North America / edited by George P. Nicholas.

p. cm. — (Interdisciplinary contributions to archaeology)

Includes bibliographies and index.

ISBN 0-306-42869-5

1. Indians of North America — Northeastern States — Antiquities. 2. Paleoecology — Northeastern States. 3. Human ecology — Northeastern States. 4. Northeastern States — Antiquities. I. Nicholas, George P. II. Series.

E78.E2H65 1988 88-11113

974′.01 — dc19 CIP

A Division of Plenum Publishing Corporation
233 Spring Street, New York, N.Y. 10013

Printed in the United States of America

To Glynn Issac,
whose work on early human ecology
set clear standards for the many disciplines involved

Contributors

Russell J. Barber • Department of Anthropology, California State University-San Bernardino, San Bernardino, California

William R. Baron • Center for Colorado Plateau Studies, Northern Arizona University, Flagstaff, Arizona, and Institute for Quaternary Studies, University of Maine, Orono, Maine

Catherine C. Carlson • Department of Anthropology, University of Massachusetts, Amherst, Massachusetts

Deborah C. Cox • Public Archaeology Laboratory, Pawtucket, Rhode Island

Denice C. Gaudreau • Geology Department, Southampton College, Southampton, New York

Arthur A. Joyce • Department of Anthropology, Rutgers University, New Brunswick, New Jersey

Douglas C. Kellogg • Institute for Quaternary Studies, University of Maine, Orono, Maine

Mitchell T. Mulholland • Department of Anthropology, University of Massachusetts, Amherst, Massachusetts

George P. Nicholas • Department of Anthropology, University of Massachusetts, Amherst, Massachusetts, and Research Department, American Indian Archaeological Institute, Washington, Connecticut

William A. Patterson III • Department of Forestry and Wildlife Management, University of Massachusetts, Amherst, Massachusetts

Kenneth E. Sassaman • Department of Anthropology, University of Massachusetts, Amherst, Massachusetts

Peter F. Thorbahn* • Hammerway Research, West Barnstable, Massachusetts

*Deceased.

Preface

Students of human behavior have always been interested in the relationship between human populations and their environment. Decades of research not only have illuminated the backdrop against which culture is viewed, but have identified many of the conditions that influence or promote technological development, social transformation, and economic reorganization. It has become increasingly evident, however, that if we are to explore more forcefully the linkages between culture and environment, a processual orientation is required. This is found in human ecology—the study of the relationship between people and the ecosystem of which they are a part.

This book is a collection of papers about the recent and distant past by scientists and humanists involved in the study of human ecology in northeastern North America. The authors critically examine the systemic interface between people and their environment first by identifying the indicators of that relationship (e.g., historical documentation, archaeological site patterning, faunal remains), then by defining the processes by which change in one part of the ecosystem affects other parts (e.g., by considering *how* an ecotonal gradient affects biotic communities over time), and finally by explicating the behavioral implications thereof.

Although the subjects of study in this collection may vary among settlement patterns, sea-level curves, land snails, and pollen records, the emphasis remains on integrative research designs that explore new ways of looking at behavioral diversity and environmental variability within an ecological context. In some cases, this involves modifying the scale at which we view elements of the human ecosystem; in others, it requires the development of analytical methods that allow us to differentiate more clearly between adaptive responses to sociocultural factors and adaptive responses to environmental factors. A number of studies examine archaeological sites in their settings, others the environmental records that document landscape development affecting prehistoric land-use behaviors; several evaluate frequently used theoretical concepts and methods to reveal and correct biases that affect our interpretations of past human lifeways. The result of this diversity is a perspective that emphasizes local adaptive responses to environmental change at a variety of temporal scales. Local perspectives are, in fact, stressed throughout not because this is the level at which most archaeology and ecology is done, but because this is where human responses to change occur.

Despite an emphasis on prehistory, this book is about Holocene human ecology and not environmental archaeology. Any differences between prehistoric and historic approaches to human ecology are more apparent than real. However, if we hope to discriminate better among the many potential sources or causes of change in cultural systems and to identify the processes of that change, prehistorians must become aware of the behavioral diversity represented in the historical record, and historians of the long-term patterns of human behaviors that occur across time and space.

The collection is organized into three parts that correspond to important aspects of human ecology. Part I, the Context of Human Adaptation, presents four very different approaches to the study of human environments, which range from the reconstruction of environmental settings using land snails (Barber), to the human perception of historic climatic change (Baron), to the nature of anadromous fish resources (Carlson), to the local and regional effects of global sea-level rise (Kellogg). Part II, People on the Landscape, offers three studies that humanize the natural history of the landscape by investigating the aboriginal use of fire as a resource management strategy (Patterson and Sassaman), the impact of horticulture on prehistoric settlement (Mulholland), and resource orientation and settlement within a small coastal estuary (Thorbahn and Cox). Part III, Long-Term Perspectives, focuses on historical developments in the human ecosystem over time, and includes a revised regional environmental reconstruction (Joyce) and studies on the processes of forest ecotone development (Gaudreau) and the importance of ecological contrasts in early prehistoric land use (Nicholas). In those cases where the authors report on long-term research projects still in progress, the real value of their contribution may not be on the data themselves, but rather on the ideas presented.

The predominance of contributions by prehistoric archaeologists in this volume is fortuitous. This collection originated in a symposium, "Modeling Cultural Responses to Environmental Change," presented at the Northeastern Anthropological Association Meetings held at Princeton University in 1982. The transformation of that original set of papers and of the ideas and data presented therein into the present volume has been long and turbulent due not only to several aborted publication attempts, but, more importantly, to the maturation of the ideas expressed by the authors. This volume includes radically revised and updated versions of the papers presented at Princeton by Carlson, Kellogg,

Mulholland, and Nicholas. The papers presented in the symposium by Dean Snow ("The Mathematical Modeling of Cultural Responses to Environmental Change") and Barbara McMillan ("Early Postglacial Adaptation in the Upper Connecticut River Valley") were unfortunately not available for incorporation here.

The breadth of this volume was increased significantly by the incorporation of six additional papers. In fact, invitations to submit material to the volume were sent out to over 50 American and Canadian scientists involved at the interface between human populations and their environments in the Northeast. These included not only prehistoric and historic archaeologists, but also dendrochronologists, palynologists, paleolimnologists, soil scientists, surficial geologists, entymologists, biological anthropologists, climatologists, and many others, including even a folklorist. Many of these individuals were unable to contribute due to other commitments. Several did submit abstracts or preliminary papers that were later withdrawn as a result of personal or publication scheduling conflicts. One of these, "A Commentary on the Relationship between Vegetation History and Cultural Change in Labrador," by William Fitzhugh and Henry Lamb, has been published in *Arctic and Alpine Research* 17:357–370 (1985), and the reader is directed to it.

Acknowledgments

The completion of this volume was possible only through the help and encouragement of many friends and colleagues. I would first like to thank both the Department of Anthropology, University of Massachusetts-Amherst, and the American Indian Archaeological Institute for their support over the last 4 years. Dena F. Dincauze (University of Massachusetts-Amherst) and W. Raymond Wood (University of Missouri-Columbia) were the critical reviewers for this volume, and contributed substantially to the improvement of its contents, as did three anonymous reviewers. Editorial advice was provided by Catherine Carlson, Dena Dincauze, and Russell Handsman. Numerous conversations with William Baron, Dena Dincauze, Russell Handsman, Howard Sargent, Brooke Thomas, W. Raymond Wood, and Martin Wobst allowed my own thinking about human ecology to develop significantly, as did one memorable meeting with Glynn Isaac. I would also like to thank Jim Chiarelli, Curley Griffith, Sissel Johannessen, Sonny Trimble, and Eric Voigt for their more distant (in time and space) contributions. Other support was provided by Irene Alm, Peter Beblowski, Roy and Maureen Carlson, Donald Foster, Betsy Guralnik, Shelly Hight, Victoria Bunker, Steven Kite, Jeffrey Maymon, Ann McMullen, Martha Pinello, and, of course, Talking Heads. I am grateful to David and Sarah Hauser for providing both encouragement and *Glen Livet* throughout the course of completing this project, and most especially to Helen Nicholas and Robert Nicholas for their long-standing support. Finally, I would like to acknowledge the important contributions of Catherine Carlson, who is an important part of my ecology, and my sons, Gordon and Graham, whose attempts at environmental disequilibrium challenged my ability to complete this book.

Contents

PART II. PEOPLE ON THE LANDSCAPE

Introduction

Human Behavior and Holocene Ecology

GEORGE P. NICHOLAS

For over 10,000 years, human populations have occupied the dynamic, ecologically diverse landscapes of northeastern North America. Throughout the course of the Holocene period, prehistoric and historic communities have each exhibited a wide array of behavioral, social, and technological adaptations in response to a diversity of environmental and social settings. The greater part of this period was dominated by various gathering–hunting–fishing strategies, and it was apparently only in the last 1,000 years that horticulture had a significant impact on prehistoric economy. By the sixteenth century, aboriginal lifeways were being regularly impacted by the early Basque fisheries; increasing contact with European colonists led initially to changes in land use in response to the fur trade, but quickly resulted in the decimation of local native populations by a series of plagues, wars, and other adverse circumstances. The historic settlement of the region was until recently agriculturally based, with broader patterns of land use focused on coastal resources, the waterways of the interior, and the timber industries. By the mid-nineteenth century, segments of the landscape were being rapidly industrialized. The Holocene period in the Northeast is thus a period of significant social transformation, favorable and unfavorable historical circumstances, and environmental variability.

The diversity of contemporaneous ecotones in the Northeast during the Holocene, and the opportunities such temporal and spatial heterogeneity may

---

GEORGE P. NICHOLAS • Department of Anthropology, University of Massachusetts, Amherst, Massachusetts 01003, and Research Department, American Indian Archaeological Institute, Washington, Connecticut 06793.

have provided, offers much to the study of a regional human ecology—one concerned as much with the processes of change as with its causes and effects. Ecology is by definition the study of dynamics (Jochim 1981:8), and change within an ecosystem is the norm. There is little that is permanent about either human lifeways or the environment of which they are a part. Human responses to internal (i.e., "social") or external (i.e., "environmental") change occur continuously and not always consciously, and may take the form of either new behaviors or technologies that strive to maintain the lifeway threatened, modifications of those lifeways to take advantage of new circumstances, or resistance to that change (e.g., Lees and Bates 1984). Similarly, members of a biotic community may respond to changes in environment by moving or expanding their range, by becoming more or less responsive to certain environmental pressures, by increasing or decreasing their population size, or by responses that eventually become visible as an evolutionary modification (e.g., Delcourt *et al.* 1983; Hardesty 1980; Krebs and Davies 1984; Macfarland 1978; Meltzer and Mead 1983). In fact, the only thing *static* about the relationship between people and their environment is the physical record available to archaeologists and paleoenvironmentalists. *Stasis* should probably be banished from our thinking about either the past or the present. Not only are the mechanisms of both change and stability (if there is a difference) dynamic by their very nature, but interaction between cultural and environmental systems may occur at different scales and rates, and it nearly always provokes new cycles of adjustment instead of equilibrium.

Researchers have traditionally focused on the *effects* of environmental change on human communities. Historians, for example, have identified change in land use within an area as a result of observed short-term (e.g., <50 years) economic shifts or social trends (e.g., Cronon 1983; Rubertone 1986; also Handsman 1981), and sometimes even to climatic episodes (e.g., Smith *et al.* 1981, 1982). Prehistorians, on the other hand, have concentrated more on long-term trends, and have correlated change in land-use patterns, social and technological developments, and other adaptive strategies with climatic or environmental conditions (e.g., Fitzhugh 1972; Fitzhugh and Lamb 1985). Despite what may be radically different scales of observation and types of data, there are no real differences between prehistoric and historic human ecology.

Human populations respond to such a wide variety of stimuli in so many remarkable ways that it is difficult to identify "what is purely cultural, or human, in human lifestyles" (Dincauze 1981:52–53; also Gunn 1975). In addition, the same environmental stimuli upon similar groups of hunter–gatherers may evoke totally different responses, not all of which may be obvious or logical. An ecological approach to human adaptation must therefore go beyond reconstructing the landscape at particular times and places, to explore the dynamic nature of human lifeways within that environment in terms of mobility, technological specialization, social differentiation, and other behavioral variables. What we are after here is a higher level of integration involving diverse classes of data at scales ranging from individual organisms through to the larger ecosystem of which they are part. People, for example, respond to localized, very short-term (i.e., daily) change in their life by using whatever adaptive strategies are readily available to them. They respond to a fresh snowfall that reveals animal tracks, for example, to the size of a day's catch, to the lack of adequate firewood nearby, or to the presence of hostile individuals in the area, by exhibiting behaviors usually appropriate to the situation. *Individuals* respond most frequently to immediate or visible threats against the status quo, or to a particular set of circumstances, and far less frequently to more abstract pressures, such as the need for long-term planning to offset the effects of some possible unexpected future event. This is the level of interaction through which change in human ecosystems is most frequently introduced. At a different scale of observation (i.e., over time), human *populations* respond to climatic fluctuations, vegetational shifts, and the influence of other groups, as well as to the feedback from their responses to these changes, through shifts in settlement and subsistence patterns, or modification of economic and social networks (see, for examples, Butzer 1982; Hodder 1986; Snow 1981; Wigley *et al.* 1981; also Vecsey and Yenables 1980). In this sense, change within either cultural or environmental systems represents a cumulative phenomenon that generally exists only at scales greater than that of the individual organisms involved.

Considerable effort is now being directed toward explaining the mechanisms by which the social, economic, or technological strategies involved may be modified by pressures external to the cultural system (e.g., *American Ethnologist*, special issue, Vol. 9, No. 2, 1982, "Economic and Ecological Processes in Society and Culture"; Braun and Plog 1982; Dean *et al.* 1985; Ellen 1982; Jochim 1981, 1984; Kirch 1980). The shift to a systemic focus is also reflected in the types of questions explored in this volume, most of which are directed to explaining, rather than describing, aspects of the complex relationship between people and their environment. How sensitive are some important paleoenvironmental indicators to ecological change? What are the linkages between a change in forest cover and a shift in land-use behaviors? What is the impact of human settlement on the landscape? How do we recognize and explain diversity in the settlement patterns of early postglacial hunter–gatherers or eighteenth century farmers? These questions are concerned with both the context and processes of human adaptation, and actively explore the middle range between archaeological patterns and the human behaviors and social relations they represent.

All of this is not to deny the importance of other environmentally oriented research in the region. There is certainly a long history of panenvironmental research in the greater Northeast, and a rich archaeological literature that extends such study back through time (see Dincauze 1981; Snow 1981; and Starna 1981 for important overviews). Correlational approaches to the relationship between cultural and environmental change are common in this literature, and probably reflect the relative ease of comparing, for example, changes in regional vegetational sequences and climate to shifts in prehistoric settlement patterns and technology. While these studies have done much to raise awareness of environmental influences upon human populations, the success of many individual projects has suffered for several reasons. First, such studies normally lack a processual component, without which it is difficult to account for, or at least discuss, the scale and dynamics of interaction between cultural and environmental systems. Second, the causal mechanisms proposed (*if* proposed) are frequently overly simplistic or deterministic. Third, such studies seldom offer independent tests of either the events involved or of the research conclusions themselves. Interdisciplinary studies (e.g., Steegman 1983; Wigley *et al.* 1981) overcome most of these problems by using multiple approaches to exploring human ecosystems through a variety of data sets, explanatory models, field methods, and research questions, and by incorporating the means of independent confirmation or adjustment of inferences within the research design (e.g., Dincauze 1981, 1987). Even at this level of integration, research must remain theoretically progressive, self-critical, and sensitive to the contributions, and limitations, of each contributing discipline.

As outlined in the Preface, the contributions contained in this volume are interdisciplinary in both content and intent. Individually, we are concerned with both the context and processes of human adaptation in the Northeast during the Holocene period, and with explaining diversity in our respective field areas. Collectively, we are interested in developing a long-term human ecology within the region.

Two broad themes are repeated throughout this collection. The first is that there is great diversity in adaptive responses to environmental change within the region during the Holocene period, and that such diversity has much to offer to the study of human ecosystem. This nonsingularity of adaptations, exhibited both by humans and the various elements of their environment, is explored in many different ways here. Ecotonal contrasts, for example, are used by Gaudreau, Thorbahn and Cox, and Nicholas to explain patterns of prehistoric land use distinct from those at other locations. From a more methodological perspective, Kellogg and Joyce illuminate diversity in the processes of landscape development in the Northeast.

The second general theme is that the human ecosystem is, not surprisingly, highly complex, and that research must be critical of differences in the scale of analysis and interpretation of human behaviors and environments. This set of papers therefore presents a variety of scalar concepts that correspond with specific points and types of interaction between human populations and their environments. The scales of observation range from short-term and local to long-term and regional, and include several permutations; interestingly, many of the papers emphasize change in the human ecosystem at the level of the individual organism. Barber and Carlson examine site environments and subsistence strategies, respectively, at a scale that is short-term, geographically focused, and occupation-specific. On the other hand, the papers of both Nicholas and of Thorban and Cox focus on the long-term record of land use at interior and coastal locations, respectively. Mulholland draws attention to the effects of the regional introduction of horticulture on local landscapes and land use, whereas Patterson and Sassaman differentiate between local, culturally induced fires and regional, natural fire histories. At an even larger scale, environmental parameters for human settlement set by broad trends in climate and environment and by regional manifestations of global events are discussed by Gaudreau, Joyce, and Kellogg. Finally, Baron demonstrates how regional climatic trends affect particular segments of a population, as evidenced in seventeenth through nineteenth century diaries.

These and related issues are addressed by the authors through approaches that range from the evaluation of the historical documentation of climate in New England, to the revision of local and regional paleoenvironmental reconstructions and corresponding models of prehistoric settlement and subsistence, to the intentional manipulation of the environment by aboriginal populations. In addition, the importance of complementarity (the application of different data sets to check research results) to research goals is emphasized in many of the papers. This addresses, in part, previous critiques of ecological research in the Northeast and elsewhere (e.g., Dincauze 1981) and the need for independent verification of the proposed behavioral/environmental interaction. For example, Barber, Carlson, Joyce, and Patterson and Sassaman all present new data sets that overturn or challenge many premature conclusions about the environmental record, whereas other authors indicate how their conclusions can be tested by future studies.

Whether the focus is on the data themselves or on the way in which we think about those data, the authors demonstrate that there is much to learn about long-term human ecology through processually based and environmentally oriented research. It is hoped that such diverse collections as this will offer new conceptual perspectives on change and stability relevant to any ecological system through time.

Acknowledgments

Catherine Carlson, Dena Dincauze, Russell Handsman, Michael Jochim, and one anonymous reviewer provided substantive comments on earlier drafts of this paper.

REFERENCES

Braun, D. P., and Plog, S., 1982, The Evolution of "Tribal" Social Networks: Theory and Prehistoric North American Evidence, *American Antiquity* 47:504–525.

Butzer, K., 1982, *Archaeology As Human Ecology: Method and Theory for a Contextual Approach,* Cambridge University Press, New York.

Cronon, W., 1983, *Changes in the Land: Indians, Colonists, and the Ecology of New England,* Hill and Wang, New York.

Dean, J. S., Euler, R. C., Gumerman, G. J., Plog, F., Hevly, R. H., and Karlstrom, T., 1985, Human Behavior, Demography, and Paleoenvironment on the Colorado Plateaus, *American Antiquity* 50:537–554.

Delcourt, H. R., Delcourt, P. A., and Webb, T., III, 1983, Dynamic Plant Ecology: The Spectrum of Vegetational Change in Space and Time, *Quaternary Science Reviews* 1:153–175.

Dincauze, D. F., 1981, Paleoenvironmental Reconstruction in the Northeast: The Art of Multidisciplinary Science, in: *Foundations of Northeast Archaeology* (D. R. Snow, ed.), Academic Press, New York, pp. 51–96.

Dincauze, D. F., 1987, Strategies for Paleoenvironmental Reconstruction in Archaeology, in: *Advances in Archaeological Method and Theory,* Vol. 11 (M. B. Schiffer, ed.), Academic Press, Orlando, Florida, pp. 255–336.

Ellen, R., 1982, *Environment, Subsistence and System: The Ecology of Small-Scale Social Formations,* Cambridge University Press, New York.

Fitzhugh, W. W., 1972, Environmental Archaeology and Cultural Systems in Hamilton Inlet, Labrador: A Survey of the Central Labrador Coast from 3000 B.C. to the Present, *Smithsonian Contributions to Knowledge,* No. 16.

Fitzhugh, W. W., and Lamb, H. F., 1985, A Commentary on the Relationship between Vegetation History and Cultural Change in Labrador, *Arctic and Alpine Research* 17:357–370

Gunn, J., 1975, An Envirotechnological System for Hugup Cave, *American Antiquity* 40:3–21.

Handsman, R. G., 1981, Early Capitalism and the Center Village of Canaan, Connecticut: A Study of Transformations and Separations, *Artifacts* (American Indian Archaeological Institute, Washington, Connecticut) IX(3):1–22.

Hardesty, D. C., 1980, The Use of General Ecological Principles in Archaeology, in: *Advances in Archaeological Method and Theory,* Vol. 3 (M. B. Schiffer, ed.), Academic Press, New York, pp. 158–187.

Hodder, I., 1986, *Reading the Past: Current Approaches to Interpretation in Archaeology,* Cambridge University Press, Cambridge.

Jochim, M. A., 1981, *Strategies for Survival: Cultural Behavior in an Ecological Context,* Academic Press, New York.

Jochim, M. A., 1984, The Ecosystem Concept in Archaeology, in: *The Ecosystem Concept in Anthropology* (E. R. Moran, ed.), AAAS Selected Symposium, No. 92, Westview Press, Boulder, Colorado, pp. 87–102.

Kirch, P. Y., 1980, The Archaeological Study of Adaptation: Theoretical and Methodological Issues, in: *Advances in Archaeological Method and Theory,* Vol. 3 (M. B. Schiffer, ed.), Academic Press, New York, pp. 101–156.

Krebs, J. R., and Davies, N. B., 1984, *Behavioural Ecology,* 2nd ed., Sinauer Press, Sunderland, Massachusetts.

Lees, S. H., and Bates, D. G., 1984, Environmental Events and the Ecology of Cumulative Change, in: *The Ecosystem Concept in Anthropology* (E. R. Moran, ed.), AAAS Selected Symposium, No. 92, Westview Press, Boulder, Colorado, pp. 133–159.

Macfarland, W. V., 1978, Determinants of Tolerance Limits in Animals and Man, in: *Biology and Quaternary Environments* (D. Walker and J. C. Guppy, eds.), Australian Academy of Science, pp. 147–168.

Meltzer, D. J., and Mead, J. I., 1983, The Timing of Late Pleistocene Mammalian Extinctions in North America, *Quaternary Research* 19:130–135.

Rubertone, P. E., 1986, Historical Landscapes: Archaeology of Place and Space, *Man in the Northeast* 31:123–138.

Smith, D. C., Borns, H. W., Jr., Baron, W. R., and Bridges, A. E., 1981, Climate Stress and Maine Agriculture: 1785–1885, in: *Climate and History* (T. M. L. Wigley, M. J. Ingram and G. Farmer, eds.), Cambridge University Press, Cambridge, pp. 404–433.

Smith, D. C., Baron, W. R., Bridges, A. E., TeBrake, J. E., and Borns, H. W., Jr., 1982, Climate Fluctuation and Agricultural Change in Southern and Central Maine, 1765–1880, *Maine Historical Society Quarterly* 21:179–200.

Snow, D. R., 1981, Approaches to Cultural Adaptation in the Northeast, in: *Foundations of Northeast Archaeology* (D. R. Snow, ed.), Academic Press, New York, pp. 97–138.

Starna, W. A., 1981, Old Data and New Models: Bridging the Gap, in: *Foundations of Northeast Archaeology* (D. R. Snow, ed.), Academic Press, New York, pp. 139–173.

Steegman, A. T. (ed.), 1983, *Boreal Forest Adaptations: The Northern Algonkians,* Plenum Press, New York.

Vecsey, C., and Venables, R. W. (eds.), 1980, *American Indian Environments: Ecological Issues in Native American History,* Syracuse University Press, Syracuse, New York.

Wigley, T. M. L., Ingram, M. J., and Farmer, G. (eds.), 1981, *Climate and History,* Cambridge University Press, Cambridge.

Part I

The Context of Human Adaptation

Environmental studies are the foundation of human ecology. Yet uncritical acceptance of apparent correlations between aspects of human behavior and environment remains more common than not. The context of human adaptation in the Northeast is explored here through four cautionary tales. These papers are critical in orientation and methodology, and are sensitive to the problems of scale and interpretation inherent in their respective data sets. As a result, the authors possess tight control over environmental variability, and thus provide more detailed reconstructions of various Holocene environments.

Barber's paper on the use of land snails as paleoenvironmental indicators builds upon a series of earlier studies on basic analytical technique and applications. Here the emphasis is on the logic and strategies of analysis. He discusses not only the biology of land snails, and why they are sensitive indicators of human habitat, but also exposes the myth of their poor preservation and limited utility in the Northeast. Land snails do occur at archaeological sites here, and are relatively sensitive indicators of highly localized (i.e., site-specific) landscape change. Some land snail species are so specialized in habitat that they may also be used to document microregional shifts in climate, and thus provide independent verification of other lines of paleoenvironmental evidence. Barber's discussion on land snail life and death assemblages emphasizes the role of the individual within the community. A critical review of the life histories of other more frequently used indicators of past environments, such as bivalves and plant macrofossils, for example, would not only expose limitations in their use by human ecologists, but also provide a more fine-grained view of conditions affecting the human ecosystem.

Land snails respond to shifts in moisture or temperature primarily by moving to a more attractive location. How do humans respond to essentially the same stimuli? Baron focuses on the relationship of weather, climate, and human populations in New England during the seventeenth and eighteenth centuries—the latter part of the Little Ice Age. During this period, interannual weather patterns in the region were highly variable, as measured along a number of climatic parameters. Baron and colleagues have developed methods that track individual climatic variables to produce very fine-grained sets of historic weather data, which can subsequently be used to separate social from environmental causes of change in the human ecosystem. The diversity of records analyzed here, which include weather records, historical documents, and diaries, indicates that some local weather and climate patterns contrasted with regional trends. The highly subjective diary entries are particularly important because they record how individuals perceived the most significant climatic event of the last several hundred years. The impact of the Little Ice Age upon human communities was significant in geographically marginal areas. In the Northeast, however, the effects were apparently so gradual that climatologists did not even record a climatic abberation; likewise, risk-lessening responses to environmental change may already have been in place, given the usual unpredictability of New England's weather.

Carlson neatly bridges the ecology of the prehistoric and historic periods in her paper on the nature of anadromous fish resources in New England and its implications for social change. This study is a good example of how a fresh look at an entrenched notion—the importance of a specific resource to aboriginal groups—provides important insights into prehistoric subsistence and social organization. Carlson utilizes detailed environmental data and life histories of salmon and other anadromous fish to evaluate historical documentation and faunal remains from both coastal and interior sites, and concludes that salmon was not an important resource. She also questions the utility of ethnographic analogs, such as those derived from the Northwest Coast, to explain prehistoric social development in the Northeast. This paper also has important implications for modern fisheries studies, providing one possible explanation as to why the restocking of New England waterways in recent years has been largely unsuccessful—that is, the regional salmon population may never have been extensive in the past.

Kellogg's study is concerned with critical differences between relative and eustatic sea-level data and how they should be utilized to reconstruct local coastal environments. Sea-level change has been perhaps the single most important factor affecting coastal environments during the early and middle Holocene periods in the Northeast in terms of environmental productivity, prehistoric land-use patterns, and archaeological site preservation. Much of the substantial body of sea-level data available for the region has been utilized by archaeologists at one time or another, resulting in interpretations that are frequently inaccurate or incorrect. Kellogg's demonstration that only data on local, relative sea-level rise are applicable to local and regional environmental reconstructions has important implications for coastal landscape history globally.

Chapter **1**

The Use of Land Snails from Prehistoric Sites for Paleoenvironmental Reconstruction

RUSSELL J. BARBER

1. INTRODUCTION

Much of recent archaeology has stressed the intimate relationship between environment and human culture and behavior, encouraging the development of new methodologies for the reconstruction of past environments. One of the findings of those engaged in developing these methodologies is that no class of remains holds a privileged position as the single data set that best can be used to infer past environmental conditions. Rather, all of the data sets—pollen, land snails, foraminifera, ostracods, macrobotanical remains, and the rest—have their strengths and weaknesses, each is better or worse for the investigation of a particular aspect of environment or for a particular set of conditions. This paper discusses the potential for using one of these classes of data, land snails, in environmental reconstruction.

Although the analysis (or at least identification) of land snail remains has

RUSSELL J. BARBER • Department of Anthropology, California State University-San Bernardino, San Bernardino, California 92407-2397.

become a standard archaeological practice in most of the world, New England has curiously lagged behind. Some of the first studies of archaeological snails were conducted by Wyman (1867), Morse (1867), and other early natural historians who pioneered New England archaeology. These initial studies were directed to the dating of sites, to the reconstruction of past environments, and to the documentation of species range shifts, and were quite sophisticated even by modern standards. By the 1870s, however, the major interest in shell middens and the great antiquity of New England had diminished, and these sophisticated analyses disappeared. With the exception of rare "laundry list" treatments (e.g., Bullen 1949), reportage of land snails is absent from New England archaeological reports until the 1970s. Even today, a full-scale analysis of land snails from a New England archaeological site is a rarity. Why?

Part of the reason for the long hiatus, of course, is the general lack of interest in New England prehistory for that considerable period when attention was focused instead on the Midwest and Southeast. Another factor is the absence of an ecological paradigm until the last few decades. The resurgence of interest in New England prehistory and cultural ecology that began in the mid-1960s, however, failed to stimulate widespread interest in incorporating land snail analysis into site reportage.

The reason may lie in beliefs held widely by New England prehistorians. While a graduate student in the 1970s, I was told that it was foolish to attempt a study of New England land snails from archaeological contexts. After all, "there really weren't any land snails to speak of in New England, and shell doesn't preserve in New England soils anyway. In addition, even if these problems didn't exist, the only methods available for the analysis of land snails give a vague, wide-area picture of past environments, at best." I hope that these erroneous conceptions no longer persist among Northeastern prehistorians, but I fear that they do. The analyses reported in this paper demonstrate that land snail assemblages in New England, although hardly as diverse or well-preserved as those on the limestone plains of Ohio or south England, are available, and not only from shell middens. Further, these assemblages are amenable to a variety of analytical approaches that can provide environmental information at a variety of levels—from that of very broad macro-climate zones (evidencing broad patterns of rainfall and temperature at a regional or larger scale) to that of narrow microenvironment habitats (evidencing such variables as local drainage, degree of shade, and vegetation type).

This paper begins with a discussion on land snail biology as it relates to paleoenvironmental reconstruction, and subsequently focuses on methodological approaches and factors that are particularly relevant to New England. Discussions of land snail assemblages analyzed by the author are included throughout. Finally, since laboratory techniques and manipulations have been discussed elsewhere (Barber 1983), this paper confines itself to the logic and strategies of analysis.

2. LAND SNAIL BIOLOGY

Some remarkably naive beliefs regarding land snails persist among archaeologists; for example, that land snails can leave their shells at will, and cannot be easily attracted in large numbers for use as food. This section dispels these and other fantasies, replacing them with well-established biological facts.

Land snails are members of the phylum Mollusca, class Gastropoda. Ancestral molluscs evolved in the sea, and the vast majority of modern molluscan species have retained their aquatic ways; some snails, however, have evolved terrestrial habits. Still, the boundary between land snails and aquatic snails is anything but clear. Within certain families, some species have terrestrial habits and others aquatic (Morton 1967:172). In some cases, the habits of a given species can encompass both terrestrial and aquatic modes.

Despite the fact that land snails do not correspond to a unitary taxon in the Linnean system, all land snails share the challenges of terrestrial living and have developed a limited suite of adaptations to it. There are five major challenges posed to a snail by terrestrial environments: temperature control, moisture control, requirements of mobility and buoyancy, reproduction in nonliquid surroundings, and respiration in air. The last two of these are of little concern to the problems at hand, but the first three are critical and are discussed in greater detail.

In an aquatic environment, temperature fluctuation is minimized, and moisture loss is obviously no threat. On land, however, moisture loss is a major risk, and the wide range of temperature variation threatens a simple organism, such as a snail, which has developed no complex mechanisms for coping with these dangers. One way of coping is to insulate the organism's body. The snail's shell accomplishes this task, but at a cost. In water, a snail's shell has greater buoyancy than in air and requires less strength to carry; in order to have mobility on land, land snail shells have been generally reduced in thickness and weight as compared to their aquatic relatives. The land snail, therefore, cannot simply increase its shell mass and insulating power to cope with land conditions, because the shell would then become too ponderous and thus limit the snail's mobility.

Land snails have been forced to more complex adaptations. All have developed a mucous secretion that covers their soft parts and impedes moisture loss. Some have become large and bulbous, maximizing the ratio of body volume to surface area. (This adaptation is analogous to the human adaptation to cold climates, which is described as Bergman's rule of zoology.) In some cases, this approach has been so successful that the external shell has been lost entirely, producing the shell-less land snails commonly known as "slugs." Other land snails have adopted approaches that modify overall shell ratios or restrict the size of the shell's aperture, hence reducing moisture loss.

Within a species, molluscs in general exhibit a high degree of phenotypic plasticity (Hubendick 1951). That is, genetically identical individuals of a species living in distinct environments will develop markedly different forms. This phenomenon has been observed in land snails repeatedly (Oughton 1948; Pilsbry 1939–1948).

Although the examination of soft body parts is critical to the biological understanding of a creature, archaeologists and paleoecologists normally have only the more durable hard parts available for study. A land snail's shell is attached to the body by muscles and the mantle, a structure that builds the shell by mineral secretion. Since land snails do not shed their shell, a shell contains a record of the individual's life, from the earliest, innermost whorls to the most recent shell at the paertural margin. Within this record may be growth checks, formed during seasonal periods of slow growth, and chemical and isotopic variations indicating environmental or feeding variations.

In addition to the external shell of most land snails, there are two other hard elements. The first, the internal shell of slugs, is really a vestigial or reduced version of the shell of other snails. These internal shells carry much the same information as external shells, and care should be exercised in recovering them, since the unwary can fail to recognize them as snail shells. The second element, the radula, is a rasping and cutting structure that is part of the feeding apparatus. It is made of chiten and is quite small, but it is potentially recoverable from archaeological and Holocene paleontological contexts. Radulae are identifiable to taxon, usually to species, and may carry much encoded information, although this latter topic has been investigated relatively little.

Despite the oft-voiced fear of archaeologists that a snail has burrowed from one level of a site to another, such activities are rare. There are few burrowing snails, and for the most part they are well recognized. Most land snails are more likely to move upward from the ground surface onto vegetation than to move downward further than the base of leaf litter. The lack of great horizontal mobility among land snails makes them the ideal subject of studies of local macrogeography. Mobility studies (e.g., Foster 1937) have shown that most land snails range over an area of less than a couple of hundred meters in diameter, and that many range over a far smaller territory (the occasional catastrophic relocation of land snails is discussed below in the section of taphonomy). Dimbleby and Evans (1974) have argued that, despite the great transport potential for pollen and the low mobility of land snails, the latter's analysis bears only on regional environment at a broad scale, and that pollen analysis better reconstructs local environments. It is more likely that their use of the species spectrum approach (discussed below) exclusively has blinded them to the considerable analytic power of other approaches.

3. APPROACHES

Ways of reconstructing past environments from biological remains can be grouped into three major approaches: species spectrum, environmental tolerance, and morph analysis. A fourth approach based on chemical and isotopic composition of species will not be treated here. Each of the three approaches is discussed here as it relates to land snail analysis.

3.1. Species Spectrum

The notion of a faunal community as a more or less stable web of interacting species has been a foundation of ecological studies for most of the period of their existence. In recent years, however, it has been increasingly recognized that this is a normative concept that carries with it an assumption of stability that many ecologists would argue is unrealistic. Despite this difficulty, the concept of community remains critical to many ecological studies, and especially to paleoecology (Laporte 1968; LaRoque 1966–1970).

If one accepts that a given ecosystem will support a given community of land snails with roughly fixed percentages of occurrence for different species, then it follows that the death assemblage produced by that community will also have more or less standard percentages of those several species. This is the basic logic underlying the *species spectrum* approach. It is not true, however, that a community consisting of 20% species A, 30% species B, and 50% species C will maintain the same percentages of those species in the death assemblage. Species A may breed rapidly and die young, whereas species C may breed rarely but survive to a much older age; under these circumstances, species percentages in the living community might also be inverted in the death assemblage. Other factors that might also lead to divergences between community composition and death assemblage composition are differential preservation, differential identifiability from fossils, differential transportability, and details of survivorship curves (see Shipman 1981 for a lucid discussion of the relationship between communities and death assemblages).

Evans (1972) is the best-known proponent of the species spectrum approach. His studies have focused on England, where descriptive knowledge of land snail communities is more advanced than anywhere else in the world. Recognizing that the composition of his death assemblages did not correspond to any modern community composition, Evans studied modern death assemblages and constructed a series of histograms representing typical species compositions of death assemblages in different environmental zones. These histograms were then used as signatures for identifying similar environments in archaeological assemblages; recognition was based on impressionistic similarity of histograms.

In Evans' hands, this approach seems to have been successful for the analysis of English land snail collections. Its application to other areas, however, is limited by two factors. First, few other parts of the world—none in New England—have received the painstaking care that Evans has lavished on the collection of information about death assemblages. In other places, one is lucky to have a good inventory of species present in the community inhabiting a given environment, let alone any quantitative summary of death assemblages. Second, recovery techniques have profound effects on species composition of death assemblages. Evans' modern death assemblages and archaeological death assemblages were collected with rigorous control for sample size and recovery procedures. This approach is exemplary, but archaeological snail assemblages in North America are usually collected in a less standardized manner; each death assemblage may be collected in a logical and rigorous manner, but there may be no modern assemblages collected with the same procedures. One is left not knowing how to correct for recovery biases.

With these two factors in mind, Evans' approach has been modified for use in New England and other areas. Essentially, Evans' histograms have been reduced to presence/absence charts or high/low/absent ranking charts; ratio measurement has been reduced to nominal or nominal-ordinal measurement. Attendant with such a reduction are increased applicability to a variety of cases and decreased ability to discriminate between environments with similar death assemblages.

Table 1 presents an example of the application of this method (Barber

Table 1. Land Snails Recovered from Lower Level Great Diamond Island Site, Maine[a]

Species	Assemblage frequency	Coniferous	Mixed	Deciduous
Cionella lubrica	L	—	L	L
Vallonia pulchella	L	—	—	L
Gastrocopta armifera	L	—	—	H
Anguispira alternata	L	L	L	L
Discus cronkhitei	L	L	L	H
Discus cronkhitei catskillsensis	L	L	L	L
Helicodiscus parallelus	H	L	L	L
Retinella binneyana	L[b]	L	L	L
Striatura ferrea	L[b]	—	L	L
Vitrina limpida	L[b]	—	?	H
Mesodon thyroidus	H	—	—	H
Triodopsis albolabris	H	L	L	H
Cepaea hortensis	H	—	L	H

[a]Assemblage frequencies and rank abundance for coniferous, mixed, and deciduous forest. $N = 1398$. H, High; L, Low; dash indicates absent.
[b]Includes proportional apportionment of unidentifiable Zontidae.

1980). The assemblage comes from the lower level of the shell midden at the Great Diamond Island site in Casco Bay, Maine. "Assemblage frequency" is a dichotomous ranking of each species of land snail occurring in this level relative to that species' general occurrence. A few members of a generally rare species may rate a high assemblage frequency, while several individuals of a generally common species may rate only a low assemblage frequency. This quantification, although based on extensive experience, is impressionistic. Unfortunately, since the rankings of abundance for the three forest types also were made impressionistically by a variety of biologists (Clarke *et al.* 1968; LaRoque 1966–1970; Moore 1963; Morse 1864; Oughton 1948), this system must serve until higher level quantification is made possible by field studies.

This chart can be looked at in two ways. First, it is a presence/absence chart, where either "L" or "H" signifies presence. Viewed in this way, there is a perfect match between the assemblage's species composition and the species composition expected from deciduous forests. In fact, several species that are rare or absent in coniferous or mixed forests. *Mesodon thyroidus, Triodopsis albolabris,* and *Cepaea hortensis,* are common in the assemblage. These data argue strongly for the presence of a deciduous forest at the time of site occupation.

The second way of looking at Table 1 is as a set of three-level rankings. From this perspective, the analyzed assemblage does not fit any of the expected patterns perfectly, although it most closely approximates the deciduous forest pattern (agreeing in 9 out of 13 species). A variety of statistical measures, including the well-known χ^2 measure, can evaluate the likelihood that this agreement is by chance only. In this case, the pattern is significant at the .05 level, using χ^2.

The key to this type of analysis is having information on modern distributions and densities of species present. Table 2 presents a case where no conclusions could be drawn using the species spectrum approach. These data come from the Late Archaic component at the Flagg Swamp Rockshelter, an inland site in Marlborough, Massachusetts (Barber 1982a). In this assemblage, small snails, almost all from the families Endodontidae and Zonitidae, dominate the assemblage. Even within those families, larger snails, like the ubiquitous *Anguispira alternata,* are rare. The predominance of small species and the virtual exclusion of all but two families suggests that environmental conditions limited species diversity. Furthermore, all of these species are catholic, frequenting a wide variety of habitats. Taken in sum, however, the assemblage is unlike any modern communities or assemblages reported in the literature, largely because of the absence of so many otherwise common species.

It seems unlikely that this is a distinctive rockshelter assemblage, limited in species diversity by edaphic factors. No specialist species seems to have adapted to this niche, leaving it open to qualified generalists. Without a series of studies of modern rockshelters, their snail communities, and their environmental differences, one cannot interpret the species composition of the assemblage.

Table 2. Assemblage Frequency of Land Snails from the Late Archaic Component, Flagg Swamp Rockshelter, Marlborough, Massachusetts[a]

Species	Assemblage frequency
Strobilops (aena?)	**
Helicodiscus parallelus	****
Helicodiscus singleyanus	**
Punctum minutissimum	****
Anguispira alternata	*
Striatura ferrea	******
Retinella sp.	******
Hawaiia miniscula	******
Zonitoides (arboreus?)	******
Mesodon thyroidus	**
Triodopsis albolabris	****

[a]Adapted from Barber 1982a. The number of stars represents relative abundance. $N = 1800$.

3.2. Environmental Tolerance

In common with most classes of organisms, land snails include both species that can survive under a wide range of conditions, and species that are specialized and have a limited set of conditions under which they can survive. *Environmental tolerance* studies take advantage only of the specialists.

If a species of land snail is known to inhabit only deciduous forests, its presence in an archaeological site should indicate that the site was in a deciduous forest zone. Of course, knowledge of the species' tolerance may be incomplete or the snail may be an accidental intrusion into the site, but these possibilities can be assessed. For example, is the species well studied? Were several specimens found? Does the species exhibit behavioral characteristics that make it likely to burrow or otherwise intrude? Are there reasons to believe that the specimen may have been transported into the site by physical processes? The answers to these and similar questions must frame one's assessment of how much faith to place in a reconstruction based on environmental tolerance.

Ultimately, of course, species spectrum analyses take advantage of the fact that certain species do not thrive in certain environments; the extreme case is when a species cannot survive in a certain environment. In this way, the presence of *Cionella lubrica* and other species unknown in coniferous forests (Table 1) can be viewed as evidence for a different vegetation type by the argument of environmental tolerance.

A more clear-cut case, however, can be made for a taxon such as *Succinea* sp., which has a highly limited environmental tolerance. *Succinea* is widely regarded as a moisture-loving species that cannot tolerate standing or running fresh water (Franzen 1982; Morse 1867:607; Shimek 1935). Field studies by the author in New England have found *Succinea* along the margins of streams and ponds and near permanent and semipermanent wood pools—always near fresh water. *Succinea* cannot tolerate salinity, and the "estuary" that Grimm (1961) reported *Succinea* from in Maryland really was the freshwater intertidal zone upriver of the estuary.

At the Wheeler's and Morrill Point sites, both shell middens in Salisbury, Massachusetts, the spatial distribution of *Succinea* bears upon local paleogeography (Barber 1979a, 1982b). A dry ravine runs between these two sites, and today there are no freshwater sources near either. The topography, however, suggests that this ravine once housed a spring-fed stream. At the Wheeler's site, a Middle Woodland site dating to around A.D. 1000, 62 *Succinea* were found. Density of this species averages 18.99 individuals per cm^3 of midden for test units located within 30 m of the ravine; for more distant units, densities averaged only 0.72 individuals per cm^3 of midden. Given the expected mobility of a species deemed virtually amphibious because of its need to return frequently to water, this distribution argues strongly for the existence of the stream at A.D. 1000. At the Morrill Point site, a Late Woodland site dating to about A.D. 1400, not a single *Succinea* was recovered despite the recovery of a sizable land snail assemblage. Taken in sum, one can argue convincingly for the drying up of the stream between A.D. 1000 and 1400 (Barber 1979a, 1982b).

While this case is especially clear-cut, the nature of environmental tolerance studies favors simple, dichotomous patterns: a species is present or absent. This is a strength, but it intensifies the need to recognize intrusive and incongruous specimens.

3.3. Morph Analysis

The phenetic plasticity of molluscs alluded to above forms the basis of *morph analysis* of land snails. If the variation in form of land snail shells is the result of environmental rather than genetic variables, then there should be a consistent relationship between the different variants (morphs) and the environments that caused them. Once these relationships have been established, the recognition of particular morphs in an archaeological assemblage can be taken as evidence for the existence of a particular environment during site formation.

For convenience, morphs are often divided into two types: discrete and metric. *Discrete morphs* are characterized by variables that have nominal values. A deer does or does not have antlers; a guinea pig is black, tan, or piebald; a cat has a bushy tail or a sleek one. *Metric morphs,* on the other hand, are distinguished by variables that have interval or ratio values, that is, "quantified" variables. One dragonfly has longer wings than another; one clam's shell is thicker than another. In reality, of course, the distinction between metric and discrete morphs is imperfect, since many apparently nominal variables can be quantified more exactly. For example, the tails of the cats mentioned above might be simply labeled "bushy" or "sleek"; alternatively, their diameters might be measured. The distinction between metric and discrete morphs lies not necessarily with the specimens under study, but with the approaches used in their study.

An excellent example of discrete morph analysis relates to *Cepaea hortensis,* a land snail with a largely European distribution. In North America, it is known virtually exclusively from coastal New England, leading early observers to believe it to be a historic introduction. It has been repeatedly found in good context from prehistoric archaeological sites, however, and is now generally accepted as native.

Cepaea hortensis has a variety of morphs, distinguished from one another by the presence or absence, number, width, and spacing of bands on their shells. The pigment forming these bands is quite persistent and the bands are easily visible under most conditions, even with heavily leached shells. Bengston *et al.* (1976) presented their own Icelandic data and summarized British studies, and made a clear case for the correlation of specific morph types with specific patterns of ground cover. In particular, they found the unbanded yellow morph to be the dominant form in open, unwooded areas, apparently due to camouflage reflectivity/temperature control factors. The author's unpublished field studies have found *C. hortensis* at 12 locations in eastern Massachusetts and New Hampshire, and have found identical patterns of morph frequency in similar habitats. The only contrary data known come from Cape Breton, Nova Scotia (Martin *et al.* 1951), where only an unbanded specimen or specimens (unquantified) were reported from deciduous woods.

Table 3 presents morph data on *C. hortensis* from three archaeological components: two in Casco Bay, Maine (Barber 1980), and one at Weymouth, Massachusetts (Barber 1979b). At all three components, the numerical dominance of the unbanded morph argues for the openness of the sites, either from clearing or natural conditions.

Table 3. Frequency of Banding Morphs of *Cepaea hortensis* at Three Archaeological Components

Component	Frequency of morph type		
	Unbanded	Three-banded	Five-banded
Great Diamond Island site			
Upper level	5	0	2
Lower level	111	1	2
Weymouth Point site	3	0	0

Cepaea hortensis is a well-studied specimen and supports useful interpretation, but its utility is limited by its own restricted distribution. The ideal candidate for discrete morph analysis not only has distinctive morphs with highly correlated environmental characteristics, but also distribution over a wide variety of environments. *Anguispira alternata* seems to fit these qualifications. It is large and has distinctive morphs based on coloration and whorl carination (Barber 1980), and there seems to be a high correlation with environmental types (Blakeslee 1945; Oughton 1948); further, it is one of the most widely distributed snails in North America, both in terms of range and environmental diversity. Its large size also makes it easy to recover from deposits and eliminates some of the difficulties of analyzing small snails. The next step in discrete morph studies is to develop reliable methods for interpreting this species and others with similar characteristics.

The phenomenon of metric morphs in gastropods appears to relate most closely to differential environmental effects on various parameters of growth and growth rates (Vermeij 1980). Accordingly, it is important to consider the effect of subadult specimens on one's analysis.

Baerreis (1974, 1980) has pioneered metric morph analysis for archaeological reconstruction, and his approach has called for the use only of adult specimens. Practically speaking, this means that species without a recognizeable terminal growth phase (a visible structure that forms only after shell growth is completed) cannot be analyzed by his methods. He has collected a large number of snail samples from localities scattered broadly over the Midwest, and has measured a variety of species. Then, using a few basic climatic variables, he has calculated regressions relating shell size to climate. To use these regressions for archaeological reconstruction, he has selected the most highly correlated variables, insantiated mean shell dimensions into the regression formula, and produced interval-level interpretations, such as centimeters of annual rainfall or number of frost-free days. The method is elegant and appears to produce useful and reliable results, but it requires a great deal of information on modern samples.

In the absence of such information from New England, I have applied metric morph analysis in a different way. From a land snail's point of view, warmth and dryness are similar, dessicating factors, and the autocorrelation of Baerreis' regressions bears this out. Warmth and dryness, then, can be combined into a single variable, as can coolness and wetness. To minimize moisture loss, a spherical shell and round aperture would be ideal; deviations from this norm would become increasingly tolerable under cooler and damper conditions.

Data on three species (*Anguispira alternata, Cepaea hortensis,* and *Mesodon thyroidus*) from both levels at both the Great Diamond Island and Moshier Island sites in Casco Bay (Barber 1980, 1988) were used to examine overall shell and apertural ratios. Within each species, there is a strong correlation of shell ratio to apertural ratio, suggesting that these ratios each relate to the same environmental variable. For both sites, there was no significant difference in metric morph distribution for any species, suggesting that overall climate changed little during the occupation of the sites.

Another application of this approach in Arizona (Barber 1984) better demonstrates its potential. In this case, overall shell length was used as an indicator of how well adapted a species was to its environment. Assuming dry-loving snails to be most successful in drier environments where their specialized adaptation could be brought to bear, and assuming the opposite for wet-loving snails, temperature/rainfall curves were constructed that spanned six phases. These curves summarized ordinal information and showed strong agreement among the three species analyzed; further, these curves agree in general form with curves derived from tree rings, pollen, and macrobotanical remains. There is no reason to believe that comparable curves cannot be derived from the study of metric morphs in New England land snails.

4. TAPHONOMIC CONSIDERATIONS

As important as analytic approaches and methods are, no analysis will be better than the sample on which it is based. Consequently, a brief discussion is included here on land snails and *taphonomy,* the study of how fossil and subfossil biological remains are modified between the death of the organisms and their subsequent recovery and study. Transportation and differential preservation of shells are the primary considerations here.

4.1. Transportation

As Grayson (1979) and others have noted, environmental reconstruction is predicated upon a death assemblage that has received minimal bias. While it will not have identical species composition to its living community, the differences should be primarily attributable to survivorship and birth rates, not transportation of remains into and out of the assemblage. The three major sources of such transportation are human processes, animal processes, and natural processes.

In most of North America, direct human interaction with land snails has always been minimal. Except for chewing slugs picked from gardens (and most slugs that eat living plants are Old World species. in North America only for one or two centuries), most North American land snails have received little or no attention from human beings. Land snails are typically small, dispersed, and time-consuming to eat, making them an inefficient food source; the only exception to this generalization appears to be *Rabdotus,* a land snail that was probably collected and consumed in prehistoric Texas (Hester and Hill 1975).

Indirect effects of human activities are less easy to assess. Human modification of living areas, such as tree clearing, could have marked effects on land snail communities. At the Calf Island site, Massachusetts (Barber 1977; Luedtke 1980), for example, the high incidence of *Heliocodiscus parallelus* and *Retinella binneyana*, species favoring habitats such as those on or beneath felled logs, suggests that clearing had had an effect on the snail community, although tree throws on rocky, windy islands cannot be discounted. This effect is really environmental modification, not transportation of shells, but it results in an assemblage that may not be identical to the assemblage accumulating off-site.

The transportation of land snails by animals is far more common. Snails are food for a variety of animals and often are collected. Rodents and birds, especially thrushes, are the most important of these predators. *Blarina brevicauda,* the short-tailed shrew, is a prominent predator on land snails. Rather than eating the snails on the spot, this shrew typically collects the snails alive and deposits them in a small burrow with an earth barrier at the mouth; the shrew then makes trips to the burrow at intervals to eat some of the snails. The shrew breaks the shells to consume the flesh, but, in the same way that squirrels often forget caches of nuts, the shrew may never return [see Banfield (1974) for details].

What has been interpreted as a shrew's cache of snails was found at the Wheeler's site (Barber 1979a, 1982b). The cross section of the burrow is shown in Figure 1, and the snails within it are listed in Table 4. This cache may have been forgotten, since all shells were intact. Alternatively, the shrew may have been robbed of its meal by *Haplotrema concavum,* a cannibalistic land snail, included in the cache.

Another rodent, probably *Tamias striatus,* the eastern chipmunk, also exploits snails. At certain localities in southwestern Vermont and east-central New York, surface concentrations of land snail shells have been observed, each chewed open on the last whorl, 90° back from the aperture. These concentrations are on and at the base of tree stumps, and are intermingled with hickory nuts,

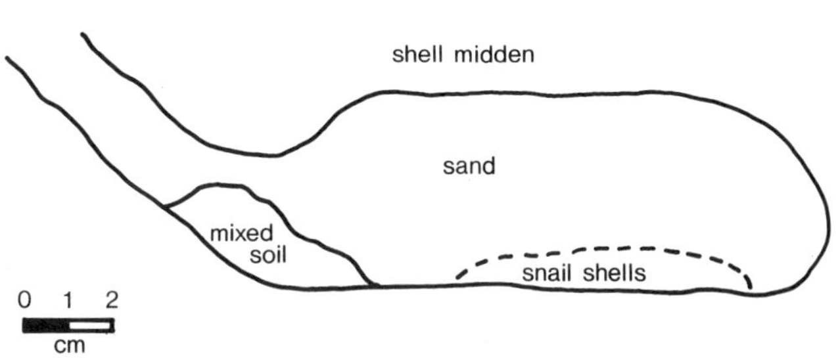

Figure 1. Cross section of burrow believed to be a snail cache deposited by a shrew, Wheeler's site, Massachusetts.

Table 4. Land Snails Recovered from a Probable Short-Tailed Shrew Cache, Wheeler's Site, Massachusetts

Species	Number
Anguispira alternata	3
Helicodiscus parallelus	2
Triodopsis albolabris	4
Haplotrema concavum	1

also chewed open. Chipmunks were observed frequenting these spots and eating nuts, although never seen eating snails. With time, these concentrations become covered with leaf litter and eventually humus.

These concentrations of shells by shrews and probably chipmunks are the major known ways that animals might bring shells *onto* a site. There are, however, a variety of animals that might bring snails *off* a site, including shrews and chipmunks. Martin *et al.* (1951) list six other mammals and birds from New England that have had snail remains found in their stomach contents; snakes and some turtles probably also consume land snails occasionally.

The final source of snail shell transportation, natural processes, is undoubtedly the most important. Land snail shells are relatively light, and some taxa, such as *Succinea* sp., have a large aperture that can act in the same way as a bucket on a water wheel; i.e., it receives the force of running water and is propelled forward. Transportation by moving water is probably the greatest source of land snail movement. In Ohio (LaRoque 1966–1970) and Arizona (Bequaert and Miller 1973), biologists have estimated that most analyzed land snail death assemblages have been deposited as river drift.

As many of New England's archaeological sites are above flood plains, the problem of drift is not so acute. Nonetheless, it is critical to examine one's data for signs of intrasite shell movement and of movement onto and off of the site. For example, one might question whether the *Succinea* sp. from the Wheeler's site, discussed earlier as indicators of an extinct stream, might not have come to their position near the ravine by downslope wash. Their lightness, great surface area, and large aperture would make them perfect for water transportation (Shipman 1981). To examine this possibility, topographic maps were constructed of both the modern and ancient ground surfaces, and it was seen that most of the site drained away from the ravine. Other evidence supporting the validity of the *Succinea* shell distribution as evidence of their distribution in life included the distribution of easily transported charcoal flecks and the irregular surface of the shell midden, which discouraged downslope wash. This kind of examination on one's data source is critical if refined analysis is to be attempted.

4.2. Differential Preservation

As with bone, the primary enemy of shell is soil acidity. Thin shells, especially, are easily attacked, and the surface sculpture or other delicate structures can be easily lost. Some pigments can be quickly leached under any conditions, and these become exceptionally fugitive in acid soils. Since many of New England's soils are highly acidic, there are obvious limitations to land snail analysis there. Many sites will have small or highly disintegrated land snail assemblages, but in cases where soils are neutral by virtue of underlying bedrock or extensive deposits of shell, significant land snail assemblages have been recovered.

Beyond the simple fact that many sites have poorly preserved assemblages, few specifics of differential preservation have been studied. It is recognized, for instance, that the apertural tooth tends to break off eroded *Mesodon thyroidus,* making them look like small *Tridopsis albolabris* (Barber 1978), and that breakage of *Succinea* sp. is about two to three times that of more robust Polygyridae shells (Barber 1979a), but the overall effects of these and other factors remain unassessed.

5. CONCLUSIONS

The analyses reported in this paper document that land snail assemblages from New England contain a considerable amount of information on past macroclimate and microenvironments. In many cases, the conclusions of these analyses have been significant in the cultural interpretation of the site being studied. It would seem that New England archaeologists would be well served to consider the benefits of analyzing their land snail assemblages.

Each of the three approaches discussed here has its advantages and cases where it will be the most productive. The species spectrum approach usually is most appropriate for well-preserved assemblages where macroclimatic information is desired, although its ratio-level form may be able to provide better resolution. The environmental tolerance approach is applicable only to that narrow set of species with known and narrow habitat requirements; as such, it can be used in New England relatively rarely. Morph analysis can provide detailed reconstructions from a wide variety of species, sometimes with ratio-level reconstructions of rainfall and temperature.

With all of these approaches, a primary limitation is the availability of detailed ecological data on modern species in New England. Without this baseline information, it can be difficult to calibrate archaeological reconstructions. To refine these reconstructions to the ratio level, analysts will have to collect data on modern species themselves.

The most important step in environmental reconstruction occurs after all the analyses have been completed. Only then can the results of analyzing pollen, land snails, and other data sets be compared and integrated into a coherent picture of past environments, recognizing the strengths and weaknesses of each data set. New England prehistorians are robbing themselves of one such data set if they ignore land snails.

ACKNOWLEDGMENTS

I am indebted to several individuals who generously gave me access to land snail collections discussed in this paper: Frederick Huntington (Flagg Swamp rockshelter; Weymouth Point site), Barbara Luedtke (Calf Island site), Lynn Teague (Arizona sites), and David Yesner (Great Diamond Island and Moshier Island sites). I also wish to thank Kenneth Boss and Ruth Turner of the Department of Mollusks, Museum of Comparative Zoology, Harvard University, for valuable assistance and advice.

6. REFERENCES

Baerreis, D. A., 1974, Molluscan Remains from 13CK405, *Journal of the Iowa Archaeological Society* 21:131–143.

Baerreis, D. A., 1980, Habitat and Climatic Interpretation from Terrestrial Gastropods at the Cherokee Site. in: *The Cherokee Excavations* (D. C. Anderson and H. A. Semken, Jr., eds.), Academic Press, New York, pp. 101–122.

Banfield, A. W. F., 1974. *The Mammals of Canada,* University of Toronto Press, Toronto, Ontario, Canada.

Barber, R. J., 1977, Report on the Land Snail Assemblage from the Calf Island Site (19-HL-23), Ms. on file, Department of Anthropology, University of Massachusetts, Boston, Massachusetts.

Barber, R. J., 1978, Report on the Molluscan Fauna from the Anderson Village Site (33-Wa-4), Ms. on file, Ohio Historical Society, Columbus, Ohio.

Barber, R. J., 1979a, *Human Ecology and the Estuarine Ecosystem,* Ph.D. Dissertation, Department of Anthropology, Harvard University, unpublished.

Barber. R. J., 1979b, Report on the Molluscan Remains Recovered from Archaeological Field Work at Weymouth Point, Massachusetts, Ms. on file, Peabody Museum, Harvard University, Cambridge, Massachusetts.

Barber, R. J., 1980, Analysis of the Gastropod Assemblage from the Great Diamond Island Site, Casco Bay, Maine, Ms. on file, Department of Anthropology and Geography, University of Southern Maine, Portland, Maine.

Barber, R. J., 1982a, Mollusc remains, in: *Preliminary Report on the Excavation of Flagg Swamp Rockshelter,* Institute for Conservation Archaeology, Peabody Museum, Harvard University, Cambridge, Massachusetts, pp. 119–126.

Barber, R. J., 1982b, *The Wheeler's Site: A Specialized Shellfish Processing Station on the Merrimack River,* Peabody Museum Monograph Series, No. 7, Harvard University, Cambridge, Massachusetts.

Barber, R. J., 1983, Basic Technique in Land Snail Analysis, *Plains Anthropologist* 28(101):219–223.

Barber, R. J., 1984 Land Snails and Environmental Reconstruction along the Salt-Gila Aqueduct, in: Hohokam Archaeology along the Salt-Gila Aqueduct, Vol. 7: Environment and Subsistence (L. S. Teague and P. L. Crown, eds.), *Archaeological Studies* (Arizona State Museum) 150(7):95–108.

Barber, R. J., 1988, Report on the Gastropod Assemblage from the Moshier Island Site, Casco Bay, Maine, In preparation.

Bengston, S.-A., Nilsson, A., Nordstrom, S., and Rundgren, S., 1976, Polymorphism in Relation to Habitat in the Snail *Cepaea hortensis* in Iceland, *Journal of Zoology* (London) 178:173–188.

Bequaert, J. C., and Miller, W. B., 1973, *The Mollusks of the Arid Southwest,* University of Arizona Press, Tucson, Arizona.

Blakeslee, C. L., 1945, Snails from the Aboriginal Deposits of Frontenac Island, New York, *The Nautilus* 58(4):109–112.

Bullen, R. P., 1949, Excavations in Northeastern Massachusetts, *Papers of the Robert S. Peabody Foundation,* Vol. 1, No. 3.

Clarke, A. H., Jr., Kelsall, J. P., and Parker, G. R., 1968, The Land Snail Fauna of Fundy National Park, New Brunswick, *National Museum of Canada, Bulletin,* No. 223.

Dimbleby, G. W., and Evans, J. G., 1974, Pollen and Land-Snail Analysis of Calcareous Soils, *Journal of Archaeological Science* 1(2):117–134.

Evans, J. G., 1972, *Land Snails in Archaeology,* Academic Press, New York.

Foster, T. D., 1937, Productivity of a Land Snail, *Polygyra thyroides* (Say), *Ecology* 18(4):545–546.

Franzen, D. S., 1982, *Succinea avara* Say from the Southern Great Plains of the United States, *The Nautilus* 96(2):82–88.

Grayson, D. K., 1979, On the Quantification of Vertebrate Archaeofauna, in: *Advances in Archaeological Method and Theory,* Vol. 6 (M. B. Schiffer, ed.), Academic Press, New York, pp. 199–237.

Grimm, F. W., 1961, Land Snails in the Upper Patuxent Estuary Margin (Maryland), *The Nautilus* 74(3):106–109.

Hester, T. R., and Hill, T. C., Jr., 1975, Eating Land Snails in Prehistoric Southern Texas: Ethnohistoric and Experimental Data. *The Nautilus* 89(2):37–38.

Hubendick, B., 1951, Recent Lymnaeidae, *Kungliga Svenska Vetenskapsakademiens Handlingar, Fjard Serien,* Vol. 3, No. 1.

Laporte, L. F., 1968, *Ancient Environments,* Prentice-Hall, Englewood Cliffs, New Jersey.

LaRoque, A., 1966–1970, Pleistocene Mollusca of Ohio, *Ohio Division of Geological Survey, Bulletin,* No. 62 (4 parts).

Luedtke, B. E., 1980, The Calf Island Site and the Late Prehistoric Period in Boston Harbor, *Man in the Northeast* 20:25–76.

Martin, A. C., Zim, H. S., and Nelson, A. L., 1951, *American Wildlife and Plants,* McGraw-Hill, New York.

Moore, R. G., 1963, Land Mollusca in the Vicinity of Wolfville, Nova Scotia, *Proceedings of the Nova Scotia Institute of Science* 25(3):187–198.

Morse, E. S., 1864, Observations on the Terrestrial Pulmonifera of Maine, *Journal of the Portland Society of Natural History,* Vol. 1, No. 1.

Morse, E. S., 1867, Land Snails of New England, *American Naturalist* 1:5–16, 95–100, 150–151, 186–188, 313–315, 411–414, 541–547, 606–609, 666–672.

Morton, J. E., 1967, *Molluscs,* rev. ed., Hutchinson University Library, London.

Oughton, J., 1948, A Zoogeographical Study of the Land Snails of Ontario, *University of Toronto, Biological Series,* No. 57.

Pilsby, H. A., 1939–1948, *Land Molluscs of North America (North of Mexico),* The Academy of Natural Sciences of Philadelphia, Monograph Series, No. 3 (2 volumes).

Schimek, B., 1935, The Habitats of Iowa *Succineas, The Nautilus* 49(1):6–10.

Shipman, P., 1981, *The Life History of a Fossil,* Harvard University Press, Cambridge, Massachusetts.

Vermeij, G. J., 1980, Gastropod Shell Growth Rate, Allometry, and Adult Size: Environmental Implications, in: *Skeletal Growth of Aquatic Organisms* (D. C. Rhoad and R. A. Lutz, eds.), Plenum Press, New York, pp. 379–394.

Wyman, J., 1867, An Account of Some Kjoekkenmoeddings or Shell-Heaps in Maine and Massachusetts, *American Naturalist* 1:561–584.

Chapter 2

Historical Climates of the Northeastern United States

Seventeenth through Nineteenth Centuries

WILLIAM R. BARON

1. INTRODUCTION

Native New Englanders have long prided themselves in their ability to earn a livelihood in a harsh, often changeable environment. The leading industries of the seventeenth through nineteenth centuries—fishing, shipping, lumbering, manufacturing, and farming—all required direct confrontation with the natural elements. One of the most important of these elements was the weather, and its fickleness is best reflected in the adage, "If you don't like the weather, just wait a minute." Early New England history is peppered with accounts of distressing droughts that withered crops and halted water-driven machinery, fierce snowstorms that cut off the countryside from the towns and cities, powerful floods that swept all before them, and raging storms that battered harbors and destroyed shipping. Upon studying the history of New England's principal crop, hay, and learning of its vulnerability to the caprice of the weather, one wonders how the region's farmers survived and even sometimes prospered. The out-migrations of the nineteenth century are evidence that some failed "to make a go of it."

That there is a relationship between weather, climate, and human activities

WILLIAM R. BARON • Center for Colorado Plateau Studies, Northern Arizona University, Flagstaff, Arizona 86011, and Institute for Quaternary Studies, University of Maine, Orono, Maine 04469.

is not a particularly new idea. For centuries, observers have speculated on the impact of certain types of climate on society. Early twentieth century environmental determinists such as Ellsworth Huntington detailed the impact of microchanges on individual human beings (Glacken 1967:551–662; Huntington 1915). Nor was the relationship believed to be unilateral. For at least 200 years, some individuals have noted mankind's impact on the climate through a number of different activities, ranging from cutting down the forests to adding carbon dioxide to the atmosphere through burning fossil fuels (Schneider and Temkin 1978; Thompson 1980).

The idea of climatic changes within historic times also is not new and has been hotly contested as late as the middle of this century. In New England and elsewhere in North America, some eighteenth and early nineteenth century observers thought they saw changes in the climate, usually in the form of milder winters and warmer summers. Their perceptions were not based on instrument records, but on ancestral and personal recollections of past weather and climate. Others contended that no changes had taken place. Weather recollections are particularly fleeting and are, at best, imprecise, and often emphasize the catastrophic (Oliver 1975). The debate could not be settled scientifically until sufficient meteorological data had been gathered. By the last third of the nineteenth century, the necessary data were available and analysis seemed to show that the climate was stable (Thompson 1981). These findings remained in vogue into the 1940s. Most observers accepted the notion that the climate had changed very little in the past 2,000 years. However, some researchers continued to support and defend the idea of climatic change (Khrgian 1970:319–329). Among this group were Huntington and Visher (1922) and C. E. P. Brooks (1970). They and their colleagues helped prepare the ground for a resurgence of the idea of climatic change that has occurred during the past 40 years.

By the 1940s, additional instrument data had begun to show that the climate was not stable after all, but, in most locations in the Northern Hemisphere, had been warming since the 1880s. This realization brought about a reevaluation of the whole idea of climatic change. Additional research to develop longer records produced further evidence that changes had taken place several times in the "historical" past (Lamb 1965; Manley 1953, 1974).

Other scholars have begun to study the relationships between changing climates and society (Lamb 1982; Maunder 1970; Tickell 1977; Parry 1978). Work on the relationship between climate and human experience has only just begun, although there is now considerable activity focusing on a number of different time periods and geographic locales. In the future, researchers will have to expend considerable effort in designing methodology to analyze climate/human relationships (Butzer 1982; Parry and Carter 1984; Warrick and Riebsame 1983; Wigley *et al.* 1985).

2. NORTHEAST CLIMATE DATA

2.1. Instrument Record Development

The debate concerning climatic change has had a definite impact on work on Northeast climate reconstructions. Early assessments of past climate were based on conjecture. Examples of these are Samuel Williams' "Change of Climate in North America" (Williams 1790) and Noah Webster's 1843 essay, "On the Supposed Change in the Temperature of Winter" (Webster 1968). These base the thrust of their arguments on memories of past weather and seasons rather than on instrumental data. Later nineteenth century evaluations, such as those by Schott (1876, 1881), were based on a relatively long series of instrument data (Thompson 1981).

Instrument records for the eighteenth century are very rare and their geographic and chronological spread is slim. The earliest of these records is one compiled for the period 1742–1779 by John Winthrop of Harvard College (Landsberg *et al.* 1968:14). Later records of varying lengths include those for Salem, Massachusetts, 1754–1829 (Holyoke 1833); New Haven, Connecticut, 1778–1795 (Landsberg 1949), Cambridge, Massachusetts, 1790–1812 (Farrar 1809); Philadelphia, Pennsylvania, 1758–1759, 1767–1777, 1798–1804 (Schott 1876:287); Morrisville, Pennsylvania, 1790–1845 (Schott 1876:287); and Nazareth, Pennsylvania, 1787–1792 (Schott 1876:287). Based on these records and other shorter record fragments, there are long reconstructions compiled for New Brunswick, New Jersey, 1738–1976 (Reiss *et al.* 1980), and for the eastern United States adjusted to Philadelphia, Pennsylvania, 1738–1967 (Landsberg *et al.* 1968).

During the early nineteenth century, records were begun at several other northeastern locations. Included among these were Massachusetts records for Williamstown commencing in 1811 (Milham 1950), for New Bedford beginning in 1813 (Rodman 1915), and for Boston starting in 1818 (Rumford Foundation 1839). In Rhode Island, record keeping began at Newport in 1817 (Gould 1857) and at Providence in 1831 (Caswell 1882). Instrumental record keeping started in Maine at Brunswick in 1807 (Cleaveland 1867), and in 1816 at both Portland (Moody 1852) and Dennysville (Lincoln 1861). Dartmouth College (1905) at Hanover, New Hampshire began a record in 1834, while in Vermont two long records commenced in 1828 for Burlington (Clayton 1927) and Randolph (Nutting 1863). In New York, there are long records for New York City, starting in 1822, and for Albany, starting in 1795 (Schott 1876:269–277). Records are also available for Belleville, New Jersey, 1803–1829 (Rutgers 1829); Germantown, Pennsylvania, 1820–1828; and Gettysburg, Pennsylvania 1839–1862 (Schott 1876:284–289).

The first meteorological network supported by an agency of the federal government was begun in 1819 by the Surgeon General's Office. Instrument readings were taken at several military posts in the Northeast (U.S. Surgeon General's Office 1855). Over the next 80 years several additional federal meteorological networks were organized to collect data. The agencies involved in this work included the Smithsonian Institute, which administered its own network from 1848 through 1872 (Langley 1894), in addition to those of the Patent Office (U.S. Patent Office, 1848–1858), the Department of Agriculture (U.S. Department of Agriculture 1863–1870), and the U.S. Army Signal Service, which took over all federally sponsored meteorological work in the early 1870s (U.S. Army 1872–1892). Finally, in 1892, federal meteorological activities were made a civilian activity under the newly created Weather Bureau. At one point in time or another, over 350 different Northeastern locations have been represented. In addition to the federal government's meteorological activities, each of the Northeast states has sponsored limited record keeping. Usually these records were connected with agricultural pursuits and were kept at state farms and, later, at state experiment stations.

Despite the large number of nineteenth century instrumental records available for the Northeast, many are unusable for climatic reconstructions because they are too short, instruments were changed or moved frequently, observation locations were poorly chosen, instruments came out of adjustment over time, and observation times were varied, sometimes on a daily basis. While not valueless, records affected by one or more of these problems do not represent mathematically discrete sets, but are actually a series of separate, short records strung together on a time line for a given location. Such a series of short records is said to be *heterogeneous*. In order to reconstruct such records into a long *homogeneous* time series or single discrete set, a climatologist must locate other long homogeneous records from locations nearby with similar geophysical features (Baron and Gordon 1985; Baron *et al.* 1980). In practice this is a very complex operation requiring a large number of records to complete a reconstruction for a single location. A heterogeneous record reconstruction can, at times, be worse than no record at all, as it is impossible to distinguish between trends and changes caused by climate or caused by problems with instrumentation and exposure. The temptation to accept at face value a long, laboriously compiled record that is the work of several lifetimes is very great. However, should one succumb, an incorrect reading of the climate record will very likely be the result.

2.2. Noninstrument Records: Sources and Methodologies

With very few instrument records available for the eighteenth century, and with the many homogenization problems of nineteenth century records, the climate historian must look elsewhere for other types of data to complete reconstructions for the seventeenth through nineteenth centuries. One type of data that has proven useful for other regions is the dendroclimatological record. It is particularly useful in areas where one climatic parameter, such as precipitation, is the predominant factor limiting tree growth. In regions such as the Northeast, where both temperature and precipitation are limiting factors, much work is being undertaken to sort out the various biological signals and produce accurate reconstructions for both parameters (Bliss 1983; Conkey 1979a,b). One recent study of the Hudson River Valley sheds some light on drought frequencies in that area (Cook and Jacoby 1977, 1979).

Another source of data that is richer in detail and has a better geographic distribution is historical documentation, such as diaries, almanacs, old newspapers, and periodicals. Surveys of over 100 archives and repositories undertaken by the author reveal over 2,500 diaries known to contain weather-related information. Of this number, at least 450 diaries have detailed, daily descriptions of the weather. A somewhat smaller number have multiple daily descriptions. The length of these diaries ranges from a few months to over 60 years. Diaries with daily entries are most frequently found for the eighteenth and nineteenth centuries. There are a few rather complete diaries for the seventeenth century, but most describe meteorological events in terms of months, seasons, or years.

Newspapers first appeared at Boston in 1704 and from their inception contained references to storms, droughts, and other climate phenomena. By the mid-nineteenth century, over 250 newspapers were being published throughout the region. Periodicals first appeared in the mid-eighteenth century. Especially useful for the researcher are those periodicals concerned with agriculture, science, and medicine, in which articles on meteorology and climatology, old weather records, and, by the very early nineteenth century, instrument records often were published.

Almanacs also are a very useful source of early weather information. The almanac was one of the first publications, after the *Bible,* to be published in New England and was so popular an item that almost every colonial family owned one (Stowell 1977:ix–xii). Owners often annotated their almanacs, keeping track of the actual daily weather and comparing it to the predictions made by the almanac's publisher. By the early nineteenth century, almanac publishers were printing tables containing descriptions of the previous year's weather.

Attempts to reconstruct New England's climate using noninstrument materials are not new. The impressionistic overviews written in the late eighteenth and early nineteenth centuries are the first examples of such work (Thompson 1981). Early in this century, Brooks (1917), Weber (1930), and Smith (1934) used frequency counts and general descriptions of colonial snowfall, rainfall, and winter weather to supplement instrument records for long time periods. Perhaps the most complete overview of Northeastern weather based on both early instrument records and published diaries, newspapers, and periodicals is that of

Ludlum (1966, 1968, 1976, 1983). While he failed to utilize existing large collections of unpublished diaries, he made a very thorough search of published materials.

To produce useful climate records from noninstrument materials requires a sometimes complex and varied methodological approach. Much recent work has been done on the methodological problems involved in the reconstruction of New England records by Baron and Gordon (1985). To reconstruct a temperature and precipitation record for the seventeenth century, they have modified a methodology used by Lamb (1965) on medieval materials. Word scales were designed that delineated temperature and precipitation departures from what was perceived by colonial diarists as "normal." The resulting records are more accurate for the second half of the century, as more descriptions from "native" observers were available. For the period 1620–1650, most of the observers evaluated the New England weather based on their English weather experience and on misconceived but generally accepted ideas concerning North American climate (Kupperman 1982).

The detailed daily diaries available for the eighteenth and nineteenth centuries were analyzed using a two-tiered content analysis methodology. Under this system, weather entries were sorted by type and were then assigned to word scales that were, in turn, related to a quantitative index. With this methodology, climate parameters such as temperature, precipitation, cloud cover, and wind direction were measured. Modifiers relating to quantity or degree and time duration also were weighted (Baron 1980, 1982a).

Records of drought frequency and growing season lengths demanded a somewhat different form of content analysis. Interpretation and analysis of data relating to these two climate parameters required a precise set of definitions for what was meant by such terms as a "white frost" or a "killing frost." The specific species of plants killed by the frost helped determine temperature limits (Baron *et al.* 1984). These data were used to determine regional dates for the beginning and end of growing seasons. The same sort of methodology was used to produce records of droughts with agricultural implications.

The methodologically simplest form of climate record reconstruction used involved making frequency counts of various weather events, such as snowfalls, storms, floods, and thunder and lightning. The occurrences of each of these phenomena were counted by month, season, and year. They were then reported as aggregates or in index form using a specified year or decade as a base period (Baron 1982b).

A number of factors can influence the analysis of noninstrument materials such as diaries. The frequency of missing daily diary entries and the amount of descriptive detail can affect the accuracy of the indices. In addition, the diarist's occupation (whether indoors or out of doors), physical condition, age, and familiarity with the geographic location can also influence the record (Baron 1982a; Baron and Gordon 1985).

2.3. Climate Record for the Northeast

Based on a large body of climate data containing both instrument records and qualitative historical materials, preliminary records for southern New England were completed for several climate parameters, including temperature, precipitation, cloud cover, wind direction, growing season parameters, floods, droughts, storm frequencies, snowfall, and thunderstorms.

A number of the records completed for temperature and precipitation are for segments of from 30 to 100 years in length. A major methodological problem has involved how to combine these various records, both instrument and noninstrument, into a homogeneous, long-time-series record for the Coastal Zone[1] spanning the period 1620–1890 (Baron and Gordon 1985). A quantified single record for this period must await the reconstruction of additional records for various time periods and locations.

Table 1, a general overview of the temperature based on a compilation of 63 records, shows that temperatures during the seventeenth and eighteenth centuries were somewhat cool, with particularly cold decades in the 1690s and 1750s. After the 1780s, a trend toward warming set in that was interrupted only in the 1810s, a decade particularly influenced by volcanic dust in the atmosphere. After the 1880s (not shown), a pronounced general warming took place, lasting until the 1940s (Baron 1982a; Baron and Gordon 1985).

Table 2 is an overview of precipitation based on a compilation of 68 records. This record shows a relatively wet period with particularly wet years in the 1630s, 1710s, and 1850s. Dry periods of note occurred in the 1760s and 1820s. Table 3, based on 143 diaries and 99 instrument records, is a reconstruction of the frequency of droughts during growing seasons and of flood frequencies. It supports some of the conclusions drawn from Table 2. The high number of droughts in the 1740s, 1760s, and 1770s correlates well with the overall picture of dryness for those decades. The relatively high number of droughts for the period 1850–1889 were offset by heavy snowpack during the winters. Evidence of heavy snowfall during this period can be seen in the increase in spring floods in northern New England (Baron and Gordon 1985).

[1]The "Coastal Zone" refers to the National Oceanic and Atmospheric Association (NOAA) climate zone that runs along the coast of the entire Northeast region. All standard climate zone maps provided by NOAA with their various publications show this zone, which covers the area from the immediate coast to 24–32 km inland, depending upon coastal topography. Elevations within the zone are usually 150 m or less, with some exceptions, such as Mount Desert, Maine.

Table 1. Relative Coolness and Warmth for Eastern Massachusetts

Decade	Cold	Cool	Average	Warm	Hot	Decade	Cold	Cool	Average	Warm	Hot
1620s			X			1760s		X			
1630s		X				1770s			X		
1640s		X				1780s		X			
1650s		X				1790s				X	
1660s			X			1800s			X		
1670s		X				1810s		X			
1680s		X				1820s				X	
1690s	X					1830s			X		
1700s			X			1840s				X	
1710s		X				1850s			X		
1720s		X				1860s				X	
1730s			X			1870s			X		
1740s			X			1880s			X		
1750s	X										

Table 2. Eastern Massachusetts Moisture Conditions

Decade	Dry	Drier	Average	Damp	Wet	Decade	Dry	Drier	Average	Damp	Wet
1620s				X		1760s	X				
1630s					X	1770s		X			
1640s				X		1780s			X		
1650s			X			1790s				X	
1660s		X				1800s		X			
1670s				X		1810s			X		
1680s			X			1820s	X				
1690s			X			1830s		X			
1700s					X	1840s			X		
1710s				X		1850s				X	
1720s			X			1860s					X
1730s				X		1870s				X	
1740s		X				1880s				X	
1750s				X							

Table 3. Dry Growing Seasons and Years with Flooding in Southern New England

Decade	Dry years	Floods	Decade	Dry years	Floods
1620s	2	0	1760s	9	1
1630s	2	0	1770s	7	6
1640s	2	0	1780s	6	8
1650s	—	—	1790s	7	5
1660s	4	2	1800s	7	3
1670s	3	1	1810s	3	7
1680s	5	1	1820s	3	5
1690s	3	1	1830s	2	3
1700s	6	3	1840s	1	3
1710s	3	2	1850s	4	3
1720s	6	4	1860s	5	2
1730s	6	2	1870s	8	3
1740s	10	2	1880s	6	2
1750s	4	5			

Decadal time frames, although helpful in reducing data from a long time series to a manageable size, obscure year-to-year and season-to-season patterns. The record of growing season lengths illustrates the great variability that existed throughout most of the eighteenth and nineteenth centuries. Only as the warming period in the 1880s through 1890s set in did the degree of variability decrease. Between 1740 and 1840, it was not uncommon for the growing season length to vary as much as 1 month from one year to the next. This record also exhibits a systematic relationship between the timing of spring and fall killing frosts (Baron *et al.* 1984).

Table 4, drawn from 38 diaries and 18 instrument records, illustrates a significant range of frequency in the number of thunderstorms from year to year. The period 1740 through the 1780s was one of increased thunderstorm frequency. From 1800 to 1849 there appears to have been a decrease in activity, but the decrease may be somewhat attributable to missing data. In the 1870s and 1880s there was a significant increase in activity, exceeded only by the 1750s (Baron and Gordon 1985).

A final reconstruction given in Table 5 concerns the percentage of each year when fair skies prevailed and is based on a compilation of nine long records. The eighteenth century was a period of relatively clear skies, with the 1760s and 1770s, two dry decades (see Table 2), the clearest of the century. During the nineteenth century and especially after the 1840s, there was a decrease in the number of fair days per year. This decrease in the number of fair days per year correlates with the decrease in the variability of the growing season.

Table 4. Mean Number of Thunderstorms per Decade for Eastern Massachusetts

Decade	Spring	Summer	Fall	Winter	Year	Range
1730s	1.1	2.8	0.2	0.1	4.2	1–7
1740s	3.4	10.3	1.1	0.2	15.0	1–28
1750s	4.4	16.6	3.3	0.6	24.9	14–32
1760s	2.7	12.8	2.6	0.2	18.3	8–24
1770s	4.1	12.9	1.6	0.2	18.8	13–22
1780s	4.0	11.8	2.7	0.1	18.6	14–24
1790s	2.6	8.4	1.5	0.2	12.7	7–17
1800s	2.5	6.8	2.4	0.3	12.0	7–18
1810s	1.5	4.4	0.8	0.1	6.8	3–12
1820s[a]	0.6	4.8	1.0	0.2	6.6	5–10
1830s[a]	1.7	5.3	0.3	0.0	7.3	5–11
1840s[a]	2.0	4.8	1.4	0.2	8.4	4–13
1850s	1.9	7.1	1.6	0.0	10.6	3–19
1860s	1.9	6.7	1.0	0.3	9.9	4–13
1870s	3.2	16.0	2.4	0.3	21.9	10–31
1880s	3.2	12.4	4.3	0.3	20.5	13–34
1730–1889	2.6	9.0	1.8	0.2	13.5	1–34

[a]Data missing for decade.

The general picture that emerges from these climate records of a cool, somewhat wet three centuries marked by occasional decades of drought and considerable interannual variability corresponds well with reconstructions for other geographic regions. According to Lamb (1977:449–473), the period from A.D. 1200–1400 to 1850–1890 was marked by a trend toward colder climates in the Northern Hemisphere. The circumpolar vortex expanded and the zones of heightened cyclonic activity shifted south. Despite the overall trends toward cooler and wetter conditions, there were intermittent episodes of warmth and drought. This climatic period, often called the Little Ice Age, continued until the middle to latter part of the nineteenth century, when a gradual warming trend began. Lamb's description is based primarily on European records, but adequately describes conditions in the northeastern United States as well.

3. CLIMATE AND SOCIETY

Now that the climate record for the historic period of New England is nearing completion, there remains a final set of questions. These concern the interrelationships of weather, climate variation, and society. There are few such

Table 5. Percentage of Fair Days per Year for Boston, Massachusetts Area

Decade	Spring	Summer	Fall	Winter	Year	Standard deviation
1730s	45.5	53.2	49.8	47.9	49.1	8.0
1740s	42.2	51.4	43.1	41.0	45.0	5.3
1750s[a]	56.4	64.4	62.7	54.9	59.4	3.9
1760s	59.5	66.2	62.3	57.8	61.5	1.8
1770s	55.5	64.1	59.2	60.8	60.2	4.1
1780s	57.3	58.5	59.7	58.9	58.7	3.6
1790s	55.7	63.1	57.5	61.7	59.4	2.7
1800s	53.6	57.8	57.8	57.1	57.3	4.1
1810s	69.0	64.0	68.6	59.5	65.3	5.4
1820s	60.8	68.0	62.0	58.4	62.4	6.8
1830s	66.0	75.7	70.3	60.5	68.3	3.0
1840s	62.4	71.9	67.3	59.6	66.0	4.7
1850s	45.7	48.4	52.6	44.6	48.1	4.5
1860s	43.0	47.4	45.8	45.1	45.3	2.3
1870s	42.7	42.6	46.9	48.5	45.1	1.9
1880s	46.3	54.0	47.3	46.8	48.7	7.0

[a]Missing data for decade.

studies for the region. Most of these concern 1816, "the year without a summer," which certainly was an aberration from the norm (Skeen 1981; Stommel and Stommel 1983). The remaining studies concern climatic impact on northern New England agriculture (Baron and Bridges 1983; Smith *et al.* 1982). A partial agenda has been suggested for eighteenth century New England (Baron 1982b), but many questions of historical importance remain unanswered.

One of the questions that demands considerable further study concerns the impact of the Little Ice Age on New Englanders. The climate records described in this paper present the first detailed picture of what the climate was like between 1600 and 1880. Its most notable characteristic was its variability in a number of climatic parameters. Growing season lengths varied greatly, often from one year to the next. Periods of drought were followed closely by periods of flooding. Hail and thunderstorms were much more frequent in some years than in others. The great detail of the climate records for time spans as small as a day allow for a number of interesting comparisons with the agricultural record. The questions that arise from a preliminary analysis of both types of records are numerous. Did the droughts and growing season length fluctuations of the 1760s and 1770s significantly reduce the New England farmer's already limited sources of ready cash by cutting short crops that normally produced a small saleable surplus? And did this loss of capital contribute to the unrest leading to the American Revolution? Was the multicropping system of New England farmers more conducive to Little Ice Age conditions than our own century's single-crop system? How was agricultural folk knowledge on when to plant and when to harvest changed by these rapid fluctuations in climate and how long did it take for the changes to take place? What were the effects of short-term climatic variation on agricultural labor practices? To answer these questions researchers should utilize research and analysis methodology developed for use in the fields of resource economics and agriculture. The research approach employed by Parry (1978) on the agriculture and settlement of geographically marginal areas of Scotland and Scandinavia can serve as an example for a similar study of the northern New England uplands. The work of Walton and Ball (1976), Rothenburg (1979), Anderson (1975), Galenson and Menard (1980), Carr and Walsh (1980), and Jones (1980) on colonial agricultural resources and productivity provide records for comparison to detailed climate/weather records and stand as examples of studies that could be conducted on similar materials for other time periods and geographic locales.

Nor is agriculture the only subject that would benefit from a close comparison of climate and societal records. There is a need for studies of the impact of storms, heavy snowfalls, and floods on the economy of businesses, towns, and the region as a whole. For example, heavy snows halt rail traffic; periods of cold weather freeze up harbors and rivers, closing them to commerce; and storms destroy wharves and shipping.

Variable weather and climate also directly affect individual members of society. For example, multilayered clothing was more common during the variable weather of the Little Ice Age. Frequent severe winters increased the demand for wood in larger towns such as Boston and Newport. Wood prices rose sharply and large areas close to the towns were deforested. People's perceptions of their environment, including the climate, need much more attention through the examination of their literature and folk art.

Before such studies can proceed, researchers must develop conceptual models of the impact of climate on humans and on society that assign various levels or orders of impact. Such models also must allow for possible feedbacks via adaptive strategies used by individuals or society as a whole. Much research has already been done in developing useful models (Kates 1985), but many of these models require considerably more testing using longer time frames. The testing of such models using historical records presents some formidable but not insoluble methodological problems. Much of this methodology has been developed by the Climatic Research Unit at the University of East Anglia (Wigley *et al.* 1985:536–544).

4. CONCLUSIONS

In the past, studies of the impact of climate change on societies have been severely limited by the absence of long, detailed climate records. The methodology described in this paper can be used for any time period and geographic locale for which there is, at a minimum, good qualitative descriptions of the weather. The development of this new methodology has broken the dependence on instrument records as a sole source of detailed climate record development. The daily climate records that result make the in-depth study of weather and short-term climate aberrations and their impact on society possible for the first time in a historical context earlier than the mid-nineteenth century. Where earlier climate/societal studies employed just a few climatic parameters such as temperature or precipitation, now it is possible to use records of the whole range of climate phenomena.

For those working on time periods and geographic areas where the type of record development used here on the Northeast is not possible, there also are some lessons to be learned. The details of the Northeast record clearly point out that it is methodologically dangerous to draw too many conclusions about climate conditions from too few data or from data derived from other climate zones. Even within the Northeast there are a number of climate records for local areas that run counter to regional trends.

Over the time span 1600–1980, there was one climate change of any significance, the change from the Little Ice Age's extreme variability to our own century's somewhat warmer, more stable climatic conditions. The two regimes seem quite different. However, the change from one to the other was not cataclysmic, but very slow, taking place over several decades, and seems to have been barely apparent to those living through it. *How else could one explain the irony of having a steady-state theory of climate in vogue during the exact time period of the only significant climatic change of the last several hundred years?*

The variety of very short-term climate aberrations, perhaps more properly called "weather aberrations," found in the Northeastern records are so small that there seems to have been little lasting impact on society as a whole. Only in areas on the geographic margins of human settlement, such as in the far north or south, the margins of deserts, or at high altitudes, would the sorts of aberrations found in the Northeastern records have a lasting effect on society. The Northeast does not fit into any of these categories. However, further study is needed on climate/societal relationships before the final chapter of the area's history can be written.

ACKNOWLEDGMENTS

I thank the members of Institute for Quaternary Studies, University of Maine at Orono, and the Center for Colorado Plateau Studies, Northern Arizona University. Much of this research was supported by National Science Foundation Grants ATM 7908415, ATM 8019514, ATM8116714, and ATM8312874.

5. REFERENCES

Anderson, T. L., 1975, *The Economic Growth of Seventeenth Century New England: A Measure of Regional Income*, Arno Press, New York.

Baron, W. R., 1980, *Tempests, Freshets and Mackerel Skies: Climatological Data from Diaries Using Content Analysis,* Ph.D. Dissertation, University of Maine, Orono, unpublished.

Baron, W. R., 1982a, The Reconstruction of Eighteenth Century Temperature Records through the Use of Content Analysis, *Climatic Change* 4:385–398.

Baron, W. R., 1982b, Eighteenth Century New England Climate Variation and Its Suggested Impact on Society, *Maine Historical Society Quarterly* 21:201–218.

Baron, W. R., and Bridges, A. E., 1983, Making Hay in Northern New England: Maine As a Case Study, 1800–1850, *Agricultural History* 57:165–180.

Baron, W. R., and Gordon, G. A., 1985, A Reconstruction of New England Climate Using Historical Materials, 1620–1980, *Syllogeus* 55:229–245.

Baron, W. R., Smith, D. C., Borns, H. W., Fastook, J., and Bridges, A. E., 1980, Long Time Series Temperature and Precipitation Records for Maine, 1808–1978, *Maine Agricultural Experiment Station Bulletin,* No. 771, University of Maine, Orono.

Baron, W. R., Gordon, G. A., Borns, H. W., and Smith, D. C., 1984, Frost-Free Season Record Reconstruction for Eastern Massachusetts, 1733–1980, *Journal of Climate and Applied Meteorology* 23:317–319.

Bliss, C. S., 1983, *The Impact of Environment on Tree-Ring Growth in One Climate Region in Maine.* M.Sc. Thesis, Institute for Quaternary Studies, University of Maine, Orono, unpublished.

Brooks, C. E. P., 1970, *Climate through the Ages,* 2nd rev. ed., Dover, New York.

Brooks, C. F., 1917, New England Snowfall, *Geographical Review* 3:222–240.

Butzer, K. W., 1982, *Archaeology As Human Ecology,* Cambridge University Press, Cambridge.

Carr, L. G., and Walsh, L. S., 1980, Inventories and Analysis of Wealth and Consumption Patterns on St. Mary's County, Maryland, 1658–1777, *Historical Methods* 13:81–104.

Caswell, A., 1882, Meteorological Observations Made at Providence, R.I. from Dec. 1831 to Dec. 1876, *Smithsonian Contributions to Knowledge* 24(443):1–31.

Clayton, H. H., 1927, Burlington, in: World Weather Records, *Smithsonian Miscellaneous Collections* 27:817–819.

Cleaveland, P., 1867, Results of Meteorological Observations made at Brunswick, Maine, between 1807 and 1859, Reduced and Discussed by C. A. Schott, *Smithsonian Contributions to Knowledge* 16(204):2–25.

Conkey, L. E., 1979a, Response of Tree-Ring Density to Climate in Maine, U.S.A., *Tree-Ring Bulletin* 39:29–38.

Conkey, L. E. 1979b, *Dendroclimatology in the Northeastern United States,* M.Sc. Thesis, Department of Geosciences, University of Arizona, Tucson, unpublished.

Cook, E. R., and Jacoby, G. C., Jr., 1977, Tree-Ring–Drought Relationships in the Hudson Valley, New York, *Science* 198(4315):399–401.

Cook, E. R., and Jacoby, G. C., Jr., 1979, Evidence for Quasi-Periodic July Drought in the Hudson Valley, New York, *Nature* 282:390–392.

Dartmouth College, 1905, Monthly, Annual and Average Temperature and Precipitation at Hanover, N.H., 1834–1904, *Climate and Crops: New England Section, Annual Summary* 1905:11.

Farrar. J., 1809, Abstract of Meterological Observations Made at Cambridge, New England, *Memoirs of the American Academy of Arts and Sciences, Old Series* 3(1):361–412.

Galenson, D. W., and Menard, R. B., 1980, Approaches to the Analysis of Economic Growth in Colonial British America, *Historical Methods* 13:3–18.

Glacken, C. J., 1967, *Traces on the Rhodian Shore,* University of California Press, Berkeley.

Gould, S., 1857, Meteorological Journal for Newport, R. I., 1817–1857, Ms. at Newport Historical Society, Newport, Rhode Island.

Holyoke, E. A., 1833, A Meteorological Journal from the Year 1786 to the Year 1829 Inclusive, *Memoirs of the American Academy of Arts and Sciences, New Series* 1:107–216.

Huntington, E., 1915, *Civilization and Climate,* Yale University Press, New Haven, Connecticut.

Huntington, E., and Visher, S. S. 1922, *Climatic Changes: Their Nature and Causes,* Yale University Press, New Haven, Connecticut.

Jones, A. H., 1980, *Wealth of a Nation To Be,* Columbia University Press, New York.

Kates, R. W., 1985, The Interaction of Climate and Society, in *Climate Impact Assessment* (R. W. Kates, J. H. Ausubel, and M. Berberian, eds.), Wiley, Chichester, pp. 3–36.

Khrgian, A. K., 1970, *Meteorology: A Historical Survey,* 2nd ed., rev., Israel Program for Scientific Translations, Jerusalem.

Kupperman, K. O., 1982, The Puzzle of the American Climate in the Early Colonial Period, *American Historical Review* 87:1262–1289.

Lamb, H. H., 1965, The Early Medieval Warm Epoch and Its Sequel, *Palaeogeography, Palaeoclimatology, Palaeoecology* 1:13–37.

Lamb, H. H., 1977, *Climate: Present, Past and Future,* Vol. 2, *Climate, History and the Future,* Methuen, London.

Lamb, H. H., 1982, *Climate, History and the Modern World,* Methuen, London.

Landsberg, H. E., 1949, Climatic Trends in the Series of Temperature Observations at New Haven, Connecticut, *Geografiska Annales* 1–2:125–132.

Landsberg, H. E., Yu, C. S., and Huang, L., 1968, Preliminary Reconstruction of a Long Time Series of Climatic Data for the Eastern United States, Technological Note B14-571, Institute for Fluid Dynamics and Applied Mathematics, University of Maryland, College Park, Maryland.

Langley, S. P., 1894, The Meteorological Work of the Smithsonian Institution, *United States Weather Bureau Bulletin* 11:216–220.

Lincoln, T., 1861, Record Book of Weather Reports for Dennysville, Maine, 1816–1861, Ms. at Maine Historical Society, Portland, Maine.

Ludlum, D. M., 1966, *Early American Winters, 1604–1820,* American Meteorological Society, Boston, Massachusetts.

Ludlum, D. M., 1968, *Early American Winters II, 1821–1870,* American Meteorological Society, Boston, Massachusetts.

Ludlum, D. M., 1976, *The Country Journal New England Weather Book,* Houghton Mifflin, Boston, Massachusetts.

Ludlum, D. M., 1983, *The New Jersey Weather Book,* Rutgers University Press, New Brunswick, New Jersey.

Manley, G., 1974, Central England Temperatures: Monthly Means 1659 to 1973, *Quarterly Journal of the Royal Meteorological Society,* 100:389–405.

Manley, G., 1953, The Mean Temperature of Central England, 1698–1952, *Quarterly Journal of the Royal Meteorological Society,* 79:242–261.

Maunder, W. J., 1970, *The Value of the Weather,* Methuen, London.

Milham, W., 1950, *Meteorology in Williams College,* McClelland Press, Williamstown, Massachusetts.

Moody, L., 1852, Thermometrical Records for Portland Observatory, Maine, 1816–1852, 2 Vols., Mss. at Maine Historical Society, Portland, Maine.

Nutting, W., 1863, Meteorological Observations for Randolph, Vt., 1828–1863, Mss. at Vermont Historical Society, Montpelier, Vermont.

Oliver, J. E., 1975, Recollection of Past Weather by the Elderly in Terre Haute, Indiana, *Weatherwise* 28:161–171.

Parry, M. L., 1978, *Climatic Change, Agriculture and Settlement,* Dawson, Folkestone, Kent.

Parry, M. L., and Carter, T. R. (eds.), 1984, *Assessing the Impact of Climatic Change in Cold Regions,* International Institute for Applied Systems Analysis, Larenburg, Austria.

Reiss, N. M., Groveman, B. S., and Scott, C. M., 1980, Seasonal Mean Temperatures for New Brunswick, N.J., *Bulletin of the New Jersey Academy of Science* 1980:1–10.

Rodman, T. R., 1905, Monthly, Annual and Average Temperature and Precipitation at New Bedford, Mass., 1813–1904, *Climate and Crops: New England Section, Annual Summary* 1905:9.

Rothenburg, W. B., 1979, A Price Index for Rural Massachusetts, 1750–1855, *Journal of Economic History* 39:975–1001.

Rumford Foundation, 1839, Meteorological Observations for Boston, 1818–1839, Mss. at Harvard University Libraries, Cambridge, Massachusetts.

Rutgers, G., 1829, Meteorological Observations, 1803–1829, Mss. at Rutgers University Library, New Brunswick, New Jersey.

Schneider, S. H., and Temkin, R. L., 1978, Climatic Changes and Human Affairs, in: *Climatic Change* (J. Gribbin, ed.), Cambridge University Press, Cambridge, pp. 132–155.

Schott, C. A., 1876, Tables, Distributions, Variations of the Atmospheric Temperature in the United States and Some Adjacent Parts of North America, *Smithsonian Contributions to Knowledge,* No. 277.

Schott, C. A., 1881, Tables and Results of the Precipitation in Rain and Snow in the United States and Some Stations in Adjacent Parts of North America, and in Central and South America, *Smithsonian Contributions to Knowledge,* 2d ed., No. 353.

Skeen, C. E., 1981, The Year Without a Summer: A Historical View, *Journal of the Early Republic* 1:51–67.

Smith, D. C., Baron, W. R., Bridges, A. E., TeBrake, J. E., and Borns, H. W., Jr., 1982, Climate Fluctuation and Agricultural Change in Southern and Central Maine, 1765–1880, *Maine Historical Society Quarterly* 21:179–200.

Smith, Jr., F. H., 1934, Some Old-Fashioned Winters in Boston, *Massachusetts Historical Society Proceedings* 65:269–305.

Stommel, H., and Stommel, E., 1983, *Volcano Weather: The Story of 1816, The Year Without a Summer,* Seven Seas Press, Newport, Rhode Island.

Stowell, M. B., 1977, *Early American Almanacs: The Colonial Weekday Bible,* Burt Franklin, New York.

Thompson, K., 1980, Forests and Climate Changes in America: Some Early Views, *Climatic Change* 3:47–64.

Thompson, K., 1981, The Question of Climatic Stability in America before 1900, *Climatic Change* 3:227–241.

Tickell, C., 1977, *Climatic Change and World Affairs,* Center for International Affairs, Harvard University, Cambridge, Massachusetts.

U.S. Army, 1872–1892, *Department of War Annual Reports,* U.S. Army, Washington, D.C.

U.S. Department of Agriculture, 1863–1870, *Annual Reports,* Department of Agriculture, Washington, D.C.

U.S. Patent Office, 1848–1858, *Annual Reports,* Patent Office, Washington, D.C.

U.S. Surgeon General's Office, 1855, *Meteorological Register,* U.S. Government, Washington, D.C.

Walton, G. M., and Ball, D. E., 1976, Agricultural Productivity Change in Eighteenth Century Pennsylvania, *Journal of Economic History* 36:102–125.

Warrick, R. A., and Riebsame, W. E., 1983, Societal Response to CO_2-Induced Climate Change: Opportunities for Research, in: *Social Science Research and Climate Change* (R. S. Chen, E. Boulding, and S. H. Schneider, eds.), D. Reidel, Dordrecht, Holland, pp. 74–102.

Weber, J. H., 1930, The Rainfall of New England: Historical Statement, *Journal of the New England Water Works Association* 44:1482–1485.

Webster, N., 1968, On the Supposed Change in the Temperature of Winter, in: *A Collection of Papers on Political, Literary and Moral Subjects,* Burt Franklin, New York [reprint of 1843 New York edition], pp. 119–162.

Wigley, T. M. L., Huckstep, N. J., Ogilvie, A. E. J., Farmer, G., Mortimer, R., and Ingram, M. J., 1985, Historical Climate Impact Assessments, in: *Climate Impact Assessment* (R. W. Kates, J. H. Ausubel, and M. Berberian, eds.), Wiley, Chichester, pp. 529–564.

Williams, S., 1790, Change of Climate in North America, in Williams Papers, Mss. at University of Vermont Special Collections Library, Burlington, Vermont.

Chapter 3

"Where's the Salmon?"

A Reevaluation of the Role of Anadromous Fisheries in Aboriginal New England

CATHERINE C. CARLSON

1. INTRODUCTION

Prehistorians have long been fascinated by anadromous fish runs. They represent a very dynamic type of resource that focuses prehistoric behavior to certain places on the landscape at particular times. Prehistoric archaeologists have seen this resource as a critical element in developing subsistence and settlement models for prehistoric populations (e.g., Binford 1964:332–333; Jochim 1979) and, in some cases, as the catalyst for social transformation (e.g., Ames 1981; Schaulk 1977). Archaeologists have been faced with the apparently perplexing problem of understanding why the majority of coastal hunter–gatherers generally appear to be more sedentary, more resource-intensive, and less egalitarian in social structure than inland hunters and gatherers.

On the Northwest Coast of North America, anadromous salmon figured importantly in both the recent and prehistoric pasts. Fladmark (1975) first suggested that the evolution of the classic Northwest Coast culture required the stabilization of the salmon streams and the subsequent climax in productivity of the salmon resources, which probably did not occur until after 5000 B.P. He proposed the idea that a middle Holocene climax productivity of salmon was the key to the existence of the Northwest Coast culture pattern, as the result of the

CATHERINE C. CARLSON • Department of Anthropology, University of Massachusetts, Amherst, Massachusetts 01003.

high energy input from this resource, which allowed for "massive cultural energy expenditures" (Fladmark 1975:94–95).

Other researchers have focused on the social implications of anadromous salmon as a major resource. For example, Schaulk (1977) developed a model for the development of complex society on the Northwest Coast that outlined how processing and storage of large amounts of salmon in a short period of time could provide the impetus for efficient labor organization and, in turn, for increasing social differentiation and specialization. Similarly, in a series of papers Ames (1979a, 1979b, 1981) argued that the specialization on salmon fishing on the Northwest Coast required coordination of task groups necessary for harvesting and processing fish, placing high value on "efficient leadership," thereby initiating a ranked social organization (Ames 1981:798). Ames further argued that increasing specialization on the salmon resource would lead to increasing sedentism and population growth, leading to competition over resources, and consequently amplifying the development of "Big Men" and of ranked social organization (Ames 1981:799).

A review of the archaeological literature for New England reveals numerous statements regarding the importance of anadromous fish, including salmon (*Salmo salar*), to prehistoric native inhabitants. Popular literature statements, such as those of Netboy (1974:521), are typical:

> New England salmon provided food for the Indians, who taught the colonists how to fish for them and prepare them in their fashion. . . The fishes, which were sometimes so thick in the rivers that they overturned small boats, were probably as vital to the aborigines as the wild turkey that has received so much publicity. To an extraordinary extent salmon served the Indians as the staple of their diet. . .

A variety of archaeological models and hypotheses have been proposed relating anadromous fish exploitation, in general, to settlement patterns, site function, and seasonality (e.g., R. Barber 1980; Bourque 1975; Brennan 1981; Brumbach 1986; Cavallo 1984; Dincauze 1976; Foster *et al.* 1982; Kenyon and McDowell 1983; Moore and Root 1979; Sanger *et al.* 1977; Spiess *et al.* 1983; Yesner *et al.* 1983). For example, Bourque (1975:42) stated that, "If we assume that fish resources were paramount in determining suitable site locations, that emphasis would have been placed upon large runs of anadromous species, the elements of a seasonal transhumance are at hand." Although there are a number of other anadromous species in New England waters, including shad, alewives, and sturgeon, it is frequently salmon that is implicitly—or explicitly—assumed to be of significance when reference is made to anadromous fish exploitation, or when the comparability of other Northeastern anadromous species to Northwest Coast utilization of salmon is implied. This is probably due to the influence of Northwest Coast ethnographies, which are often used as analogs for New England where comparable ethnographic records do not exist. For example, Spiess *et al.* (1983:98–104) state that:

> The April/May/June period coincides with the heaviest anadromous fish runs in Maine rivers (principally shad, then alewife, then Atlantic salmon). . . The numerous estuarine and interior camps seem to be oriented primarily to anadromous fishing. . . In comparing cultural development in Maine and the Northwest Coast some general similarities and a rather large number of specific differences appear to be present. General similarities may include economies focused on marine resources and anadromous fish. . . . Although early depletion of salmon stocks in Maine makes reconstruction of pre-contact conditions difficult, Maine anadromous fish runs (salmon, alewifes, shad, sturgeon) are primarily spring and early summer phenomena.

Likewise, Yesner *et al.* note:

> Whether land-locked or sea-run, salmon were obviously an important resource to Middle Archaic populations [in Maine]. This is reflected in archaeological site distributions in the Sebago Lake region. . . The fact that all known Middle Archaic sites (other than those at the headwaters of the Presumpscot River) are located within a few kilometers of streams or lake outlets flowing into Sebago Lake therefore suggests that salmon fishing must have been critical to the subsistence of these populations, or at least their decision for locating their settlements. (Yesner *et al.* 1983:316)

Cole and Mulholland (1983:4), in presenting historic and prehistoric patterns of water use in the Connecticut River valley, outline a model of Archaic hunting and gathering, c. 8000–1000 B.C., that consists of a "strong dependence upon salmon and shad fishing".

The question addressed in this paper is whether these models are appropriate for aboriginal New England. Certainly, a number of comparisons have been made between the Northwest Coast and the Northeast Coast in terms of cultural development and aboriginal utilization of fish resources (e.g., Nash 1983; Rostlund 1952), although Spiess *et al.* (1983:105) concede that "we lack data to compare the relative numbers of available salmon in Maine and the Northwest Coast rivers during the pre-Contact period." Suggestions have been made that the cultural groups on the Northeast Coast may have fostered an equally complex chiefdom level of social organization (Miller 1983), which the fragmentary ethnohistorical accounts of the seventeenth and eighteenth centuries had failed to record before the demise of the traditional aboriginal way of life. More importantly, models for anadromous salmon and other fish utilization have been used in New England to assign functionally and seasonally specific interpretations to various site locales based on the model that people were exhibiting a coastal–inland transhumant settlement pattern; that is, that groups were moving from the coast to the interior at different seasons of the year to exploit anadromous fish runs at special locales. For example, Sanger *et al.* (1977:469) suggested that for northern New England, "The research at Hirundo has contributed towards the elaboration of a theoretical model that sees the interior as an extension of the coastal zone in that the carrying capacity of the inland areas for hominids is largely a function of fish originating in the marine environment."

The most immediate problem with the assumption of anadromous salmon utilization and its importance to prehistoric New England is that the archaeological data base thus far does not reflect the heavy reliance on anadromous salmon described in the literature, while at the same time suggests that fishing for other, nonanadromous species was of considerable importance. The purpose of this paper, therefore, is to review the archaeological evidence of fishing in New England, to reconstruct the basic structure of the salmon resource during pre-Contact times through historical accounts, and to discuss the probable relationship of salmon to prehistoric aboriginal populations. The ultimate goal here is to explain why the evidence for aboriginal salmon exploitation in New England is almost totally nonexistent, while questioning the entrenched notion that salmon in New England was as important as salmon on the Northwest Coast. A discussion of "anadromy" in general, and of the life cycle patterns of salmon and other anadromous fish in New England in relation to those of the Pacific salmon, illustrates the profound differences in the species, and suggests that models based on Northwest Coast salmon exploitation are inappropriate for the Northeast on this basis as well.

By critically examining one of the numerous myths that have become ingrained in the environmental archaeological literature, that of the relationship between prehistoric populations and salmon in New England, a more precise understanding of the mechanisms and parameters of prehistoric culture and the aquatic environment may result, and may reveal otherwise obscured aspects of the relationship between fish resources and aboriginal social organization. When one reviews the data base for fisheries exploitation, it becomes apparent that alternative explanations are necessary in order to understand the prehistoric record. Little attention has been paid to the heuristic utility of examining prehistoric social organization which has developed due to a focus on the sea as a subsistence source; hypotheses derived from models of maritime social organization among hunter–gatherers in New England based upon marine fish exploitation must be tested (Carlson 1986a). The way in which commitment to marine fishing practices affected social organization, sedentism, and resource intensification is discussed in the concluding section as an alternative model for aboriginal New England.

2. SEVENTEENTH AND EIGHTEENTH CENTURY HISTORICAL ACCOUNTS OF SALMON IN NEW ENGLAND

When Europeans first arrived in northeastern North America, the Atlantic salmon (*Salmo salar*) was reportedly found in every river not barred by impassable falls, from northeastern Labrador to the Housatonic River emptying into Long Island Sound (Figure 1). In New England, salmon runs were reported for the St. John, St. Croix, Dennys, Orange, East Machias, Pleasant, Narraguagus, Union, Penobscot, St. George, Medomak, Sheepscot, Androscoggin, Kennebec, Royal, Presumpscot, Saco, Mausam, Piscataqua, Merrimack, Connecticut, and Housatonic Rivers (Atkins 1887; Bigelow and Schroeder 1953:125). The building of dams and pollution along these rivers reportedly caused the almost total extinction of the Atlantic salmon by the early 1800s. For example, James Birket wrote during his voyage to North America in 1750–1751:

> Formerly this river [Piscataqua] was well stored with Salmon which they took in plenty but of Late they have quite forsaken this river, Occasion'd it's believed, from the number of Saw mills on the different branches of P. [Piscataqua] river, the weirs of which running crop the same And the sawdust from the pine, which it's believed is very ungratefull to the fish. (Birket 1916:7)

Prior to that time, salmon was described as an abundant species, although data on actual catch statistics are unavailable for the early Contact period. It is difficult to make resource biomass comparisons between Atlantic salmon and Pacific salmon in terms of their availability and abundance during the prehistoric period, because there is no numerical record. A review of the historical accounts and anecdotal material relating to Atlantic salmon, however, can provide a baseline for discussion on what the salmon resource may have provided to pre-Contact Native Americans, and what its importance and potential might have been in influencing aboriginal subsistence and settlement patterns. This review emphasizes seventeenth and eighteenth century accounts to provide the most pristine descriptions possible of the fisheries resources before extensive dam construction, and hence is not exhaustive in terms of later nineteenth century ethnohistorical accounts. As actual fisheries statistics on numbers of salmon are almost nonexistent, it is particularly useful to compare the statements regarding salmon abundance with those of other fish species in order to achieve an impression of *relative* abundance.

The implications of the data contained in these accounts for the regional anadromous fish resource, their applicability to the issues of prehistoric resource exploitation, and the biases that may be contained in historical accounts are addressed in the discussion section below.

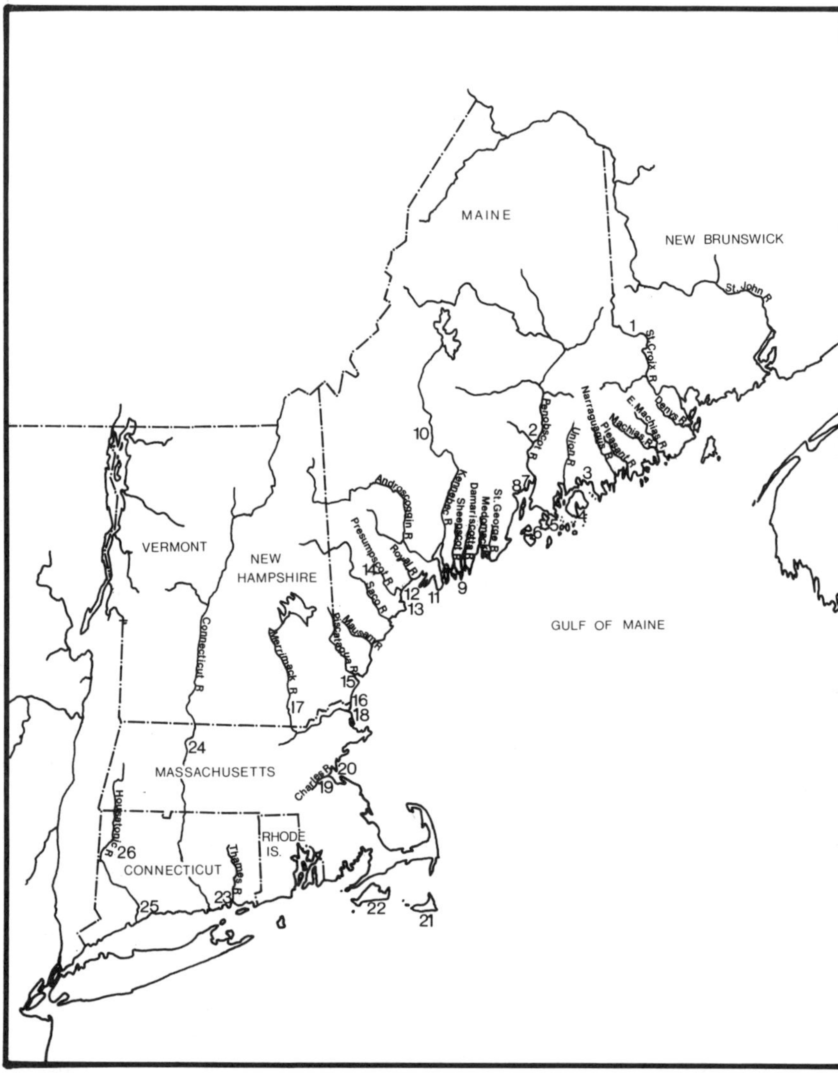

Figure 1. Major river drainages and archaeological sites mentioned in the text. 1. Mud Lake Stream site. 2. Hirundo/Young sites. 3. Frazer Point site. 4. Fernald Point site. 5. Goddard site. 6. Turner Farm site. 7. Sears Island site. 8. Kidder Point site. 9. Boothbay, Sheepscot, and Damariscotta sites. 10. Hodgdon site. 11. Basin site. 12. Great Diamond Island site. 13. Mosher Island site. 14. Linquist site. 15. Wadleigh Falls site. 16. Seabrook Marsh site. 17. Neville/Smyth sites. 18. Buswell and Wheeler's sites. 19. Boylston Street fish weir. 20. Calf Island and Long Island sites. 21. Nantucket Island sites. 22. Martha's Vinyard sites. 23. Fort Shantok site. 24. Riverside/Gill sites. 25. Seaside Indian Village site. 26. Woodruff Rockshelter.

2.1. Historical Accounts

One of the first mentions of salmon in the historical literature is by Champlain in 1604; he states that the harbors in Passamaquoddy Bay, "abound with fish in their season, such as codfish, salmon, bass, herring, halibut, and others in great numbers" (Champlain 1922:270). In this instance, salmon is listed second after the cod, and may imply a lesser importance, which would be expected, given the marine environment to which Champlain is referring. In the bay in front of Plymouth harbor, Champlain describes Indians in canoes who "approached us on their way back from fishing for cod and other fish, which are plentiful thereabouts" (Champlain 1922:343).

Captain John Smith's first voyage to the New England coast in 1614 describes the abundance of fish from Monhegan Island on the Maine coast to Cape Cod:

> In March, April, May and half of June, here is cod in abundance; in May, June, July and August, mullet, and sturgeon, whose roes do make caviar and puttargo. Herring, if any desire them, I have taken many out of the bellies of cods, some in nets; but the savages compare their store in the sea to the hairs of their heads: and surely there are an incredible abundance upon this coast. In the end of August, September, October, and November, you have cod again, to make cor fish . . . (Smith 1884:198). Much salmon some have found up the Rivers as they have passed . . . (Smith 1884:199). Scarce any place, but cod, cusk, halibut, mackerel, skate or such like, a man may take with a hook or line what he will. And in diverse sandy bays, a man may draw with a net great store of mullets, basses, and diverse other sorts of such excellent fish, as many as his net can draw on shore. no river where there is not plenty of sturgeon, or salmon, or both. (Smith 1884:207–208)

The Reverend Francis Higgeson wrote in 1629, upon arriving to the Massachusetts Bay Colony at Salem, that "The aboundance of sea-fish are almost beyond believing." He describes the great abundance of mackerel, cod, bass, skate, thornback, herring, turbut, sturgeon, cuskes, haddock, mullets, eels, and shellfish. Regarding bass, he states that it is "altogether as good as our fresh sammon" (Higgeson 1792:119–120), which implies that salmon were not available there. Thomas Morton, also an early Massachusetts colonist, wrote on the abundance, variety, and quality of food fishes. He describes in length the multitude of bass:

> There are such multitudes that I have seen stopped into the [Neponset] river . . . at one tide so many as will load a ship of 100 tons. Other places have greater quantities insomuch as wagers have been layed that one should not throw a stone in the water but that he should hit a fish. I, myself, at the turning of the tide, have seen such multitude pass out of a pound that it seemed to me that one might go over their backs dry-shod. (Morton 1637:222)

Similarly, he writes at length about the great abundance of mackerel: "so many thousands have shot themselves ashore with the surf of the sea that whole hogsheads have been taken up on the sands"; of sturgeon, "Every man in New England may catch what he will. There are multitudes of them"; of smelts, "there is such abundance that the savages take them up rivers with baskets, like sieves"; of shad, "taken in such multitudes in every river that has a pond at the end that the inhabitants dung their ground with them"; and about other fish, such as the halibut, herring, and eels. Interestingly, he makes only one short statement about salmon, that "there is great abundance, and these may be allowed for a commodity" (Morton 1637:223–228). William Wood wrote in 1634 that "The salmon is as good as it is in England and in great plenty," but, like Morton, writes considerably more about the abundance and quality of bass, eels, sturgeon, cod, mackerel, and other fish (Wood 1634:37).

An account by an Englishman, John Josselyn, who made two trips to the Maine/Massachusetts coast in 1638 and 1663, states during the second voyage that, "The salmon likewise is a sea-fish, but, as the bass, comes into rivers to spawn," and that, "There are another sort of salmon [that] frequent in those parts called white salmon" (Josselyn 1675:83–85). Josselyn does not give any idea of the abundance of salmon in this account, even though other types of fish are described in great quantities. It is also interesting that Josselyn refers to a fish called the "white salmon," which undoubtedly is the anadromous shad. In his account of a voyage of 1605, Rozier (1906:376) also used the term "white salmon" when he stated that the Indians showed him "another kind of great scaly fish, broiled on the coals, much like white salmon, which the Frenchmen call aloza"—aloza is French for shad. This suggests that there may be ambiguity in the use of the term *salmon* elsewhere in the early accounts, i.e., certain references may be to the shad rather than the Atlantic salmon even when the name salmon is being used.

In a general history of New England written in 1682 by Hubbard, it is stated that, "Up higher, at the falls of the great rivers, they [the Indians] used to take salmon, shad, alewives, that use in great quantities, more than cart loads, in the spring, to pass up into the fresh water ponds and lakes, therein to spawne, of all which they, with there wares, used to take great store for their use." (Hubbard 1815:31). This history, however, does not appear to be a first hand account.

James Birket described the Merrimack River during his 1750–1751 voyage as "the most remarkable for salmon of any in those parts" (Birket 1916:14). Peters (1781) wrote of the Connecticut River that "The shad, bass, and salmon more than half support the province," (Peters 1877:189) again ranking shad and bass before salmon. In reference to the Connecticut and Merrimack rivers, Douglass wrote in 1749–1751 that:

> Salmon and shad come up this river [Connecticut] to spawn, but in smaller quantities and later, and continue a shorter time [about three weeks in May] than in Merrimack River (Douglass 1972 vol 1:461) . . . Salmon are a high-

> latitute fish. They are not to be found south of New England; the farther south, the later they set in and continue a shorter time. For instance, in Connecticut River they set in the beginning of April, to spawn, and lie in the deep cold brooks until September and October, then silently (so as not to be observed) and with despatch, they return to the sea. (Douglass 1972 vol 2:212)

This account implies that the timing of the salmon runs is variable, depending on latitude (i.e., water temperature), that the runs are short in duration, and that the salmon are somewhat of an elusive fish, perhaps suggesting difficulty in capture.

In 1749, Douglass also wrote that,

> Salmon are plenty in all the British North America rivers from Newfoundland to about N. lat 41 D. They set in to Massachusetts Bay about the middle of April; they do not choose warm weather, therefore do not continue there long after having spawned; farther north they continue many months. This salmon is not of a good quality and is not so good quantity and is not so good for a market as the salmon of Great Britain and Ireland. . . Many fish go up the rivers and into ponds, earlier or later in the spring to spawn, viz. salmon, shad, alewives, tom-cod, smelts, etc. (Douglass 1972 vol 1:304)

Douglass' account is important because he states that salmon was not of such a great quality and quantity as others, such as Wood (1634), had implied. The quality of the fish may refer to the fact that the meat of spawning salmon is not as good as those caught at sea before spawning. His reference to other types of anadromous fish is also important.

Reference to the Indians fishing up river in Maine for salmon and "sundry other fish" is found in a letter from Thomas Westbrook to William Dummer, May 21, 1725:

> We judge that the greatest part of the enemy [Indians] are drawn some distance back on the great rivers, this being their time to fish for salmon and sundry other fish up the fresh rivers on which the Indians yearly make a fishing voyage. Our winter scouts discovered sundry of their fishing places on Saco, Pesomscott and Amuscoggin Rivers where they made large quantities last summer. (Baxter 1907)

In the 1785 diary of Matthew Patten of Bedford, New Hampshire, fishing at Amoskeag Falls on the Merrimack is described, an account that suggests that shad and eels were more abundant than salmon:

> June 7, 1785: I went to the falls toward the middle of the day and I fished at the upper setting place. I got a salmon of nine pounds, and I sold it for one-half a dollar. I changed nine shad for half a bushel of salt and I sold two shad for 8 pence. I had three shad left. Alexander got six shad on the Island. I fished eels the night following and got forty-nine, three of which I gave to Eliphant Emery and Mr. Wyman. On the 8th I got five shad and a nine pound salmon. Alexander got four shad and a twenty-pound salmon on the

> Island. I came home and brought fifteen shad, forty-seven eels, and Alexander's salmon and mine to Major Goffes' field in the canoe. (Gordon and Marston 1940:4)

In a history of Newbury, Massachusetts, Coffin (1845:189) cites references to salmon, which imply that they were highly valued as food, but a rare commodity:

> 1718. The town (Newbury) granted to Moses Chase . . . eighty rods of the flats above Holt's rocks to fish on, on condition they pay as an acknowledgement to the town two salmon per year; one to Mr. Toppan, the other to Mr. Tufts, *if they catch them* [emphasis his].

In a letter to Anthony Morse from H. Whitton, Boston, March 21, 1728, the value of salmon at this time is shown:

> 1728. Mr. Anthony Morse: This is to desire the favor of you to get me one, two, or three or more of the first salmon *if it can be had this year*. I am willing to give a good price and a great price rather than not have it . . . H. Whitton, Boston [emphasis mine]. (Coffin 1845:189)

A British visitor to Boston in 1740, Mr. Bennett, compared the cost and abundance of fresh fish, and from this note, salmon is evidently at least six times as expensive as cod, again suggesting its lesser abundance:

> They sell a fine fresh cod, that will weigh a dozen pounds or more, just taken out of the sea, which are generally alive, for about two-pence sterling. They have smelts, too, which they sell as cheap as sprats are in London. Salmon they have, too, in great plenty, which is as fine as any I ever eat of anywhere in my life; and those they will sell for about a shilling a-piece, which will weigh fourteen or fifteen pounds. They have flounders and plaice too, and eels, and likewise mackerel in their season, and several other sorts of fish not known in England,—all of which are good and cheap. (Winsor 1881:464)

In the *Niles National Register,* April 25, 1840, it was written that "The first salmon caught in the Kennebec River this year was sold in Boston on Friday last for forty-two dollars" (Craigie and Hulbart 1943 vol. 4:2007), implying again that salmon was a rare and costly commodity at that time.

3. ARCHAEOLOGICAL EVIDENCE OF PREHISTORIC FISHING

Definitive evidence for prehistoric fish exploitation is derived from the faunal record in sites where bone is preserved. On the Northwest Coast, faunal evidence for anadromous salmon exploitation is found in archaeological sites that are located in coastal, estuarine, and riverine settings (Carlson 1982a). The largest proportion of these sites is from the coastal and estuarine zones because shell middens are ideal for bone preservation, while interior upriver sites, such as the Dalles on the Columbia River or Yale on the Fraser River, provide evidence of strictly riverine exploitation of salmon. The estuaries of salmon spawning rivers have often been regarded as optimal settings for anadromous fish exploitation due to the biotic richness of estuaries in general and to the scaling of rivers and streams within a drainage, which logistically suggests that the densest concentrations of fish runs are to be located at the mouths of rivers, dispersing into side streams as they travel further up the drainages (Baumhoff 1963; Moore and Root 1979; Perlman 1980; Schaulk 1977; Sneed 1971).

If a comparable situation to the Northwest Coast for aboriginal salmon exploitation existed in New England, then one would also expect to find evidence there of salmon remains in sites located in the coastal, estuarine, and riverine areas. Faunal remains have been recovered in a number of New England sites from all three of these environmental zones. These data are reviewed for evidence of fishing and species exploitation. Information pertinent to this discussion from each site is presented below. Table 1 lists the common and scientific names of all fish species identified from New England archaeological sites that are referred to in the text by their common name.

3.1. Sites in Coastal Areas

Most of the sites from coastal areas in New England that have fish faunal information available are located on the Maine coast, with the exception of two sites located in Boston Harbor, five sites on Martha's Vinyard, six sites on Nantucket Island, and one in coastal Connecticut. Other excavated coastal sites, such as those on Cape Cod (McManamon 1984; Fred Dunford, personal communication 1985), have not had the fish remains analyzed and reported to date. From the following presentation, it is evident that salmon is virtually nonexistent in sites located in the coastal zones of New England.

3.1.1. Goddard Site

The Goddard site, located on the central Maine coast east of Penobscot Bay, has fish remains that are predominantly sturgeon, with some sculpin, flounder, and bluefish identified (Bourque and Cox 1981:18).

3.1.2. Fernald Point Site

The Fernald Point shell midden is located on Mount Desert Island, Maine, and the fish remains comprise 78% of the faunal sample. Longhorn sculpin, winter flounder, and cod are the predominant species; no salmon was identified (Carlson 1980; Sanger 1980, 1982).

Table 1. Common and Scientific Names of Fish Identified from Prehistoric Sites in New England

Common name	Scientific name
Marine species	
Alewife	*Alosa pseudoharengus*[a]
Atlantic cod	*Gadus morhua*
Atlantic salmon	*Salmo salar*[a]
Bluefish	*Pomatomus saltatrix*
Cunner	*Tautogolabrus adspersus*
Cusk	*Brosmium brosme*
Eelpout	*Macrozoarces americanus*
Goosefish	*Lophius americanus*
Haddock	*Melanogrammus aeglefinus*
Hake	*Urophycis* sp.
Herring	*Clupea harengus*
Longhorn sculpin	*Myoxocephalus octodecemspinosus*
Mackerel	*Scomber scombrus*
Northern barracuda	*Sphyraena borealis*
Pollock	*Pollachius virens*
Ray	*Dasyatis americana*
Sand shark	*Carcharias taurus*
Sea catfish	*Galeichthys felis*
Sea robin	*Hemitripterus americanus*
Scup	*Stenotomus chrysops*
Shad	*Alosa sapidissima*[a]
Shorthorn sculpin	*Myoxocephalus scorpius*
Smooth flounder	*Liopsetta putnami*
Spiney dogfish	*Squalus acanthias*
Striped bass	*Morone saxatilis*
Sturgeon	*Acipenser* sp.[a]
Swordfish	*Xiphius gladius*
Tautog	*Tautoga onitis*
Tomcod	*Microgadus tomcod*[a]
Weakfish	*Cynoscion regalis*
White shark	*Carcharodon carcharias*
Winter flounder	*Pseudopleuronectes americanus*
Wolffish	*Anarhichas* sp.
Freshwater species	
Bluegill/Sunfish	*Percimorph* sp.
Bullhead catfish	*Ictalurus nebulosus*
Catfish/Bullhead	*Dermethmoid* sp.
Minnow/Carp	*Cyprinidae* sp.
Perch	*Percidae* sp.

[a]Anadromous.

3.1.3. Frazer Point Site

The Frazer Point shell midden is located in Frenchman Bay, Maine, and 64% of the faunal assemblage is fish bone. Of the 1,668 bone fragments identifiable to species, 91% was made up of winter flounder, longhorn sculpin, and Atlantic cod, with the remaining 9% comprised of wolffish, sturgeon, cusk, striped bass, cunner, and mackerel. Two perforated fish vertebrae that *may* be salmon (*Salmo salar*) were also recovered (Carlson 1981; Sanger 1982).

3.1.4. Turner Farm Site

The Turner Farm site, located in Penobscot Bay, Maine, has had the fish remains reported in Morse (1975), Bourque (1975), and Spiess *et al.* (1983). No salmon has been identified from the site. The fish remains identified consist of swordfish, cod, sturgeon, tomcod, sculpin, striped bass, flounder, mackerel, and goosefish.

3.1.5. Sears Island Site

The Sears Island site is a shell midden located in Penboscot Bay, Maine. Only 12 identifiable elements of fish bone were recovered from the site, and these were identified as sturgeon, flounder, cod, and sculpin (Spiess and Hedden 1983:177).

3.1.6. Kidder Point Site

The Kidder Point site is a shell midden located in Penobscot Bay in close proximity to the Sears Island site. Fish is a minor component in the assemblage, of which there were 38 identifiable fragments. Of these, one is a salmon vertebra; the remainder consists of cod, cunner, flounder, and possibly swordfish (Spiess and Heddon 1983).

3.1.7. Boothbay Harbor Region Sites

In the marine zone of the Boothbay region of Maine (i.e., in Sheepscot, Linekin, and Johns Bays), 12 shell midden sites were analyzed for fish material. From these sites, 6,621 fish bone fragments were identifiable to species. Atlantic cod represents the greatest proportion, although tomcod, haddock, winter flounder, smooth flounder, longhorn sculpin, shorthorn sculpin, sea robin, alewife, striped bass, sturgeon, dogfish, eelpout, and wolffish were also identified. No salmon was identified from any of these sites (Carlson 1986a).

3.1.8. Basin Site

This Woodland period shell midden is located within a few feet of the high tide line at Phippsburg, Maine. Sturgeon and striped bass have been identified from the faunal remains (Waters 1965).

3.1.9. Great Diamond Island Site

The Great Diamond Island site is a shell midden located in Casco Bay, Maine. From a sample of 950 fish bone fragments, 83% were Atlantic cod, the remainder being pollock, sculpin, winter flounder, sturgeon, alewife, swordfish, and cunner (Yesner 1984).

3.1.10. Mosher Island Site

The Mosher Island site is a shell midden located in Casco Bay, Maine. This site has a similar assemblage to Great Diamond Island in that cod, pollock, sculpin, winter flounder, sturgeon, alewife, swordfish and cunner were identified. In addition, tomcod, striped bass, and haddock were also identified (Yesner 1984).

3.1.11. Calf Island Site

The Calf Island site is located in Boston Harbor, and from this site Atlantic cod was the predominant species, and tautog, cusk, spiney dogfish, goosefish, cunner, and bass were also identified (Luedtke 1980).

3.1.12. Long Island Site

Fish bone collected by Barbara Luedtke from this site in Boston Harbor, identified by Carlson (1986b), consisted of cod and tomcod.

3.1.13. Hornblower II Site

This shell midden site is located on the southwest point of the island of Martha's Vineyard on the north shore of Squibnocket Pond. It is separated from open access to the sea only by a narrow sand bar and dune, which did not always form a barrier in the past. It was postulated that at the time of occupation the site had access to annual spring runs of herring or alewives and other fish (Ritchie 1969:10). Of 1,093 elements of fish bone recovered from the site, scup was identified as the most abundant species, followed by sturgeon, tautog, goosefish, cod, striped bass, and bluefish (Ritchie 1969:41–44).

3.1.14. Pratt Site

This site is a shell midden located on a small pond 0.4 km inland from Vineyard Haven Harbor on the north shore of Martha's Vineyard. At this site there were very few fish remains; the minimum number of three individuals identified represents striped bass, tautog, and sturgeon (Ritchie 1969:81).

3.1.15. Cunningham Site

This shell midden site is located on Lagoon Pond in Vineyard Haven Harbor, Martha's Vineyard. Of a minimum number of 33 individual fish identified, 27 were scup, 4 were tautog, and 2 were striped bass (Ritchie 1969:115).

3.1.16. Vincent Site

This shell midden site is located on a small, brackish swamp adjoining Lagoon Pond in Vineyard Haven, Martha's Vineyard. Of greater than 270 identifiable fragments of fish bone recovered, 200 fragments were scup, 67 were striped bass, and 4 were sturgeon (Ritchie 1969:155).

3.1.17. Peterson Site

This shell midden is located on Squibnocket Pond, Martha's Vineyard, near the Hornblower II site. A minimum number of 21 fish were identified, of which there were six scup, six striped bass, five goosefish, two sea robin, one sturgeon and one tautog (Ritchie 1969:190).

3.1.18. Nantucket Island Sites

Six shell midden sites on Nantucket Island off the coast of Massachusetts have had fish remains reported. Collectively, the fish species identified are sturgeon, sculpin, sand shark, sea robin, ray, cod and spiney dogfish (Waters 1965; Little 1985). Recent analysis of the Quidnet site also includes the identification of bluefish (Carlson 1987).

3.1.19. Seaside Indian Village Site

This shell midden, located at the west entrance to Milford Harbor on the Connecticut coast, at the mouths of the Wopowaug and Indian Rivers, contains abundant faunal remains, which were minimally reported on by Coffin (1963). Of the unquantified species Coffin lists (1963:14), the only fish identified were sturgeon and shad.

3.2. Sites in Estuarine Areas

Despite the fact that estuaries have been singled out as being particularly productive environments for resource exploitation, very few sites to date have been excavated and analysed for faunal material in New England estuaries. What few have been done show no evidence whatsoever of prehistoric salmon exploitation.

3.2.1. Boothbay Region: Sheepscot and Damariscotta Estuarine Sites

In the Sheepscot and Damariscotta river estuaries of the Boothbay region of Maine, nine sites were tested and analyzed for faunal remains. Five sites were analyzed in the Damariscotta estuary and four in the Sheepscot estuary, and these provided a combined faunal sample of fish remains of 5,978 identifiable elements. The species identified from these sites were, in order of abundance, Atlantic cod, tomcod, flounder, sturgeon, longhorn sculpin, sea robin, dogfish, haddock, alewife, and striped bass (Carlson 1986a).

3.2.2. Seabrook Marsh Site

This site is located in the Seabrook/Hampton estuary about 1.6 km inland of the New Hampshire coast. Faunal preservation was good, providing numerous fish remains that included cod, cunner, winter flounder, spiney dogfish, swordfish, haddock, hake, shad, striped bass, wolffish, herring, sturgeon, tomcod, weakfish, bluefish, and bullhead catfish (Robinson 1985).

3.2.3. Buswell Site

The Buswell site is located on the upper reaches of the Merrimack River estuary. Sturgeon and two unidentified fish species were recovered from this site. This site has been interpreted as representative of the pattern of anadromous fish exploitation during spring spawning runs. This is suggested on the basis of faunal remains of sturgeon, a possible fish weir adjacent to the site, and the environmental location of site on the upper reaches of the estuary (Barber 1980).

3.2.4. Wheeler's Site

The Wheeler's site is located on the Merrimack River estuary, and from a sample of 3,227 fish bone fragments, Atlantic cod was identified as predominant, with lesser amounts of haddock, pollock, sturgeon, and possible herring (Barber 1982).

3.2.5. Boylston Street Fish Weir

The Boylston Street fish weir is located under several meters of fill in the city of Boston, in an area that was once part of the middle reaches of the Charles River estuary. Detailed analyses of the faunal material from the sediments surrounding the fish weir stakes were undertaken; no fish remains were found (Barghoorn *et al.* 1949; Johnson 1942). Additional excavation and faunal analysis are presently being undertaken, and while a very small number of fish remains have been recovered, they have not yet been received for analysis by the author, who is a member of the project research team.

3.2.6. Fort Shantok Site

Fort Shantok is a post contact site occupied between 1600 and 1750, and located 24 km inland from the Connecticut coast on the Thames River estuary. At this site white shark, sea sturgeon, northern barracuda, bluefish, striped bass, tautog and sea catfish were identified (Waters 1965).

3.3. Sites in Interior Riverine Areas

Several sites located up river drainages have produced assemblages of faunal remains that have been analyzed and reported on, despite the fact that bone does not preserve well in inland sites due to the acidic soils of New England. Fish remains have been recovered at all of the sites reported below, although the evidence for salmon utilization is almost totally absent.

3.3.1. Mud Lake Stream Site

This site is located on Spednik Lake, within the Chiputneticook–St. Croix drainage between New Brunswick and Maine. Fish bones from the site have been identified as alewife and/or shad (Deal 1986).

3.3.2. Hirundo and Young Sites

The Hirundo and Young sites are situated directly across from each other on Pushaw Stream, a tributary of the Penobscot River, about 20 km north of Bangor, Maine (Borstel 1982; Sanger *et al.* 1977). Originally it was hypothesized that this site locality "coincides closely with a stretch of rapids, the only such feature on the stream. It would appear, therefore, that the sites were situated for the purpose of exploiting the rapids for fishing, especially anadromous species" (Sanger *et al.* 1977:463). The more than 30 identifiable fish vertebrae recovered from the Young site were identified to the *Percidae* (perch) family (Borstel 1982). Calcined bone fragments were also recovered from the Hirundo site, and although fish remains were not abundant, some anadromous sturgeon was identified (Knight 1985).

3.3.3. Hodgdon Site

The Hodgdon site is an interior riverine site on the Kennebec River, Maine, where "an abundant faunal sample" was preserved, but included no fish. Despite this absence, it is implied that the site is a fishing station because it is suggested that "anadromous species (salmon, alewives) could have run this far upriver despite very severe falls", and the petroglyphs associated with the site are suggested to have been drawn by fishermen with "time on their hands" (Lahti *et al.* 1981)—perhaps because the fishing was so poor?

3.3.4. Linquist Site

The Linquist site is located in the Presumpscot drainage on Sebago Lake, Maine. Yesner *et al.* (1983:329) state that the "data from the Sebago Lake region of Southwestern Maine suggest the intensive development of an aquatically-based subsistence regime beginning in Middle Archaic times, focused on anadromous fish such as salmon, shad, and alewives." Faunal data from the Linquist site are derived from a sample of 162 identifiable calcined bone fragments, of which only 1.8% was fish; within this sample, "only one [bone] appears to be a salmon vertebra" (Yesner *et al.* 1983:327). It is important to note that this one bone *is not even positively identified as Atlantic salmon,* i.e., it could be trout or char. The remainder of the faunal sample consisted of 58.6% turtle, 21% deer, 9.3% small mammal, 6.2% reptile/amphibian, and 3.1% bird.

3.3.5. Wadleigh Falls Site

The Wadleigh Falls site is located below the upper falls of the Lamprey River approximately 8 km inland from Great Bay on the New Hampshire coast. The faunal remains from the site have been identified by Arthur Spiess; there are eight identifiable fish bones, all representing shad, from a total of approximately 150 identifiable bone fragments (Maymon and Bolian 1985).

3.3.6. Neville and Smyth Sites

The Neville and Smyth sites are located on the Merrimack River at Amoskeag Falls, on terraces above what is considered a "classic" fishing locale. At the

Neville site (Dincauze 1976:97), no faunal remains were recovered, although mercury measured in the soil is suggested to be the result of anadromous fish, such as salmon or shad. As with Neville, the Smyth site is predicted to have functioned, in part, as a station for the exploitation of anadromous fish runs (Foster *et al.* 1982). Of the 1,383 fragments of calcined bone recovered from the Smyth site, only two were identifiable fish vertebrae, the other identifiable elements being other classes of fauna. Neither of these fish vertebra were identified as salmon or any other anadromous species (Carlson 1982b).

3.3.7. Riverside/Gill Sites

The Riverside Archaeological District is located on the Connecticut River at Turner's Falls, Gill, Massachusetts. From the WMECO site, Thomas (1980) reported on shad remains along with other species of vertebrates. At the Walnut Street locality, Curran and Thomas (1979) recovered 590 fragments of fish bone from 3 m^2 of midden, which comprise 87% of the faunal sample. The fish remains were identified as shad and alewife.

3.3.8. Woodruff Rockshelter

This site, located near the outlet of Lake Waramaug in northwestern Connecticut, overlooks the Aspetuck River, a tertiary tributary of the Houstatonic River (Swigart 1986). Of the 149 fragments of fish bone recovered, 141 were *Cyprinidae* sp. (minnows and carps), 7 were *Dermethoid* sp. (catfish and bullhead), and 1 was a *Percimorph* sp. (Bluegills and sunfish), all presumed to have come from Lake Waramaug. This lake is not a location where spawning salmon would be expected, although the Aspetuck River supports a resident salmonid (i.e., brook trout) population.

4. DISCUSSION

Clearly, the archaeological faunal evidence for fishing in New England does not suggest a heavy utilization of Atlantic salmon (*Salmo salar*) in either coastal, estuarine, or riverine settings. To the contrary, all available evidence indicates that salmon was an extremely minor component of the prehistoric resource base, having been positively identified *at only one site* (Kidder Point; one vertebra), and *possibly* identified at two others (Linquist and Frazer Point; total of three vertebrae). Even at "classic" fishing locales, such as the falls at Riverside on the Connecticut River, Amoskeag Falls on the Merrimack River, and Pushaw Stream off the Penobscot River, where one might expect evidence of salmon utilization given the historical accounts, there is none. Furthermore, fishing for salmon at coastal sites, or fishing for them upriver and bringing them back to coastal sites, is also not evident.

The lack of salmon in New England sites can be attributed to two possible factors: (1) the bones do not survive the archaeological record; or (2) salmon were not being fished either because they were not abundant or were technologically too difficult to catch. Despite the fact that the former is implicitly an entrenched notion among many New England archaeologists, I would argue that it is not justifiable, for a number of reasons. First, other types of fish are preserved in a wide variety of archaeological deposits in New England, including such species as shad and alewives, which have very thin, fragile skeletons. The vertebrae of salmon are, in fact, quite dense and durable relative to many species of fish. Notably, even the semicalcified vertebrae of elasmobranchs (sharks and rays) have survived the archaeological record (Follett, cited in Casteel 1976:90; Carlson 1986a). Furthermore, even in a fragmented state, Pacific salmon vertebrae (albeit a different species) have been shown by faunal analysts (e.g., Casteel 1976:91–92) to be easily identifiable. As a typical example of this preservation myth about salmonids, Swigart (1986:16) states, "In addition salmonoid sp. (i.e., trout) may not have been found on the site because of the extremely poor preservation possibilities of the softer fish bone" despite the fact that bone remains of minnows/carps, catfish/bullhead, and bluegills/sunfish had been preserved.

Casteel (1976:88–92) addressed the issue of salmon bone preservation in an earlier attempt to dispel, and provide an alternate explanation for, the commonly held misconception that salmon should not be expected to occur in archaeological sites due to poor preservation. Casteel (1976:89–90) states that, "In a rather unfortunate series of published reports a number of North American investigators have been suggesting that salmonids should not be expected to occur in archaeological sites." Apparently, the source for this view is a report by Bryan (1963:8), in which it is stated that, "salmon bones are cartilagenous and therefore do not preserve."

In fact, salmonids *are not* cartilaginous, being members of the class Osteichthyes, or bony fishes (Bigelow and Schroeder 1953). Moreover, Casteel (1976:90) cites numerous reports of salmonids (i.e., *Salmo* and *Oncorhynchus* genera) in the archaeological literature indicating that "remains of salmonids, including their scales, are rather abundant in North American sites. Thus, it appears that salmonid remains are not 'invisible' archaeologically. Indeed, they seem to be present over long periods of time and wide geographic areas." Furthermore, Atlantic salmon vertebrae are large enough that they would not suffer from a sampling bias due to the generally employed 7mm mesh screens that are used in archaeological excavations in New England. This is in comparison to other small species, such as alewives, herring, tomcod, lake trout, or pickerel, to name some that fall through that mesh size (Carlson 1986a).

Ruling out the possibility, then, that Atlantic salmon would not preserve in the archaeological record in New England puts the weight on the other explanation for its lack of evidence, i.e., that they were not fished. The question at hand, therefore, becomes why was the salmon resource not utilized extensively in aboriginal New England? Furthermore, given that salmon could not, in all probability, have been the tremendous catalyst that is often involved in models for prehistoric New England settlement and subsistence patterns, how does this reflect on the models of anadromous fish exploitation in New England in general? Are models based on anadromous salmon exploitation on the Northwest Coast appropriate for prehistoric New England? And finally, how should the historical accounts of salmon be evaluated? These questions are addressed in detail below.

The explanation for why salmon were not extensively fished in aboriginal New England lies either with biological or cultural reasons. The most likely explanation is biological, i.e., that salmon were never as extremely abundant in New England as has been supposed. This is not to imply that salmon was not present, only that its natural abundances were more limited than hitherto supposed. If salmon runs were as large and predictable as those on the Northwest Coast, then it seems likely that the Native Americans would have spent considerable energy exploiting them; it is unlikely that the lack of evidence of salmon utilization in the archaeological record is attributable to a lack of technology or interest on the part of the prehistoric populations. I would suggest that fisheries biologists take heed of the archaeological data because it may give a clearer and longer term picture of the pre-Contact fisheries than the historical accounts, which are probably unreliable, or the paleontological evidence, which is virtually nonexistent. Salmon in New England certainly never was the great commercial industry that cod fishing was and still is, or that salmon is in the Northwest. The generally disappointing results of the modern salmon enhancement programs in New England may be due more to the fact that salmon is not naturally abundant in these waters than to historical and modern dams and pollution. It is likely that the most productive areas for Atlantic salmon in the Northeast are much farther north, in Newfoundland and Labrador, where the salmon fishery was historically, and is at present, a viable commercial industry (Taylor 1985).

4.1. Evaluation of the Historical Accounts

In view of the preceding discussion, how then should the historical accounts be interpreted? I would suggest that the historical accounts of salmon in New England be viewed with considerably more caution than has been the common practice. First, it must be understood that the earliest settlers would probably be biased in presenting a considerably brighter picture of life in New England to the folks back in the old country than was necessarily the case; in this regard, since salmon was a much esteemed fish at home, it faired well to mention it. An emic

European perspective on salmon was undoubtedly presented in the historical accounts, rather than an accurate depiction of the natural abundance of such species or their significance to the native diet. It is probably appropriate to classify Atlantic salmon as a status fish to the English, and due to its scarcity and final eradication in the Thames River by 1833, any amount of salmon found in New England would be praised and embellished upon. As stated by Scott and Crossman :

> Few animals, whether fish, bird, or beast, have attracted as much attention through the ages as has the Atlantic salmon. Prized by the Gauls, then by the Romans, an abundant commercial fish in the British Isles, mentioned in the Magna Charta, revered by the sportsman and esteemed by gourmets, its relation with man has been truly unique. After centuries of exploitation, Yarrell (1836) told us that the last salmon was caught in the Thames in 1833. (Scott and Crossman 1973:196)

I am also not convinced that the historical accounts of salmon, as presented in the previous section, really support the notion of abundant salmon runs in New England, in the way that they are often made out to do. For instance, while none of the accounts of fish explicitly states that species are ranked in terms of their abundance or importance, the order of their presentation could be interpreted as an inherent ranking. Given this, salmon is presented with considerably lesser significance than several other species of fish in New England, on the basis of its order of presentation in the accounts.

Other interesting historical comments concerning the biology of those species that may have been significant to Native American populations include that of Douglass (1760) cited above, which implies that the timing of the salmon runs in New England was variable and short in duration, and certainly not the lengthy and predictable resource considered significant for those models of social complexity and subsistence specialization based on anadromous salmon exploitation. Notably, Schaulk (1977:222) states that, "temporal compression of the resource also implies increasingly formidable conditions for processing and storing fish." Moreover, lack of stability in a resource would argue against specialization on that resource (Schaulk 1977:228). Furthermore, Douglass (1760) suggests that the Atlantic salmon was somewhat of an elusive fish, lying "deep in cold brooks," then "silently" returning to the sea, suggesting a lack of overt visibility. In addition, the fact that Douglass writes that, "This salmon is not of a good quality and is not so good quantity and is not so good for a market as the salmon of Great Britain and Ireland" may reveal a more honest appraisal of the salmon resource in New England than other early writers.

Additional areas for caution with the historical accounts lies in the naming of fish. Rozier in 1605 (Rozier 1906) and Josselyn (1675) describe "white salmon," which is interpreted here to mean shad because it is a white-meat fish that is anadromous in habit and can attain sizes roughly comparable to that of Atlantic salmon. This suggests that there may be ambiguity in the use of the term *salmon*. In other words, it may be possible that some accounts describing salmon actually may be referring to shad, thereby artificially inflating the importance of salmon in the historical accounts, but there is no way to ascertain this.

4.2. The Geographic and Biological Baseline

The next question is whether or not the ethnographic models based on anadromous salmon exploitation from the Northwest Coast are appropriate for prehistoric New England. Obviously they are not in terms of the salmon resource *per se*. However, there are several other species of anadromous fish in New England, such as shad, alewives, and sturgeon, and perhaps these species could substitute well into the anadromous salmon models. For example, Burley (1983:171) has argued for the Micmac that "a Northwest Coast like specialized adaptation is viable in the Maritimes. However, rather than an intensive concentration on a single resource, it would have to be developed on the total range of anadromous, partially anadromous and catadromous species." Are shad, alewives, and/or sturgeon *combined* the "salmon" of prehistoric New England? Unequivocally, these three species are evident in the archaeological faunal record (see previous section). A discussion of the biology of these species and the Atlantic salmon in comparison to those of the Pacific salmon, of the habits of their "anadromy," and of the implications for social behavior in prehistoric New England is presented here.

The anadromous life cycle pattern exhibited by Pacific and Atlantic salmons and other species means that they are born in fresh water, migrate to the sea until they attain reproductive age, and then return to fresh water to spawn. The difference in this life-cycle pattern between the Atlantic salmon and the group of Pacific salmons is that the former does not necessarily die after it has spawned, whereas the latter unquestionably does. Other anadromous species in New England, such as the shad, alewives, and sturgeon, also do not die after spawning. A brief description of the important anadromous species and their habits is presented below.

The shad (*Alosa sapidissima*) is a medium-sized fish, ranging in size from 0.7 to 3.6 kg, occasionally reaching sizes of up to 6.5 kg (Bigelow and Schroeder 1953:108). They run up rivers to spawn in New England during May and June, depending upon the temperature of the rivers, and begin their return journey to the sea immediately after spawning. They often ran extremely far up some of the larger rivers in great abundance before dams prevented their migration.

The sturgeon in New England consist of two species, the Atlantic sturgeon (*Acipenser sturio*) and the short-nose sturgeon (*Acipenser brevirostrum*), the latter being a much smaller fish, and not as common. These fish also run up rivers to spawn in New England, usually to about the reach of tide, in May and June, and sometimes as late as July, returning to the sea occasionally as late as September (Bigelow and Schroeder 1953:82). They are also known to spawn in the estuaries in brackish water. Young sturgeon may live several years in the lower tidal reaches of rivers before migrating to the sea. These habits indicate that young sturgeon may be caught in the estuaries year-round, and the adults in the estuaries and rivers from May through September. This suggests that sturgeon found in archaeological sites are not a temporally restricted indicator of seasonality, and also that, due to the lack of intensive, temporally focused runs, there would have been no impetus for coordinated, specialized fishing efforts aboriginally, a feature prerequisite to most anadromous fishing models. Sturgeon are not particularly plentiful in New England waters today, although they were once quite abundant in most good-sized streams entering into the Gulf of Maine, of which the Merrimack, St. John, Penobscot, and the Kennebec Rivers are notable. Atlantic sturgeon are very large fish, the average size at one time in the Kennebec being 54 kg, with reports of some attaining weights of 270 kg (Bigelow and Schroeder 1953:81–82), suggesting that the catch of even one of these large fish would have provided a considerable food resource.

Alewives (*Alosa pseudoharengus*), another anadromous species in New England, are a small fish of the herring family, averaging about 25–28 cm in length and 0.22–0.25 kg in weight. Alewives run up rivers and streams in April to spawn in ponds and sluggish stretches of streams. They run indifferently up either very large rivers for considerable distances or extremely small coastal streams. The spent fish return downriver shortly after spawning, so "that many of them pass others coming up" (Bigelow and Schroeder 1953:103). Alewives were once extremely abundant in most waters of New England, although, as stated by Bigelow and Schroeder (1953:106), "Though alewives may seem almost incredibly numerous when crowding into some stream, they made but a sparse population, even in their days of greatest plenty, when spread over our coastal waters of the Gulf." Alewives also tend to remain in the general vicinity of the coast and estuaries, and therefore were easily captured in abundance in weirs throughout most of the year (Bigelow and Schroeder 1953:104), indicating that their presence on an archaeological site does not necessarily suggest an April occupation of the site.

In comparison with these three species, Atlantic salmon (*Salmo salar*) begin making their annual spawning migrations upriver as early as March, with runs continuing irregularly into August, even though spawning does not occur until October. Bigelow and Schroeder (1953:129) state that, "Every salmon fisherman is familiar with the fact that salmon enter in runs that are spaced irregularly in time, and that vary in date from year to year, depending on the height of water in the river and on the strength of the current." This would suggest a certain amount of unpredictability in the resource structure. It appears that the fish generally stay upriver until the fall, although, as noted by Bigelow and Schroeder:

> A good deal of discussion has centered about the question whether the earliest fish stay in fresh water from then until spawning time (a matter of 6 months) or whether there is more or less movement in and out of the river mouths at the beginning of the season. The latter view may be correct for the small streams, but it seems safe to say that after the run is well under way in late May or early June no fish return from fresh to salt water until autumn. (Bigelow and Schroeder 1953:129)

Furthermore, Bigelow and Schroeder (1953:129) state that when the fish are in the rivers they are quiescent and do not feed, which "is the chief reason why salmon angling is so uncertain a sport, even in the best of rivers." This would concur with the statements made by Douglass (1760) cited above. Atlantic salmon in the Gulf of Maine currently attain an average size of 4.5 kg, with sizes of up to 18 kg recorded, but very rare (Bigelow and Schroeder 1953:122).

Salmon on the Northwest Coast belong to an entirely different genus than Atlantic salmon, although both are members of the same family, *Salmonidae*. There are five species of Pacific salmon, of the genus *Oncorhynchus,* which begin their annual spawning migrations upriver from August through November, depending upon the species (Hart 1973). Certain species make very short trips upriver to their spawning streams, while others, such as the sockeye salmon, travel incredible distances; for example, the spawning beds in the Yukon are 2,000 miles from the sea, requiring that the fish travel at great speeds to reach the beds before winter ice develops (Curtis 1949:188).

The relationship between the various salmonids in the Pacific Northwest and the Atlantic Northeast is poorly understood from biological, paleontological, zoogeographic, and evolutionary perspectives. While both Pacific and Atlantic salmon are of the family *Salmonidae,* they are of separate genera, namely *Oncorhynchus* and *Salmo,* respectively. The genus *Salmo* includes, besides the Atlantic salmon, the rainbow/steelhead and cutthroat trouts of the Pacific Northwest, and the brown trout. The genus *Salvelinus* (eastern brook trout and Pacific Dolly Varden) also belongs to this family. The rainbow/steelhead trout (*Salmo gairdnerii*) of the Pacific has an anadromous ocean-going form called the steelhead and a "landlocked" form called the rainbow, which is taxonomically most related to the anadromous and landlocked forms of the Atlantic salmon. The only anatomical difference between the two species is "the shape of one small bone lying deep within the skull" (Curtis 1949:179). Biologically, therefore, it is to the steelhead trout that we should be making comparisons with the Atlantic salmon, and not to the five species of salmon of the *Oncorhynchus* genus. A factor noteworthy in this regard, and which may have some bearing on the lack of salmon in New England sites, is that the anadromous steelhead trout in the Pacific Northwest is an extremely difficult fish to catch, as any avid fisherman will attest, being considered the king of game fishing today (Christopher Carlson, personal communication, 1985). As stated by Scott and Crossman (1973:190), "The difficulties of landing a hooked steelhead in a swift, rocky river in winter are legendary and the subject of many books." The steelhead is also not an overly abundant species, particularly when compared to the Pacific salmon (as evidenced by the "catch and release" conservation policy for steelhead), suggesting that it is a more suitable biological analog for Atlantic salmon than for Pacific salmon, and perhaps indicating how we should be looking at the Atlantic salmon as a resource prehistorically.

The evolutionary relationships between members of the Salmonidae family are poorly understood due to a lack of paleontological materials. Studies on the systematics and evolution of salmon species suggest that the development of races within species may have been the result of the complex environments around glacial margins during the Pleistocene (Barber 1978; Behnke 1972; McPhail and Lindsey 1970). The work of Hoar (1976) supports the probability that *Oncorhynchus* evolved in the North Pacific after the closure of the Bering Strait during glacial periods. Behnke (1972) supports this hypothesis, and suggests that the genus *Oncorhynchus* may have evolved from the char, *Salvelinus alpinus,* as an adaptation to being trapped in the layer of freshwater under an annual ice cover during glaciation, which caused transport to the saltwater environment of the Bering Sea. The presence of landlocked sockeye salmon called the "Kokanee" in interior lakes of British Columbia can only be explained be postglacial changes in sea levels and river gradients, which isolated those populations that had migrated up rivers to lakes during glacial periods of raised sea level or glacial marine transgression (Behnke 1972). A similar situation exists with the landlocked form of the Atlantic salmon found in some lakes in the interior of Maine (Atkins 1884). As summarized by Curtis:

> The most widely accepted theory about the salmonids holds that they originated in the Arctic as migratory fish. When the ice came down across the parts of the world in which we now live, they came with it. When the ice retired and the seas grew warm again, colonies of them which had become fresh-water dwellers were trapped in the cool streams they had entered—some as far south as the Atlas Mountains in North Africa—and have remained there ever since. Cut off from each other in these streams, the differing conditions brought about different hereditary tendencies. And from this arose the bewildering number of species and subspecies and races and varieties of trout which make life difficult for the taxonomist. (Curtis 1949:175)

From the above descriptions of the anadromous habits of the Atlantic Salmon and of the three other species of anadromous fish found in New England, it is evident that the resource structures are entirely different from that of the five species of Pacific salmon. Not only do they run up the rivers in the spring and summer instead of the fall, but they descend the rivers after spawning, which the Pacific salmon do not. The timing of the runs may be the most important factor affecting aboriginal exploitation in the Northwest and New England coasts for several reasons. First, spring and summer are times of increasing abundance and availability of various other types of plant and animal resources. This includes numerous marine and lake fish species, which are exploitable from a local home base without having to move to a special fishing locality far upstream in order to catch shad, alewives, or sturgeon in any great quantity; during spawning these species, including the salmon, could only be caught in any great abundance at upstream locations such as falls or river constrictions, due to the fact that they do not feed during spawning, and therefore would not be easily captured with baited hooks or in intertidal weirs, such as marine species are. Given this situation, one could question what the incentive would be to move upstream to exploit a run of one or the other of these kinds of fish when other resources were abundant in the local vicinity, particularly when inter- or intragroup competition over space at the fishing locales would make them relatively crowded.

Second, one of the critical factors about the Pacific salmon runs is that they occur in the fall, before the onset of the winter season, a time when resources are lean. The timing of these runs is opportune for preparation and storage of fish for the winter ahead. This would encourage or support large, sedentary aggregations of people in the winter village settlement pattern (Fladmark 1975). The spring runs in New England do not have the advantage of this timing, and hence are not amenable to winter storage and the concomitant sedentary winter village settlement pattern. If storage is not a prime factor with the anadromous fish runs in New England, then the social attributes of coordinated group work effort and storage of the Northwest Coast models (Schaulk 1977) should not be expected either. If this is the case, then the applicability of the model for New England is questionable. In fact, there is little evidence in the ethnohistorical accounts for New England of extensive fish storage and preservation technology (Carlson 1986a).

Finally, the constraints imposed by the greatly differing geographic and hydrological structures of the river drainages in the Pacific Northwest, which are fast-moving streams of water flowing out of the steep coastal mountain range, as opposed to the gently flowing and relatively short drainages of New England, require detailed investigation and comparison between the two regions if the fish resources are to be compared. This is beyond the scope of this paper, however, but it is important to understand that great differences do exist, making analogous situations improbable.

In conclusion, given the above archaeological data, historical biases, and structure of the anadromous fish resources in New England, I would suggest that models that narrowly assign the function of "spring anadromous fishing camps" to specific site locales are questionable, particularly for the exploitation of salmon. Furthermore, I would argue that there would have been little adaptive advantage in a prehistoric New England settlement system for seasonal transhumancy up rivers from the coast, given that abundant fish and other resources exist year-round in the coastal zones, as evident from the marine faunal record. I would postulate that groups of people inhabiting the inland zones, but who were part of separate inland adaptive systems and ethnic identities, would be more likely to have taken advantage of the upriver anadromous fish runs than did coastal populations. Nevertheless, I would still feel uncomfortable about assigning site function to particular riverine locales such as great falls as spring fishing stations in this scenario because falls or other peculiar geographic features along river drainages are natural landmarks for site locations regardless of the spring fish runs. It seem logical that people inhabiting the inland regions would orient their settlement patterns to the river drainages, both because of the resource ecotone effect and the advantageous travel routes afforded by the rivers. In this regard, then, sites at falls may be interpreted as major village locations that could have been occupied at various times of the year depending upon the social behavior of the group, and therefore where a wide range of subsistence activities could have been taking place (i.e., not just spring fishing for anadromous fish runs). In fact, the diverse range of faunal remains found at these site locales, such as Riverside on the Connecticut River or Amoskeag on the Merrimack River, supports this statement.

5. ALTERNATIVE MODELS FOR SOCIAL ORGANIZATION BASED ON FISH RESOURCES

In comparing the archaeology of the New England coast to that of the Northwest Coast and the basically maritime orientation inherent in both, I would suggest that the marine fish resources, *as opposed to the anadromous riverine fish resources,* played a greater role in New England social organization and settlement pattern. While more salmon remains will undoubtedly be recovered at archaeological sites in New England, it is clear from substantial evidence that their exploitation was not a focal point of prehistoric lifeways. The data on fish exploitation in New England do not fit the anadromous salmon models from the Northwest Coast. Probably what we are seeing in New England is an overwhelming adaptation to sea fishing, particularly for the Atlantic cod, a species that largely requires hook and line fishing from ocean-going canoes. This appears to have been augmented by possible weir fishing of flounders and sculpins, with some sturgeon, alewives, and tomcod, depending upon the season, in northern New England (Carlson 1986a). In southern New England, the majority of fish remains that have been reported to date come from Martha's Vineyard, where the fishing pattern may be somewhat atypical; there, the fish remains are dominated by scup, a small species that infrequently ranges north of Cape Cod, and is easily caught with baited hook and line when the fish move close inshore in schools from May through October (Bigelow and Schroeder 1953:413–416). In Boston Harbor, the pattern of cod fishing is predominant, augmented by probably incidental catches of other bottom-feeding fish, similar to the northern New England situation. Notably, further north, in the Canadian Maritimes, faunal analysis of sites in Passamaquoddy Bay, from Cape Breton Island, Nova Scotia, and from the Miramichi River estuary also shows no evidence of salmon utilization, although sturgeon and striped bass are reported (Stewart 1982). Even today, the cod fisheries form the basis of the commercial industry in the New England–Maritimes region, just as the salmon form the basis in the Pacific Northwest ("Some things never change, they just don't stay the same" [Anonymous]).

In this fishing pattern for prehistoric coastal New England, anadromous species do not appear to have been most important. Even the presence of a few remains of anadromous sturgeon, alewives, or tomcod does not necessarily imply specialization on spawning "runs," since these species are generally available in the estuaries and close to shore outside of their spawning seasons. Given these circumstances, a model that examines the conditions under which commitments are made to marine fishing practices and the ways in which social organizational and technological demands of marine fishing may have affected social structure may be more appropriate. For example, what are the implications for social organization and production given fishing technologies such as boats, weirs, or large nets, in terms of group size, number of people required to produce, amount of cooperation demanded, amount of work effort, degree of division of labor, and task specialization? The use of a seine net, for example, implies the use of a boat and a certain amount of teamwork (Clark 1952:45).

Other questions and variables useful for modeling the structure of marine fish resources and human populations include what the degree of access to technology, such as boats or weirs, would suggest about social hierarchies. Access might imply why one would go heavily into difficult technology or travel far out to sea in the first place (e.g., for control or status), or to enter into the social order certain exotic trade or status items, which may eventually lead to social ranking. Finally, how is the degree of sedentism and population size affected by the commitment to marine fishing, particularly if the sea is viewed as an inexhaustible "catchment area" where resource depletion is unlikely even if a permanent land base is established? In conclusion, I would suggest that these are the kinds of questions and hypotheses that should be asked and investigated when attempting to model prehistoric social organization and settlement patterns along the New England coast. Emphasis on the runs of anadromous fish in these waters, and most of all on salmon, should be relegated to the proper position of a minor component with insufficient similarities to Northwest Coast models to warrant cross-cultural comparisons.

Acknowledgments

The individual who deserves my heartfelt thanks is George Nicholas, without whose encouragement, editorial abilities, persistence, and invitation to present a paper in the first place I may never have undertaken the task. I would also like to thank Dena Dincauze for the several stimulating discussions about prehistoric anadromous fish that I have enjoyed with her, and also for her review comments on an earlier draft of this paper. Finally, I thank those individuals who provided information, unpublished reports, or useful discussion necessary to seeing various aspects of this paper to completion: Marie Bourassa, Bruce Bourque, Christopher Carlson, Roy Carlson, Mary Lou Curran, Frederick Dunford, Knut Fladmark, Elizabeth Little, Barbara Luedtke, Lucinda McWeeney, Jeffrey Maymon, Harold Prins, David Sanger, Peter Thomas, James Wherry, and Martin Wobst.

The responsibility for interpretations or errors is solely my own.

6. REFERENCES

Ames, K., 1979a, Stable and Resilient Systems along the Skeena River: The Gitskam/Carrier Boundary, in: *Skeena River Prehistory* (R. Inglis and G. MacDonald, eds.), Mercury Series, No. 87, National Museums of Canada, Ottawa, pp. 219–243.

Ames, K., 1979b, The Evolution of Ranked Societies in the Northwest Coast of North America, Paper presented at the 44th Annual Meeting of the Society of American Archaeology and Canadian Archaeological Association, Vancouver, Canada.

Ames, K., 1981, The Evolution of Social Ranking on the Northwest Coast of North America, *American Antiquity* 46:789–805.

Atkins, C. G., 1884, Notes on Land-Locked Salmon, *Transactions of the American Fish Cultural Association* 13:40–56.

Atkins, C. G., 1887, The River Fisheries of Maine, in: *The Fisheries and Fishery Industries of the United States,* Section 5, Vol. 1 (G. B. Goode, ed.), U.S. Commission of Fish and Fisheries, Washington, D.C., pp. 673–728.

Barber, F. G., 1978, On the Origin of *Oncorhynchus gorbuscha, Fisheries Bulletin of the Fish and Wildlife Service,* No. 53, Washington, D.C.

Barber, R., 1980, Post-Pleistocene Anadromous Fish Exploitation at the Buswell Site, Northeastern Massachusetts, in: *Early and Middle Archaic Cultures in the Northeast* (D. R. Starbuck and C. E. Bolian, eds.), Occasional Publications in Northeastern Archaeology, No. 7, pp. 97–114.

Barber, R., 1982, *The Wheeler's Site: A Specialized Shellfish Processing Station on the Merrimack River,* Peabody Museum Monographs, No. 7, Harvard University, Cambridge, Massachusetts.

Barghoorn, E., Conger, P. S., Judson. S., Phleger, F. B., and Wilson, L. R., 1949, The Boylston Street Fishweir II: A Study of the Geology, Palaeobotany, and Biology of a Site on Stuart Street in the Back Bay District of Boston, Massachusetts (F. Johnson, ed.), *Papers of the Robert S. Peabody Foundation for Archaeology,* No. 4, Part 1.

Baumhoff, M., 1963, Ecological Determinants of Aboriginal California Populations, *University of California Publications in American Archaeology and Ethnology* 49:155–236.

Baxter, J. P. (ed.), 1907, Letter from Thomas Westbrook to William Drummer, May 21, 1725, *Documentary History of Maine,* Vol. 10.

Behnke, R., 1972, The Systematics of Salmonid Fishes of Recently Glaciated Lakes, *Journal of the Fisheries Research Board of Canada* 29:639–671.

Bigelow, H. B., and Schroeder, W. C., 1953, Fishes of the Gulf of Maine, *Bulletin of the U.S. Department of Fisheries,* No. 40, Washington, D.C.

Binford, L. R., 1964, Post-Pleistocene Adaptations, in: *New Perspectives in Archaeology* (S. R. Binford and L. R. Binford, eds.), University of Chicago Press, pp. 313–341.

Birket, J., 1916, *Some Cursory Remarks Made by James Birket in his Voyage to North America, 1750–1751,* Yale University Press, New Haven, Connecticut.

Borstel, C., 1982, *Archaeological Investigations at the Young Site, Alton, Maine,* Occasional Publications in Maine Archaeology, No. 2, Maine Historic Preservation Commission, Augusta, Maine.

Bourque, B. J., 1975, Comments on the Late Archaic Populations of Central Maine: The View from Turner Farm, *Arctic Anthropology* 12(2):35–45.

Bourque, B. J., and Cox, S. L., 1981, Maine State Museum Investigations of the Goddard Site, 1979, *Man in the Northeast* 22:3–27.

Brennan, L., 1981, Pick-up Tools, Food, Bones, and Inferences on Lifeway Function of Shell Heaps along the Lower Hudson, Archaeology of Eastern North America 9:42–51.

Brumbach, H. J., 1986, Anadromous Fish and Fishing: A Synthesis of Data from the Hudson River Drainage, *Man in the Northeast* 32:35–66.

Bryan, A. L., 1963, An Archaeological Survey of Northern Puget Sound, *Idaho State University Museum, Occasional Papers,* No. 11.

Burley, D. V., 1983, Cultural Complexity and Evolution in the Development of Coastal Adaptations among the Micmac and Coast Salish, in: *The Evolution of Maritime Cultures on the Northeast and Northwest Coasts of North America* (R. J. Nash, ed.), Department of Archaeology Publication, No. 11, Simon Fraser University Burnaby, British Columbia, Canada, pp. 157–172.

Carlson, C. C., 1980, Report on the Analysis of the Fernald Point Fish Remains, Ms. on file, Department of Anthropology, University of Maine, Orono, Maine.

Carlson, C. C., 1981, Report on the Analysis of the Frazer Point Site Fish Remains, Ms. on file, Department of Anthropology, University of Maine, Orono, Maine.

Carlson, C. C., 1982a, A Reevaluation of the Role of Salmon to the Evolution of Ranked Society on the Northwest Coast, Paper presented at the 15th Annual Meeting of the Canadian Archaeological Society, Hamilton, Ontario, Canada.

Carlson, C. C., 1982b, Report of the Faunal Analysis of the 1967 Excavation of the Smyth Site, New Hampshire, *New Hampshire Archeologist* 23:91–102.

Carlson, C. C., 1986a, *Maritime Catchment Areas: An Analysis of Prehistoric Fishing Strategies in the Boothbay Region of Maine,* M.Sc. Thesis, Institute for Quaternary Studies, University of Maine, Orono, Maine.

Carlson, C. C., 1986b, Long Island Fish Identification, Unpublished data on file with Barbara Luedtke, University of Massachusetts, Boston.

Carlson, C. C., 1987, Report on the Fish Remains from Locus 6, Site M52/65, Quidnet, Nantucket, Massachusetts, Archaeology Department, Nantucket Historical Association, Nantucket, Massachusetts, unpublished.

Casteel, R, 1976, *Fish Remains in Archaeology and Paleo-environmental Studies,* Academic Press, New York.

Cavallo, J., 1984, Fish, Fires and Foresight: Middle Woodland Economic Adaptions in the Abbott Farm National Landmark, *North American Archaeologist* 5(2):111–138.

Champlain, Samuel de, [1613] 1922, Les Voyages, in: *The Works of Samuel de Champlain,* Vol. 1, *1599–1607* (H. P. Biggar, ed.), The Champlain Society, Toronto, Ontario, Canada, pp. 206–459.

Clark, J. G. D., 1952, *Prehistoric Europe: The Economic Basis,* Stanford University Press, Stanford, California.

Coffin, C. C., 1963, Seaside Indian Village Site, Fort Trumbull Beach, Milford, Connecticut, *Bulletin of the Archaeological Society of Connecticut* 32:11–25.

Coffin, J., 1845, *A Sketch of Newbury, Newburyport, and West Newbury from 1635 to 1845,* S. G. Drake, Boston, Massachusetts.

Cole, J. R., Mulholland, M. T., 1983, *Historic and Prehistoric Patterns of Water Use in the Connecticut River Valley: Implications for Water Resources Planning,* Water Resources Research Center, University of Massachusetts, Amherst.

Craigie, W. A., and Hulbart, J. R. (eds.), 1943, *A Dictionary of American English on Historical Principles,* Vol. 4, University of Chicago, Chicago, Illinois.

Curran, M. L., and Thomas, P. A., 1979, Phase III—Data Recovery: Wastewater Treatment System in the Riverside Archaeological District of Gill, Massachusetts, Department of Anthropology Report, No. 19, University of Vermont, Burlington, Vermont.

Curtis, B., 1949, *The Life Story of the Fish: His Manners and Morals,* General Publishing Company, Toronto [reprint, Dover, New York, 1961].

Deal, M., 1986, Late Archaic and Ceramic Period Utilization of the Mud Lake Stream Site, Southwestern New Brunswick, *Man in the Northeast* 32:67–94.

Dincauze, D. F., 1976, *The Neville Site: 8000 Years at Amoskeag Falls,* Peabody Museum Monographs, No. 4., Harvard University, Cambridge, Massachusetts.

Douglass, W., 1972, *A Summary, Historical and Political, of the First Planting, Progressive Improvements, and Present State of the British Settlements in North America,* Vol. 1 (1749), Vol. 2 (1971), Rogers and Fowle, Boston [reprinted, Arno Press, New York, 1972].

Fladmark, K., 1975, *A Paleoecological Model for Northwest Coast Prehistory,* Mercury Series, No. 43, National Museums of Canada, Ottawa, Canada.

Foster, D., Kenyon, V. B., and Nicholas, G. P., 1982, Prehistoric Lifeways at the Smyth Site, NH 38-4, *New Hampshire Archeologist* 22(2):1–91.

Gordon, M., and Marston, P. M., 1940, The Diary of Matthew Patton of Bedford, N.H., 1785, *New England Naturalist* 8:4.

Hart, J. L., 1973, Pacific Fishes of Canada, *Fisheries Research Board of Canada Bulletin,* No. 180.

Higgeson, F., 1792, New Englands Plantation, a Description of New England in the Yeare 1629, *Collections of the Massachusetts Historical Society for the Year 1792* 1:117–124.

Hoar, W. S., 1976, Smolt Transformation: Evolution, Behavior and Physiology, *Journal of the Fisheries Research Board of Canada* 33:1233–1252.

Hubbard, W., 1815, *A General History of New England from the Discovery to 1680,* Massachusetts Historical Society, Cambridge, Massachusetts.

Jochim, M. A., 1979, Catches and Caches: Ethnographic Alternatives for Prehistory, in: *Ethnoarchaeology, Implications of Ethnography for Archaeology* (C. Kramer, ed.), Columbia University Press, N.Y., pp. 219–246.

Johnson, F., 1942 (ed.), *The Boylston Street Fishweir,* Paper No. 2, Robert S. Peabody Foundation for Archaeology, Phillips Academy, Andover. Massachusetts.

Josselyn, J., 1675, *An Account of Two Voyages to New England,* G. Widdowes, London, reprinted (1865), William Veazie, Boston, Massachusetts.

Kenyon, V., and P. McDowell, 1983, Environmental Setting of Merrimack River Valley Prehistoric Sites, *Man in the Northeast* 25:2–23.

Knight, J., 1985, *Differential Preservation of Calcined Bone at the Hirundo Site, Alton, Maine,* M.Sc. Thesis, Institute for Quaternary Studies, University of Maine, Orono, Maine.

Lahti, E., Spiess, A., Hedden, M., Bradley, R., and Faulkner, A., 1981, Test Excavations at the Hodgdon Site, *Man in the Northeast* 21:19–36.

Little, E. A., 1985, Prehistoric diet at Nantucket Island, Massachusetts, Nantucket Archaeological

Studies, No. 6, Archaeology Department, Nantucket Historical Association, Nantucket, Massachusetts.

Luedtke, B. E., 1980, The Calf Island Site and the Late Prehistoric Period in Boston Harbor, *Man in the Northeast* 20:25–74.

Maymon, J. H., and Bolian, C. E., 1985, The Wadleigh Falls Site: An Early Holocene Site in Southeastern New Hampshire, unpublished.

McManamon, F., 1984, *Chapters in the Archaeology of Cape Cod,* I: *Results of the Cape Cod National Seashore Archeological Survey, 1979–1981,* Vols. 1 and 2, Cultural Resources Management Study, No. 8, Division of Cultural Resources, North Atlantic Regional Office, National Park Serice, U.S. Department of the Interior.

McPhail, J. D., and Lindsey, C. C., 1970, Freshwater Fishes of Northwestern Canada and Alaska, *Fisheries Research Board of Canada Bulletin,* No. 173.

Miller, V. P., 1983, Social and Political Complexity on the East Coast: The Micmac Case, in: *The Evolution of Maritime Cultures on the Northeast and Northwest Coasts of America* (R. Nash, ed.), Department of Archaeology Publication, No. 11, Simon Fraser University, Burnaby, British Columbia, Canada, pp. 41–56.

Moore, J., and Root, D., 1979, Anadromous Fish, Stream Ranking and Settlement, in: *Ecological Anthropology of the Middle Connecticut River Valley* (R. Paynter, ed.), Research Report 18, University of Massachusetts, Amherst, pp. 27–44.

Morse, D. R., 1975, *Prehistoric Subsistence of Coastal Maine: A Zooarchaeological Study of the Turner Farm,* M.A. Thesis, Ball State University, Muncie, Indiana, unpublished.

Morton, T., 1637, *New English Canaan or New Canaan* J. F. Stam, Amsterdam, 2nd printing (1883) Prince Society, Boston, Massachusetts.

Nash, R. J. (ed.), 1983, *The Evolution of Maritime Cultures on the Northeast and Northwest Coasts of America,* Department of Archaeology Publication, No. 11, Simon Fraser University, Burnaby, British Columbia, Canada.

Perlman, S. J., 1980, An Optimum Diet Model, Coastal Variability and Hunter–Gatherer Behavior, in: *Advances in Archaeological Method and Theory,* Vol. 3 (M. B. Schiffer, ed.), Academic Press, New York, pp. 257–310.

Peters, S. A., 1781, *General History of Connecticut,* London, 2nd printing (1877) D. Appleton, New York.

Ritchie, W. A., 1969, *The Archaeology of Martha's Vineyard,* Natural History Press, New York.

Robinson, B., 1985, The Nelson Island and Seabrook Marsh Sites: Late Archaic, Marine Oriented People on the Central New England Coast, *Occasional Publications in Northeastern Anthropology* 9:1–107.

Rostlund, E., 1952, *Freshwater Fish and Fishing in Native North America,* University of California Publications in Geography, Vol. 9, University of California Press, Berkeley, California.

Rozier, J., [1605] 1906, A True Relation of Captaine George Weymouth his Voyage, made this Present Yeere 1605; in the Discouerie of the North Part of Virginia, in: *Early English and French Voyages, 1534–1608,* Scribners, New York, pp. 353–394.

Sanger, D., 1980, Archaeological Salvage and Test Excavations, Fernald Point, Acadia National Park, Maine, Report prepared for the U.S. National Park Service, Contract No. 6-0022-77-03, U.S. Department of the Interior, National Park Service, Denver, Colorado.

Sanger, D., 1982, Changing Views of Aboriginal Seasonality and Settlement in the Gulf of Maine, *Canadian Journal of Anthropology* 2(2):195–203.

Sanger, D., Davis, R., MacKay, R., and Borns, H., Jr., 1977, The Hirundo Archaeological Project: An Interdisciplinary Approach to Central Maine Prehistory, *Annals of the New York Academy of Science* 288:457–471.

Schaulk, R., 1977, The Structure of an Anadromous Fish Resource, in: *For Theory Building in Archaeology* (L. R. Binford, ed.), Academic Press, New York, pp. 207–250.

Scott, W., and Crossman, E., 1973, Freshwater Fishes of Canada, *Fisheries Research Board of Canada Bulletin,* No. 184, Ottawa.

Smith, J., 1884, *Capt. John Smith, Works, 1608–1631* (E. Arbor, ed.), Birmingham, England.

Sneed, P., 1971, Of Salmon and Men: An Investigation of Ecological Determinants and Aboriginal Man in the Canadian Plateau, in: *Aboriginal Man and Environments on the Plateau of Northwest America* (A. Stryd and R. Smith, eds.), University of Calgary, Alberta, pp. 229–242.

Spiess, A., and Hedden, M., 1983, *Kidder Point and Sears Island in Prehistory,* Occasional Publications in Maine Archaeology, No. 3, Maine Historic Preservation Commission, Augusta, Maine.

Spiess, A., Bourque, B. J., and Cox, S. L., 1983, Cultural Complexity in Maritime Cultures: Evidence from Penobscot Bay, Maine, in: *The Evolution of Maritime Cultures on the Northeast and Northwest Coasts of America* (R. Nash, ed.), Department of Archaeology Publication, No. 11, Simon Fraser University, Burnaby, British Columbia, Canada, pp. 91–108.

Stewart, F., 1982, Seasonal Movements of Indians in Acadia as Evidenced by Faunal Remains and Historical Documents, Paper presented at the 15th Annual Meeting of the Canadian Archaeological Association, Hamilton, Ontario, Canada.

Swigart, E. K., 1986, The Woodruff Rock Shelter Faunal Report, American Indian Archaeological Institute, Washington, Connecticut, unpublished.

Taylor, V. R., 1985, *The Early Atlantic Salmon Fishery in Newfoundland and Labrador,* Canadian Special Publication of the Fisheries and Aquatic Sciences, No. 76, Department of Fisheries and Oceans, Ottawa.

Thomas, P. A., 1980, The Riverside District, The WMECO Site, and Suggestions for Archeological Modeling, in: *Early and Middle Archaic Cultures in the Northeast* (D. R. Starbuck and C. E. Bolian, eds.), Occasional Publications in Northeastern Anthropology, No. 7, pp. 73–96.

Waters, J. H., 1965, Animal Remains from Some New England Woodland Sites, *Bulletin of the Archaeological Society of Connecticut* 33:5–12.

Winsor, J., 1881, *Notes of Mr. Bennet, 1740 (Boston 1862)* [Reprinted in *Memorial History of Boston 1630–1880,* Vol. 2, Ticknor, Boston, Massachusetts].

Wood, W., 1634, *New England's Prospect,* London, reprinted (1865) Prince Society, Boston.

Yesner, D., 1984, L'Évolution de la Pêche Préhistorique dans la Région Sud-Ouest du Maine, *Recherches Amérindiennes au Québec* 14(1):34–44.

Yesner, D., Hamilton, N., and Doyle, R., Jr., 1983, "Landlocked" Salmon and Early Holocene Lacustrine Adaptations in Southwestern Maine, *North American Archaeologist* 4:307–333.

Chapter 4

Problems in the Use of Sea-Level Data for Archaeological Reconstructions

DOUGLAS C. KELLOGG

1. INTRODUCTION

Accurate environmental reconstructions are essential for modeling cultural responses to changing environments and testing archaeological hypotheses of cultural adaptation. The character of the paleoenvironmental record and assumptions inherent in analyzing and interpreting data must therefore be appreciated by archaeologists (Dincauze 1978, 1981). Fluctuating sea levels are dominant in determining the character and development of Holocene coastal environments (Fladmark 1983; Perlman 1980; Richardson 1981; Salwen 1975). This paper considers problems in the use of sea-level data and in constructing sea-level curves. The emphasis is on sources of data, not on actual techniques of reconstructing paleoenvironments. The goal is to provide the reader with a critical introduction to sea-level studies and to the sea-level literature. An extensive bibliography includes the classic references and current reviews of Holocene sea-level data, where more detail can be found on particular topics. The history of sea level in the Gulf of Maine is summarized to illustrate the complexities that may be encountered. Although the focus is on the Atlantic coast of North America, the issues discussed are global.

DOUGLAS C. KELLOGG • Institute for Quaternary Studies, University of Maine, Orono, Maine 04469.

2. SEA-LEVEL CURVES

Although sea-level curves have been the subject of much discussion recently (Inman 1983; Marcus and Newman 1983; Kraft 1985), the details of sea level are worth review, and many factors have been ignored in the archaeological literature. Two basic types of curves are common: eustatic and relative. Eustatic originally referred to changes in "mean sea level" that occur simultaneously worldwide. "Relative" curves apply to particular locations. Much of the earlier literature is concerned with eustatic sea level. In the 1970s, the focus shifted to local, relative sea level because, as more sea-level data became available, the problems involved in determining a eustatic curve were realized.

2.1. Eustatic Sea Level

In theory, mean sea level is "the surface level that the oceans and seas of the world would adopt if tidal forces ceased to operate" (Jardine 1975:163). In reality, mean sea level is "the average level of the sea calculated from a large number of observations" (Jardine 1975:163; see also Hicks 1978). Since the melting of the Late Wisconsin ice sheets, beginning around 16,000 years ago, mean sea level has risen. The amount of rise, however, is the subject of active debate because the magnitude of sea-level drop at glacial maximum is not known (Andrews 1982; Bloom 1983; Kidson 1982; Morner 1976).

Estimates have been attempted by several methods. For example, theoretical ice volumes have been used (e.g., Clark and Lingle 1979; Donn *et al.* 1962; Hughes *et al.* 1981:308–311; Mix and Ruddiman 1984; Paterson 1972), but several competing models of ice sheet character and dynamics are current (Andrews 1982; Andrews and Miller 1985; Denton *et al.* 1986; Mayewski *et al.* 1981:67–178; Ruddiman and Duplessy 1985). The various models are based on identical or overlapping data, but each implies different results for the magnitude, rate, and character of eustatic sea-level rise, especially in the early postglacial period. The controversy over ice sheet reconstructions extends to the basic assumptions of how the earth's climate system operates and to philosophical problems of scientific method.

Oxygen isotope measurements of foraminiferal assemblages from deep ocean cores also yield estimates of ice volumes over the long term, on the order of hundreds of thousands of years (e.g., Emiliani and Shackleton 1974; Ericson and Wollin 1968; Imbrie *et al.* 1973; Shackleton 1967; Shackelton and Opdyke 1973). Temperature, however, also exerts an influence on oxygen isotope values. Other questions, such as nonequilibrium deposition of oxygen isotopes in foraminifera shells (e.g., Fairbanks *et al.* 1980; Hecht and Savin 1970, 1972; Schiffelbein and Hills 1984; Williams *et al.* 1979) and postdepositional changes (e.g., Killingley 1983), have been raised, so the exact relationship between the oxygen isotope record and sea-level fluctuations remains problematic (Andrews 1982; Imbrie *et al.* 1973; Mix and Ruddiman 1984). Furthermore, such sea-level curves do not apply to specific coastlines and lack the precision required for specific archaeological applications, such as reconstructing shorelines. On the other hand, the oxygen isotope record is unquestionably vital for paleo-oceanographic and paleoclimatic studies (Imbrie and Imbrie 1986).

Recently, global sea-level curves have been produced based on seismic stratigraphy of continental margins (Haq *et al.* 1987; Poag and Ward 1987; Vail *et al.* 1977). These curves show fluctuations of sea level over the last 250 million years, but are somewhat controversial (Bond 1978; Watts 1982), and are very generalized for the Quaternary time period.

Estimates of the *minimum* value of the eustatic sea-level drop at ice maximum have been derived from submerged terrestrial and littoral deposits on the continental margins (e.g., Dillon and Oldale 1978; Emery and Garrison 1967; Emery *et al.* 1965, 1967; Merrill *et al.* 1965; Oldale and O'Hara 1980; Oldale *et al.* 1983; Shepard 1964). This type of data, however, is ill-suited for constructing sea-level curves because the depth and context of the dated material are often unknown or subject to large possible errors (Macintyre *et al.* 1978; Milliman and Emery 1968). Often only one date at one depth is obtained. To construct a curve, data points from different depths at different localities must be combined. Such a practice can be imprecise.

Except for those interested in the possibilities of Paleoindian occupations on continental margins (e.g., Carter 1964; Edwards and Merrill 1977; Emery and Edwards 1966; Masters and Flemming 1983; Oldale 1985; Pearson *et al.* 1986; Roberts 1979; Stright 1986), the rate and character of postglacial sea-level rise are more important than its exact magnitude. There have been many attempts to derive a curve showing eustatic sea-level rise (e.g., Blackwelder *et al.* 1979; Coleman and Smith 1964; Curray 1965; Fairbridge 1958, 1960; Goodwin *et al.* 1958; Kazanskiy 1985; Milliman and Emery 1968; Morner 1971; Redfield 1967; Scholl and Stuiver 1967; Scholl *et al.* 1969; Shepard 1964; Shepard and Suess 1956). Presumably such a curve could be applied to the topography of an area and maps drawn showing its morphology at dates in the past. This is the technique employed in early paleogeographic reconstructions (e.g., Salwen 1962). At the time, Salwen's (1962) study was based on the best available information. Ignoring the important processes of sedimentation and erosion (Kraft 1976, 1985; Kraft *et al.* 1975, 1977; van Andel and Lianos 1984), the major drawback of this approach is in assuming that no crustal movements have affected the region of interest.

The weight of Pleistocene ice sheets deformed the earth's crust, depressing it hundreds of meters, so that mantle material was forced out horizontally (Flint 1971:343–366). Forebulges may have extended out beyond the ice (Bloom 1967; Clark *et al.* 1978; Walcott 1972a,b), and the shapes and volumes of the oceans basins may have changed as well (Bloom 1967; Walcott 1972a,b). The extent of these *isostatic* changes in the crust depends on the nature and thickness of the material that forms the mantle (Walcott 1980; Weertman 1978; Wolf 1985). Mathematical models provide theoretical maximum and minimum estimates (Clark *et al.* 1978; Hughes *et al.* 1981; T. S. Hughes, personal communication). Because the mantle cannot be studied directly, hypotheses derived from mathematical models have been tested against the observed tilts of proglacial lakes (Bloom 1967; Broecker 1966; Wolf 1985) and rebounding ocean shorelines (e.g., Bloom 1967; Dillon and Oldale 1978; McGinnis 1968; Morner 1971; Newman and Marsh 1968; Peltier 1984; Quinlan and Beaumont 1981, 1982). Estimates of ice volumes and eustatic sea-level rise either enter into or are used to test these calculations; thus, a danger of circularity is introduced into the study of both isostasy and eustacy. These topics are subjects of active controversy in geophysics and glaciology (Andrews 1982; Morner 1976, 1980; Peltier *et al.* 1978; Wolf 1985) and are part of the controversy over ice sheet reconstructions discussed earlier.

Crustal movements can take place independently of glacial activity; therefore, the local tectonic history of a region must also be evaluated (Curray 1965). No place on earth is completely stable (Clark *et al.* 1978). Thus, sea-level curves apply only to a particular location from which the original data were drawn (Kraft and John 1978:42–43; Marcus and Newman 1983).

Further complicating the concepts of "eustatic" and "mean" sea level are irregularities in the earth's gravitational field. The *geoid*—defined as the equipotential surface of the earth's gravitational field—is not smooth and flat, but is distorted by rises and depressions (Morner 1976). For example, there is a 180 m difference in the elevation of the sea surface between a hump at New Guinea and a depression at the Maldive Islands in the Pacific Ocean (Morner 1976). The concept of a worldwide sea level is no longer accepted and curves that compile worldwide data, such as Fairbridge (1958), are only of historical interest (Kidson 1982). Morner (1976) defines eustatic sea level specifically as water volumes added to the surface of the oceans. Eustatic waters spread over and conform to the geoid, contributing to "geoidal sea level." Geoidal sea level is the water level that affects the shores of the continents. Eustatic sea level is an outmoded concept (Kidson 1982) that Farrell and Clark (1976:649) find ambiguous in modern usage. Geoidal perturbations are considered in recent discussions and models of the earth's interior composition and of isostacy and in ice sheet modeling (Wolf 1985).

2.2. Relative Sea-Level Curves

Eustatic sea-level curves are useful to archaeologists only for the most general purposes (e.g., Binford 1968; Edwards and Merrill 1977; Emery and Edwards

1966; Perlman 1980; van Andel and Shackelton 1982). Relative sea-level curves, based on local data, are much more valuable to archaeologists (Inman 1983; Kraft 1985), although they are not without problems. Among these is the reference point for sea level itself. In some cases, datum is taken as mean high water (e.g., Redfield and Rubin 1962) because that is the critical elevation for floral and faunal indicators of past sea level. Others, however, have used mean low water as datum for similar reasons (Belknap and Kraft 1977; Fairbridge 1958; Zeigler *et al.* 1965). Technically, such curves record changes in tidal level through time, which may or may not directly reflect sea level (Jardine 1975).

We usually take sea level for granted, but determining the exact reference point for sea level can be confusing (Shalowitz 1964). Mean sea level may be different from *local* mean sea level, which varies from place to place, depending on local tidal fluctuations. Maps and charts add to the possible confusion. National Ocean Survey navigational charts use *local* mean low water as a reference for depth soundings, but elevations are given relative to mean high water. Mean sea level, however, is not indicated. On U.S. Geological Survey quadrangle maps depths are also relative to local mean low water, but elevations are relative to mean sea level. Theoretically it is possible to correlate these water levels to the National Geodetic Vertical Datum of 1929 (NGVD) as defined by the National Oceanic Survey, Department of Commerce, Washington, D.C. (e.g., Hicks 1978); however, this datum is not always readily determined. Moreover, due to local tectonics, the survey benchmarks on which NGVD was defined have moved in relation to each other (e.g., Anderson *et al.* 1984).

Relative sea-level curves have been constructed from many different sources of data. The most common are:

1. Salt marsh cores (e.g., Anderson and Race 1981; Bloom and Stuiver 1963; Keene 1971; Nelson 1979; Newman and Rusnak 1965; Newman *et al.* 1980; Redfield and Rubin 1962; Stuiver and Daddario 1963; Thompson 1973).
2. Drowned forests (e.g., Bloom 1963; Bradley 1953; Frankel and Crow 1961; Fujii *et al.* 1987; Harrison and Lyons 1963; Hussey 1959; Lyons and Goldthwait 1934; Redfield and Rubin 1962).
3. Shoreline features (e.g., Andrews 1980; Fairbridge 1958; Grant 1980; Macintyre *et al.* 1975; McMaster and Garrison 1967; Oldale 1985, 1986; Oldale *et al.* 1983; Schnitker 1974; Stephens and Synge 1966).
4. Fossil shells (e.g., Andrews 1980; Macintyre *et al.* 1975; Stuiver and Borns 1975).
5. Archaeological and historical sites (e.g., Anderson *et al.* 1984; Andrews *et al.* 1971; Barghoorn 1953; Brooks *et al.* 1979; Colquhoun and Brooks 1986; Colquhoun *et al.* 1981; DePratter and Howard 1981, 1983; Fairbridge 1976; Flemming 1969; Goldthwait 1935; Grant 1970; Holmes

and Trickey 1974; McFarland 1961; MacIntyre 1971; Ronen 1983; Sanger 1985b,c; Sanger and Sanger 1986; Smith 1979; Smith and Bridges 1981a,b).

6. Tide gauge records (e.g., Emery 1980; Gornitz *et al.* 1982; Hicks 1972, 1978; Hicks and Crosby 1974; Lisitzin 1974:165–183).

2.2.1. Salt Marshes

Salt marshes may build up peat deposits as sea level rises. Sea-level curves have been constructed either by dating sections of a single core at different depths or by dating the basal sections of several cores along a transect up the back slope of the marsh to dry land. The single-core technique is fraught with difficulties. Salt marshes are dynamic environments. Tides continually carry organic material (such as drift wood) from unknown contexts into the marsh; storms can deposit foreign material on and in the marsh; and ice movement can displace blocks of marsh peat. Tidal channels meander continually, eroding and redepositing material (Deevey 1948:334–335; Letzsch and Frey 1980). As the marsh grows in depth with sea-level rise, the whole column "autocompacts" so that its original relationship to sea level is distorted (Kaye and Barghoorn 1964). Finally, some sampling techniques can displace or compact peat in the coring process (Belknap *et al.* 1987; Belknap and Kraft 1977; Nelson 1979; Redfield and Rubin 1962). "Vibracores" (Lanesky *et al.* 1979) are preferred because a large-diameter, continuous record can be obtained with a minimum of compaction or distortion of the sediments.

The transect technique is less susceptible to errors. The best technique is to date rhizomes and roots of salt-sensitive, high-marsh species recovered from above an incompactable substrate at the base of a large-diameter (>5 cm) core (Nelson 1979; Redfield and Rubin 1962). Care must be taken, however, to avoid dating humic material from pre-marsh terrestrial soils (Belknap *et al.* 1987; Nelson 1979). Recent work has used marsh foraminifera species to determine more precisely marsh environments and associated sea levels (Scott *et al.* 1984; Smith *et al.* 1984). Few sea-level curves have been produced that pay adequate attention to all of the problems that can arise in dealing with salt marsh peats.

2.2.2. Drowned Forests

Drowned forests can only yield *minimum* dates for sea-level positions because it is difficult to prove that trees died due to sea-level rise. Salt tolerances of various tree species are not well known, so that the relationship between killing water level and sea level cannot be clearly defined. There may also be an unknown time lag between the death of a tree and inundation (Thompson 1973).

2.2.3. Shoreline Features

Submerged shoreline features are difficult to identify and often impossible to date directly. Assumptions must be made about the context of dated material and the relationship of the shoreline feature to sea level. For example, storms can produce beach deposits containing dateable material of unknown provenance well above the associated sea level. Emerged shoreline features are easier to study, but again the context of dateable material in relation to contemporaneous sea level is difficult to establish.

2.2.4. Fossil Shells

The use of fossil shells is susceptible to error because assumptions must be made about their relationship to sea level (I. G. Macintyre *et al.* 1978). Shellfish species grow over different ranges of nearshore and intertidal environments. Elevational relationships to sea level may vary widely with local conditions. In addition, carbon dates on shells may not be strictly comparable to dates on other material (Berger *et al.* 1966; Mangerud 1972; Mangerud and Gulliksen 1975; Stuiver and Borns 1975; Stuiver *et al.* 1986; Tanaka *et al.* 1986) because shellfish incorporate carbon from sea water, not the atmosphere, and may produce metabolic carbonate. Thus, several of the asumptions of the radiocarbon dating technique (Terasmae 1984) are violated.

2.2.5. Archaeological Sites

In order to date sea-level positions using archaeological data, assumptions must be made about cultural interactions with the marine environment. The relationships between cultural occupation, site formation processes, discard behavior (Gluckman 1982), and contemporaneous sea level are subject to interpretation (Belknap and Hine 1983; Ceci 1984; Sanger and Kellogg, In press; Scarre 1984). Because such relationships are often the subject of study for archaeologists, sea-level curves based on archaeological data should be approached with extreme caution because circular reasoning may be involved.

Archaeological data, for example, are central to some controversial studies of sea level along the South Carolina and Georgia coasts (Brooks *et al.* 1979; Belknap and Hine 1983; Colquhoun and Brooks 1986; Colquhoun *et al.* 1981; DePratter and Howard 1981, 1983). Brooks *et al.* (1979) interpret changes in upland settlement patterns as independant evidence for sea-level fluctuations inferred from geological studies of coastal marshes. The settlement pattern models make many assumptions about cultural behavior.

Where historic structures are present, maximum and minimum values for sea-level positions can sometimes be obtained (e.g., Anderson *et al.* 1984; Flemming 1969; Grant 1970; Sanger 1985c; Smith and Bridges 1981a,b). Site presence gives only a *minimum* age for a shoreline if it can be established that activities represented by the archaeological deposits were, in fact, carried out above the sea level (i.e., tidal) datum (Barghoorn 1953; Kellogg 1985b; MacIntyre 1971; Sanger 1985b).

2.2.6. Tide Gauge Records

Tide gauge data are available for the very recent past. Regression analysis is used to obtain the trend of mean sea level from seasonal and other variations in the tides. Local subsidence is difficult to separate from eustatic contributions to tidal trends (Lisitzin 1974:165–183). Although such data are of little value for paleoenvironmental reconstructions, they can be important for those concerned with modern impacts on sites in the coastal zone. Coastal erosion is a particularly serious problem affecting many areas (e.g., Garrett 1983; Kellogg 1982; 1984; 1985a; Luedke 1980:25; Moran 1980; Sanger 1980; Simonsen 1978; Spiess 1981).

Present-day tidal variations are often ignored in compiling data for sea-level curves (Kidson 1982; Tooley 1985), but should be carefully considered in light of the potential for introducing errors (Belknap and Hine 1983).

2.3. Constructing Sea-Level Curves

Interpreting dates and constructing sea-level curves usually involves selecting the "best" dates from a study location and then fitting a regression line to them. Some studies apply tree ring calibration corrections to the radiocarbon dates (Berger 1983; Ralph *et al.* 1974; Stuiver and Kra 1986) to obtain more accurate rates of rise or sedimentation. Decisions as to which dates are anomalous, or what type of regression to use, are commonly justified on the basis of agreement with previous curves. Thus, original errors or oversimplifications can be propagated. The magnitude of possible errors is often not emphasized, especially in older studies. Another common practice is to adjust a eustatic curve or a local curve from a neighboring region on the basis of limited local data (e.g., Perlman 1979; van Andel *et al.* 1980). The resulting curve may bear little relationship to the actual sea-level history of the region of interest.

A realistic curve can only be constructed with data from one area and dates on comparable material. Errors in both dates and datum should be clearly indicated (Kidson 1982). Any line through the points on the curve is, of course, an abstraction. Because no data exist between the points on the line, the behavior of sea level between the points is technically unknown. Furthermore, it must be emphasized that each data point is relative to some tidal datum, such as mean high or mean low water. *Sea-level curves show changes in local water level, with respect to a tidal datum, through time.* Finally, some (e.g., Kidson 1982) have argued that the practice of plotting sea-level curves be discontinued in favor of probability envelopes which indicate the uncertainties in both datum and dating.

A further problem is the lack of continuous data at any one locale for the entire Holocene (at least along the eastern Atlantic coast of the United States). Data are usually available up to 6,000 years ago and for the period 13,000 to about 8,500 years ago. The preservation of nearshore and intertidal environments that yield evidence of former sea levels requires unusual conditions (Belknap and Kraft 1981, 1985; Kraft *et al.* 1983).

3. SEA LEVEL IN THE GULF OF MAINE

As an example of the complexities arising in the use of sea level data, the postglacial history of the Gulf of Maine (Figure 1) is considered here. The follow-

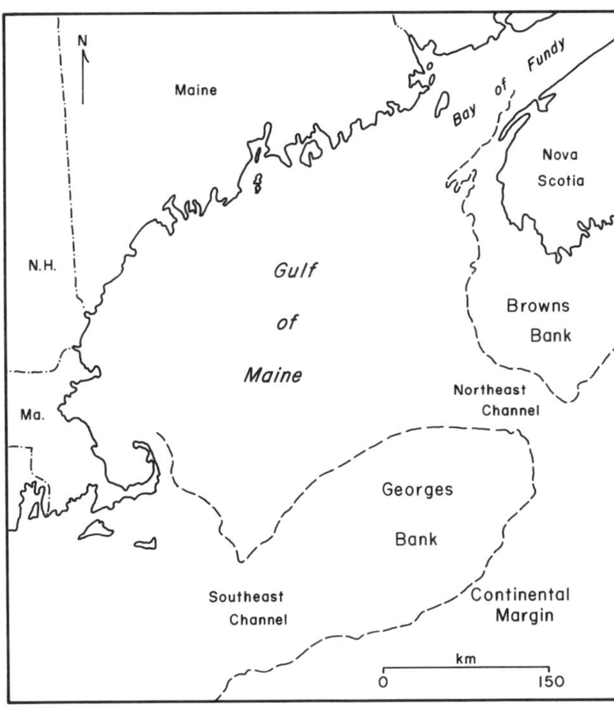

Figure 1. The Gulf of Maine region of the northeastern United States and adjacent Canada. The margins of Georges and Browns Banks have been drawn at 100 m depth. Places on Georges Bank, however, are as little as 10 m in depth.

ing is the standard late Quaternary history of the central Gulf of Maine based on Borns (1973), Hughes *et al.* (1985), Mayewski *et al.* (1981), and Schafer and Hartshorn (1965), with specific references as indicated. A critical examination of the data follows.

3.1. The Late Quaternary History of the Gulf of Maine

At glacial maximum, approximately 18,000–22,000 years ago (B.P.), Laurentide ice extended onto the outer banks (Georges and Browns Banks) to the edge of the Continental Shelf (Pratt and Schlee 1969). Recession began about 14,000 B.P. By 13,500 B.P., the ice had reached the position of the present coastline of Maine (Schnitker 1974). As the ice continued to recede, marine waters covered the land that had been depressed beneath the ice up to a maximum elevation of about 135 m above the present level (Bloom 1963; Schnitker 1974; Smith 1985; Stuiver and Borns 1975). By 12,000 B.P., the rate of isostatic rebound outpaced eustatic sea-level rise, so that marine waters receded past the present coastline. Recession continued until some time between 8000 and 9000 B.P. At that time, relative sea level stood about 65 m below its present level on the central Maine coast (Schnitker 1974). The Gulf of Maine would have been an inland sea connected to the Atlantic Ocean only by the narrow Northeast and shallower Southeast channels.

Tidal range in the early Gulf was probably minimal because it was cut off from the North Atlantic tidal system. After the outer banks became submerged between 4000 and 7000 B.P. (Grant 1970) due to relative sea-level rise (e.g., Belknap *et al.* 1987; Bloom 1963; Bradley 1953; Harrison and Lyons 1963; Johnson 1942, 1949; Keene 1971; Nelson 1979; Thompson 1973), the present meso- and macrotidal regime of the Gulf of Maine developed (Grant 1970). The natural frequency of the Gulf of Maine, determined by its length, width, and depth, is very close to the 12.4-hour period of North Atlantic tidal forcing. Thus, tides have increased in the Gulf due to resonance as Atlantic waters deepened over the outer banks. Small sea-level changes continue, as indicated by tide gauge records (Hicks and Crosby 1974).

Cultural responses to changes in the marine environment, such as the development of tidal flats, marsh systems, and upwelling, have been discussed extensively (e.g., Barber 1979, 1982; Borns 1971; Braun 1974; Dincauze 1973, 1974; Fladmark 1983; Johnson 1942, 1949; McManamon 1984; Newman 1974; Oldale 1985; Robinson 1985a,b; Sanger 1975, 1982, 1985a; Sanger and Belknap, 1987; Sanger and Sanger 1986; Snow 1972, 1974; Spiess *et al.* 1983; Tuck 1975; Yesner 1980, 1983).

3.2. A Critical Discussion

As a general outline of the past 14,000 years in the Gulf of Maine, the scenario outlined above is reasonable, and the basic sequence still stands. On close examination, however, all of the problems with sea level discussed earlier contribute uncertainties to the reconstruction.

First, there are competing ice sheet models for northern New England and maritime Canada (see Grant 1977; Hughes *et al.* 1985; Mayewski *et al.* 1981), which imply different positions for the Laurentide ice on the continental margins. As a consequence, the isostatic reaction of the whole region remains problematic (e.g., Quinlan and Beaumont 1981, 1982). It is especially difficult to relate early postglacial sea-level data from one area of the Gulf of Maine to another because depression and rebound varied widely (Oldale 1986). The problem has been approached by iterative modeling of ice sheet behavior and earth rheology, using relative sea-level data to constrain the mathematical models. The sea-level data are interpreted based on the glacial and earth models to be constrained (e.g., Quinlan and Beaumont 1981, 1982). Circular reasoning is almost impossible to avoid in this situation. Furthermore, there are few data on the isostatic behavior of the outer banks of the Gulf of Maine.

Fixed points do exist for some early postglacial events based on shells found in delta deposits above present sea level (Anderson *et al.* 1984; Bloom 1963; Crossen 1985; Stuiver and Borns 1975; Swift and Borns 1967). The dates given by Schnitker (1974), however, are extrapolated from the eustatic sea-level curve of Milliman and Emery (1968). The −65 m position for sea level at 8000–9000 B.P. is based on interpreting seismic profiles as showing a beach. Support for that interpretation comes from Belknap's (1985) report of locating the paleodelta of the Kennebec River at −65 m depth. The paleodelta of the Merrimack River (Oldale 1986; Oldale *et al.* 1983) has been defined by seismic profiles at −47 m depth. On the other hand, Knebel (1986) places the low stand at −40 m in Penobscot Bay. Similar features have been reported from seismic profiling elsewhere along the coast (e.g., Belknap *et al.* 1987; Kelley *et al.* 1984; Knebel and Scanlon 1985). The dating of the low stand has not improved, however, and the seismic interpretations have not been verified by coring.

Grant's (1970) modeling of tidal amplification in the Gulf of Maine was very simplistic and was also based on the Milliman and Emery (1968) eustatic sea-level curve. Recently, Scott and Greenberg (1983) applied a more detailed model to estimate the tidal behavior of the Gulf of Maine and Bay of Fundy systems. In contrast to Grant's linear increase in tidal range from 6000 B.P. to the present, they found that most of the increase in tidal ranges occurred earlier. The tidal model is very sensitive to the placement of the boundary to the Gulf of Maine on the outer banks (Greenberg 1979). The placement is adjusted according to the model's ability to reproduce *modern* tidal conditions. Although Scott and Greenberg (1983) used local sea-level data instead of an eustatic curve, their curve was a compilation of dates on a variety of materials from several locations in the Fundy/Scotia region, and is therefore somewhat questionable.

The issue of tidal amplification in the Gulf of Maine is of considerable importance to archaeologists in the region. Sanger (1975), based on Grant

(1970), noted that tidal amplification and cultural adaptation to the Gulf of Maine may be correlated. The development of intertidal resources—such as the soft shell clam, *Mya arenaria,* of obvious importance to the prehistoric inhabitants of the region, as indicated by abundant shell middens—may be closely tied to increasing tidal range. Tidal variations are important not only at the regional scale, but also at the local scale. The size, shape, and depth of a basin affects the tidal range so that it varies around the basin (Jardine 1975). For a dramatic example, tidal range at the mouth of the Bay of Fundy is about 5 m, whereas at the head of the bay it is as much as 15 m. "Sea level," though, is the same at both places.

Amos (1978) placed the onset of tidal amplification at about 6300 B.P. based on sedimentological studies of cores from Minas Bay at the head of the Bay of Fundy. By relating particle sizes to theoretical current velocities, Amos attempted to separate tidal amplification from relative sea-level rise. The dating is based on the total organic carbon of sediments in sections of the core from critical depths—a technique that is subject to possible errors because the provenance of dated carbon is unknown and may not be contemporaneous with the sediments. In addition, there may be other explanations for changes in current velocities.

Increasing tidal ranges after about 6000 B.P. could distort the rate of sea-level rise recorded by salt marsh peats. If tidal range is increasing while sea level remains constant, then salt marsh peat may reflect the accelerated tidal range rather than sea-level rise because salt marsh grasses will grow to the new tidal levels. For this reason, applying a curve from one locale to estimate the sea-level history of another or combining data from several places are questionable practices.

This may be part of the problem with comparing sea-level curves from around the Gulf of Maine region (e.g., Oldale 1986). Grant (1970) originally suggested tidal amplification to account for anomalous sea-level data from the Bay of Fundy. Sea-level curves from most of the East Coast showed a slowing rate of rise in the last 2,000–6,000 years, while the Bay of Fundy yielded increasing rates. Curves from Addison, Maine (Thompson 1973) and Seabrook Marsh, New Hampshire (Keene 1971) show significantly different relative sea-level rise rates, possibly due to differences in tidal range. Tidal range at Addison is 3.6 m, while tidal range at Seabrook is 2.5 m. Theoretically, the difference could be normalized to NGVD for present sea level, but not for the past, because differences in tidal range are not known. While fluctuations in tidal range may not be important for establishing shoreline positions, they are critical for reconstructing the extent of marshes or mudflats (e.g., Dincauze 1973; Perlman 1979). This problem is exaggerated in the Gulf of Maine, but should not be ignored in other areas (Belknap and Hine 1983). It should also be noted that Oldale (1986) redrafted the sea-level curves compared in his review paper. The original data are preferable for any specific studies.

The sea-level curves produced by Thompson (1973) and Keene (1971) were criticized by Nelson (1979:99–101) for use of small-diameter cores and failure to emphasize the magnitude of possible errors in datum. Furthermore, pre-marsh forest litter, not salt marsh peat, was dated at Seabrook (Keene 1971). The curve from Addison, Maine (Thompson 1973) may be affected by erosion and redeposition, as it was produced by the single-core method. Addison Marsh is currently undergoing intensive reevaluation to determine its bearing on local, relative sea level (Belknap *et al.*, 1987).

To further complicate matters, the coast of Maine is presently subject to differential downwarping (Anderson *et al.* 1984; Tyler *et al.* 1979). Therefore, separate sea-level histories may pertain to each area of the coast regardless of local or regional tidal variations. This problem is not limited to the Gulf of Maine region. Differential tectonic activity has been recorded for other areas of the Atlantic coast as well (e.g., Brown 1978; Grant 1980; Holdahl and Morrison 1974; Kranck 1972; Newman *et al.* 1980; Vanicek and Hamilton 1972).

4. CONCLUSIONS

Given the many and varied problems with sea-level data, it can be extremely difficult to reconstruct the marine environment adjacent to an archaeological site at specific times in the past, especially on the basis of sea-level data alone (Kraft *et al.* 1975). The same observation applies to all scales of interest—from a single cove to an ocean basin. Most of the difficulties, however, can be minimized by carefully considering the available information. Eustatic sea-level curves are useless for either local or regional reconstructions. Only *local, relative* sea-level rise data are applicable.

If this discussion has been overly cynical, it is because the literature on sea level is voluminous and often confusing or contradictory. Scientists in many different fields are involved in sea-level research (Lisitzin 1974:1). Margins of error and needs for precision vary widely, depending on the purposes of a particular discussion. For example, an error of 2 m means little in reconstructing Pleistocene ice sheets and would represent less than a 2% error in estimating the low stand of sea level at glacial maximum, but a 2 m difference could radically alter a local geomorphological reconstruction developed for archaeological purposes, and could represent a 100% error in sea level over the last few thousand years. Sea-level curves are rarely produced for their own sake, but are constructed for a specific reason, and thus may be totally inappropriate for application to another problem. If paleoenvironmental data are to contribute to archaeological studies, they must be critically applied, keeping in mind the problems, assumptions, and inherent limitations involved. This is especially true when dealing with sea-level data.

Acknowledgments

I thank George Nicholas for the invitation to write this paper, his editorial commitment and perseverance, and emergency road services. David Sanger provided guidance and editorial comments. I also thank Sharon L. Pekrul, James B. Oliver, William R. Belcher, Daniel F. Belknap, R. Craig Shipp, and Margaret P. O'Connell for reading and commenting on various drafts of this paper. My knowledge of sea-level data has benefited from many discussions with R. Scott Anderson, Daniel F. Belknap, Harold W. Borns, Jr., L. Kenneth Fink, Jr., Joseph T. Kelley, David Sanger, and R. Craig Shipp. Some bibliographic research was supported by a Research Assistanceship from the Graduate School of the University of Maine. Steve Bicknell drafted the figure.

5. REFERENCES

Amos, C. L., 1978, The Postglacial Evolution of the Minas Basin, Nova Scotia: A Sedimentological Interpretation, *Journal of Sedimentary Petrology* 48:965–982.

Anderson, R. S., and Race, C. D., 1981 Investigation of Salt Marsh Stratigraphy As an Indicator of Sea-Level Rise in Coastal Maine, in: *New England Seismotectonic Study Activities in Maine During Fiscal Year 1980* (W. B. Thompson, ed.), Maine Geological Survey, Augusta, Maine, pp. 79–96.

Anderson, W. A., Kelley, J. T., Thompson, W. B., Borns, H. W., Jr., Sanger, D., Smith, D. C., Tyler, D. A.. Anderson, R. S., Bridges, A. E., Crossen, K. I., Ladd, J. W., Anderson, B. J., and Lee, F. T., 1984, Crustal Warping in Coastal Maine, *Geology* 12:677–680.

Andrews, J. T., 1980, Progress in Relative Sea Level and Ice Sheet Reconstructions, Baffin Island, N. W. T. for the Last 125,000 Years, in: *Earth Rheology, Isostasy, and Eustacy* (N. A. Morner, ed.), Wiley-Interscience, New York, pp. 175–200.

Andrews, J. T., 1982, On the Reconstruction of Pleistocene Ice Sheets: A Review, *Quaternary Science Reviews* 1:1–30.

Andrews, J. T., and Miller, G. H., 1985, Discussion on the Paper by Denton and Hughes 1983, *Quaternary Research* 24:361–363.

Andrews, J. T., McGhee, R., and McKenzie-Pollock, L., 1971, Comparison of Elevations of Archaeological Sites and Calculated Sea Levels in Arctic Canada, *Arctic* 24:210–228.

Barber, R. J., 1979, *Human Ecology and the Estuarine Ecosystem: Prehistoric Exploitation in the Merrimack Valley,* Ph.D. Dissertation, Harvard University, Cambridge, Massachusetts, unpublished.

Barber, R. J., 1982, *The Wheeler Site: A Specialized Shellfish Processing Station on the Merrimack River,* Peabody Museum Monographs, No. 7, Harvard University, Cambridge, Massachusetts.

Barghoorn, E. S., 1953, Recent Changes in Sea Level along the New England Coast: New Archaeological Evidence, *Science* 117:597–598.

Belknap, D. F., 1985, The Submerged Glaciofluvial Paleodelta of the Kennebec River, West-Central Maine Coast, *Geological Society of America Abstracts with Programs* 17:4.

Belknap, D. F., and Hine, A. C., 1983, Evidence for a Sea-Level Lowstand between 4,500 and 2,400 Years B.P. on the Southeast Coast of the United States—Discussion, *Journal of Sedimentary Petrology* 53:680–682.

Belknap, D. F., and Kraft, J. C., 1977, Holocene Relative Sea Level Change and Coastal Stratigraphic Units on the Northwestern Flank of the Baltimore Canyon Trough Geosyncline, *Journal of Sedimentary Petrology* 47:610–629.

Belknap, D. F., and Kraft, J. C., 1981, Preservation Potential of Transgressive Coastal Lithosomes on the U.S. Atlantic Shelf, *Marine Geology* 42:429–442.

Belknap, D. F., and Kraft, J. C., 1985, Chapter VI. Barrier Platforms. Influence of Antecedent Geology on Stratigraphic Preservation Potential and Evolution of Delaware's Barrier Systems, *Marine Geology* 63:235–262.

Belknap, D. F., Andersen, B. J., Anderson, R. S., Anderson, W. A, Borns, H. W., Jr., Jacobson, G. L., Jr., Kelley, J. T., Shipp, R. C., Smith, D. C., Stuckenrath, R., Jr., Thompson, W. B., and Tyler, D. A., 1987, Late Quaternary Sea-Level Changes in Maine, in: *Sea-Level Change and Coastal Depositional Systems Architecture* (D. Nummedal, Q. H. Pilkey, Jr., and J. D. Howard, eds.), Society of Economic Paleontologists and Mineralogists Special Publication, pp. 65–79.

Berger, R., 1983, Sea Levels and Tree-Ring Calibrated Radiocarbon Dates, in: *Quaternary Coastlines and Marine Archaeology* (P. M. Masters and N. C. Flemming, eds.), Academic Press, New York, pp. 51–61.

Berger, R., Taylor, R. E., and Libby, W. F., 1966, Radiocarbon Content of Marine Shells from the California and Mexican West Coast, *Science* 153:864–866.

Binford, L. R., 1968, Post-Pleistocene Adaptations, in: *New Perspectives in Archaeology* (L. R. Binford and S. Binford, eds.), Aldine, Chicago, Illinois, pp. 313–341.

Blackwelder, B. W., Pilkey, O. H., and Howard, J. D., 1979, Late Wisconsinan Sea Levels on the Southeast U.S. Atlantic Shelf Based on In-Place Shoreline Indicators, *Science* 204:618–620.

Bloom, A. L., 1963, Late-Pleistocene Fluctuations of Sea Level and Postglacial Crustal Rebound in Coastal Maine, *American Journal of Science* 261:862–879.

Bloom, A. L., 1967, Pleistocene Shorelines: A New Test of Isostasy, *Geological Society of America Bulletin* 78:1477–1494.

Bloom, A. L., 1983, Sea Level and Coastal Morphology of the United States through the Late Wisconsin Maximum, in: *Late Quaternary Environments of the United States* (S. C. Porter, ed.), University of Minnesota, Minneapolis, pp. 215–229.

Bloom, A. L., and Stuiver, M., 1963, Submergence of the Connecticut Coast, *Science* 139:332–334.

Bond, G., 1978, Speculations on Real Sea-Level Changes and Vertical Motions of Continents at Selected Times in the Cretaceous and Tertiary Periods, *Geology* 6:247–250.

Borns, H. W., Jr., 1971, Possible Paleoindian Migration Routes in Northeastern North America, *Maine Archaeological Society Bulletin* 11(1):33–39.

Borns, H. W., Jr., 1973, Late Wisconsin Fluctuations of the Laurentide Ice Sheet in Southern and Eastern New England, *Geological Society of America Memoir* 136:37–45.

Bradley, W. H., 1953, Age of Intertidal Tree Stump at Robinhood, Maine, *American Journal of Science* 251:543–546.

Braun, D. P., 1974, Explanatory Models for the Evolution of Coastal Adaptation in Prehistoric New England, *American Antiquity* 39:582–596.

Broecker, W. S., 1966, Glacial Rebound and the Deformation of the Shorelines of Proglacial Lakes, *Journal of Geophysical Research* 71:4777–4783.

Brooks, M. J., Colquhoun, D. J., Pardi, R. R., Newman, W. S., and Abbott, W. H., 1979, Preliminary Archaeological and Geological Evidence for Holocene Sea-Level Fluctuations in the Lower Cooper River Valley, South Carolina, *The Florida Anthropologist* 32:85–103.

Brown, L. D., 1978, Recent Vertical Crustal Movement along the East Coast of the United States, *Tectonophysics* 44:205–231.

Carter, G. F., 1964, Some Thoughts on Coastal Archaeology, *Quarterly Bulletin of the Archaeological Society of Virginia* 18(3):46–55.

Ceci, L., 1984, Shell Midden Deposits As Coastal Resources, *World Archaeology* 16:62–74.

Clark, J. A., and Lingle, C. L., 1979, Predicted Relative Sea-Level Changes (18,000 Years B.P. to Present) Caused by Late-Glacial Retreat of the Antarctic Ice Sheet, *Quaternary Research* 11:279–298.

Clark, J. A., Farrell, W. E., and Peltier, W. R., 1978, Global Changes in Postglacial Sea Level, *Quaternary Research* 9:265–287.

Coleman, J. M., and Smith, W. G., 1964, Late Recent Rise of Sea Level, *Geological Society of America Bulletin* 75:833–840.

Colquhoun, D. J., and Brooks, M. J., 1986, New Evidence from the Southeastern U.S. for Eustatic Components in the Late Holocene Sea Levels, *Geoarchaeology* 1:275–291.

Colquhoun, D. J., Brooks, M. J., Michie, J., Abbott, W. B., Stapor, F. W., Newman, W., and Pardi, R. R., 1981, Location of Archaeological Sites with Respect to Sea Level in the Southeastern United States, *Striae* 14:144–150.

Crossen, K. J., 1985, *Glaciomarine Deltas in Southwestern Maine: Formation and Vertical Movements,* M.S. Thesis, Institute for Quaternary Studies, University of Maine, Orono, Maine, unpublished.

Curray, J. R., 1965, Late Quaternary History: Continental Shelves of the United States, in *The Quaternary of the United States* (H. E. Wright and D. G. Frey, eds.), Princeton University, Princeton, New Jersey, pp. 723–736.

Deevey, E. S., Jr., 1948, On the Date of the Last Rise of Sea Level in Southern New England, with Remarks on the Grassy Island Site, *American Journal of Science* 246:329–352.

Denton, G. H., Hughes, T. J., and Karlen, W., 1986, Global Ice-Sheet System Interlocked by Sea Level, *Quaternary Research* 26:3–26.

DePratter, C. B., and Howard, J. D., 1981, Evidence for a Sea-Level Lowstand between 4,500 and 2,400 Years B.P. on the Southeast Coast of the United States, *Journal of Sedimentary Petrology* 51:1287–1295.

DePratter, C. B., and Howard, J. D., 1983, Evidence for a Sea-Level Lowstand between 4,500 and 2,400 Years B.P. on the Southeast Coast of the United States—Reply, *Journal of Sedimentary Petrology* 53:682–685.

Dillon, W. P., and Oldale, R. N., 1978, Late Quaternary Sea-Level Curve: Reinterpretation Based on Glacio-Tectonic Influence, *Geology* 6:56–60.

Dincauze, D. F., 1973, Prehistoric Occupation of the Charles River Estuary: A Paleogeographic Study, *Bulletin of the Archaeological Society of Connecticut* 38:25–39.

Dincauze, D. F., 1974, An Introduction to the Archaeology of the Greater Boston Area, *Archaeology of Eastern North America* 2:39–66.

Dincauze, D. F., 1978, Open Season on Sacred Cows, *Archaeology of Eastern North America* 6:82–87.

Dincauze, D. F., 1981, Paleoenvironmental Reconstructions in the Northeast: The Art of Multidisciplinary Science, in: *Foundations in Northeast Archaeology* (D. R. Snow, ed.), Academic Press, New York, pp. 51–96.

Donn, W. L., Farrand, W. R., and Ewing, M., 1962, Pleistocene Ice Volumes and Sea-Level Lowering, *Journal of Geology* 70:206–214.

Edwards, R. L., and Merrill, A. S., 1977, A Reconstruction of the Continental Shelf Areas of Eastern North America for the Times 9,500 B.P. and 12,500 B.P., *Archaeology of Eastern North America* 5:1–43.

Emery, K. O., 1980, Relative Sea Level from Tide Gauge Records, *Proceedings of the National Academy of Sciences of the USA* 77(12):6968–6972.

Emery, K. O., and Edwards, R. L., 1966, Archaeological Potential of the Atlantic Continental Shelf, *American Antiquity* 31:733–737.

Emery, K. O., and Garrison, L. E., 1967, Sea Level 7,000 to 20,000 Years Ago, *Science* 157:684–687.

Emery, K. O., Wigley, R. L., and Rubin, M., 1965, A Submerged Peat Deposit off the Atlantic Coast of the United States, *Limnology and Oceanography* 10:R97–R102.

Emery, K. O., Wigley, R. J., Bartlett, A. S., Rubin, M., and Barghoorn, E. S., 1967, Freshwater Peat on the Continental Shelf, *Science* 158:1301–1307.

Emiliani, C., and Shackleton, N. J., 1974, The Brunhes Epoch: Paleotemperatures and Geochronology, *Science* 183:511–514.

Ericson, D. B., and Wollin, G., 1968, Pleistocene Climates and Chronology in Deep-Sea Sediments, *Science* 162:1227–1234.

Fairbanks, R. G., Weibe, P. H., and Be, A. W. H., 1980, Vertical Distribution and Isotopic Composition of Living Planktonic Foraminifera in the Western North Atlantic, *Science* 207:61–63.

Fairbridge, R. W., 1958, Dating the Latest Movements of the Quaternary Sea Level, *Transactions of the New York Academy of Sciences, Series II* 20:471–482.

Fairbridge, R. W., 1960, Recent World-Wide Sea-Level Changes, *Bulletin of the Massachusetts Archaeological Society* 21(3):49–51.

Fairbridge, R. W., 1976, Shellfish-Eating Preceramic Indians in Coastal Brazil, *Science* 191:353–359.

Farrell, W. E., and Clark, J. A., 1976, On Postglacial Sea Level, *Geophysical Journal of the Royal Astronomical Society* 46:646–667.

Fladmark, K. R., 1983, A Comparison of Sea Level and Prehistoric Cultural Development on the East and West Coasts of Canada, in: *Cultural Complexity and Cultural Development on the Northeast and Northwest Coasts* (R. Nash, ed.), Department of Archaeology, Simon Fraser University, Burnaby, British Columbia, Canada, pp. 65–75.

Flemming, N. C., 1969, *Archaeological Evidence for Eustatic Change of Sea Level and Earth Movements in the Western Mediterranean during the Last 2,000 Years,* Geological Society of America Special Paper No. 109, Boulder, Colorado.

Flint, R. F., 1971, *Glacial and Quaternary Geology,* Wiley, New York.

Frankel, L., and Crow, G. H., 1961, Drowned Forests along the Eastern Coast of Prince Edwards Island, *Canadian Journal of Geology* 69:352–357.

Fujii, S., Nasu, N., Smith A. J., Fuji, N., Mizutani, Y., Shimakura, M., Konishi, K., Igarashi, C., Muramoto, J., Takemura, T., Shimoda, T., Boggs, S., Jr., Fujioka, K., Mappa, H., Kawahata, H., Kong, Y. S., and Tanaka, T., 1987, Submerged Forest off Nyuzen, Kurobgawa Alluvial Fan, Toyama Bay, Central Japan, *Boreas* 15:265–277.

Garrett, S. E., 1983, *Coastal Erosion and Archaeological Resources on National Wildlife Refuges in the Southeast,* Southeast Regional Office, National Park Service, Atlanta, Georgia.

Gluckman, S. J., 1982, Underwater Sites in South Florida: A Preliminary Predictive Model, *The Florida Anthropologist* 35:93–99.

Goldthwait, R. P., 1935, The Damariscotta Shell Heaps and Coastal Stability, *American Journal of Science* 230:1–13.

Goodwin, H., Suggate, P., and Willis, E. J., 1958, Radiocarbon Dating the Eustatic Rise in Ocean Sea Level, *Nature* 181:1518–1519.

Gornitz, V., Lebedeff, S., and Hansen, J., 1982, Global Sea-Level Trend in the Past Century, *Science* 215:1611–1614.

Grant, D. R., 1970, Recent Coastal Submergence of the Maritime Provinces, *Canadian Journal of Earth Sciences* 7:676–689.

Grant, D. R., 1977, Glacial Styles and Ice Limits, the Quaternary Stratigraphic Record, and Changes of Land and Ocean Level in the Atlantic Provinces, Canada, *Geographie Physique et Quaternaire* 31(3–4):247–260.

Grant, D. R., 1980, Quaternary Sea-Level Change in Atlantic Canada As Indication of Crustal Delevelling, in: *Earth Rheology, Isostasy, and Eustacy* (N. A. Morner, ed.), Wiley-Interscience, New York, pp. 201–214.

Greenberg, D. A., 1979, A Numerical Model Investigation of Tidal Phenomena in the Bay of Fundy and Gulf of Maine, *Marine Geodesy* 2:161–187.

Haq, B. U., Hardenbol, J., and Vail, P. R., 1987, Chronology of Fluctuating Sea Levels Since the Triassic, *Science* 235:1156–1167.

Harrison, W., and Lyons, C. J., 1963, Sea Level and Crustal Movements along the New England–Acadian Shore 4,500–3,000 B.P., *Journal of Geology* 71:96–108.

Hecht, A. D., and Savin, S. M., 1970, Oxygen-18 Studies of Recent Planktonic Foraminifera: Comparisons of Phenotypes and of Test Parts, *Science* 170:69–71.

Hecht, A. D., and Savin, S. M., 1972, Phenotypic Variation and Oxygen Isotope Ratios in Recent Planktonic Foraminifera, *Journal of Foraminiferal Research* 2:55–67.

Hicks, S. D., 1972, Vertical Crustal Movements from Sea-Level Measurements along the East Coast of the United States, *Journal of Geophysical Research* 77:5930–5934.

Hicks, S. D., 1978, An Average Geopotential Sea-Level Series for the United States, *Journal of Geophysical Research* 83:1377–1379.

Hicks, S. D., and Crosby, J. E., 1974, *Trends and Variability of Yearly Mean Sea Level, 1893–1972,* National Oceanic and Atmospheric Administration Technical Memo NOS13, Department of Commerce, Washington, D.C.

Holdahl, S. R., and Morrison, N. L., 1974, Regional Investigations of Vertical Crustal Movements in the U.S., Using Precise Relevelings and Mareograph Data, *Tectonophysics* 23:373–390.

Holmes, N. H., Jr., and Trickey, E. B., 1974, Late Holocene Sea-Level Oscillations in Mobile Bay, *American Antiquity* 39:122–124.

Hughes, T. J., Denton, G. H., Anderson, B. G., Schilling, D. H., Fastook, J. L., and Lingle, C. S., 1981, The Last Great Ice Sheets: A Global View, in: *The Last Great Ice Sheets* (G. H. Denton and T. J. Hughes, eds.), Wiley-Interscience, New York, pp. 257–317.

Hughes, T., Borns, H. W., Jr., Fastook, J. L., Hyland, M. R., Kite, J. S., and Lowell, T. V., 1985, Models of Glacial Reconstruction and Deglaciation Applied to Maritime Canada and New England, *Geological Society of America Special Paper* 197:139–150.

Hussey, A. M., II, 1959, Age of Intertidal Tree Stumps at Wells Beach and Kennebunk Beach, Maine, *Journal of Sedimentary Petrology* 29:464–465.

Imbrie, J., and Imbrie, K. P., 1986, *Ice Ages: Solving the Mystery,* 2nd ed., Harvard University Press, Cambridge, Massachusetts.

Imbrie, J., van Donk, J., and Kipp, N. G., 1973, Paleoclimatic Investigation of a Late Pleistocene Caribbean Deep-Sea Core: Comparison of Isotopic and Faunal Methods, *Quaternary Research* 3:10–38.

Inman, D. L., 1983, Application of Coastal Dynamics to the Reconstruction of Paleocoastlines in the Vicinity of La Jolla, California, in: *Quaternary Coastlines and Marine Archaeology* (P. M. Masters and N. C. Flemming, eds.), Academic Press, New York, pp. 1–49.

Jardine, W. G., 1975, The Determination of Former Sea Level in Areas of Large Tidal Range, in: Quaternary Studies: Selected Papers from the IX INQUA Congress (R. P. Suggate and M. M. Cresswall, eds.), *Royal Society of New Zealand Bulletin* 13: 163–168.

Johnson, F. (ed.), 1942, *The Boylston Street Fishweir,* Paper No. 2, R. S. Peabody Foundation for Archaeology, Phillips Academy, Andover, Massachusetts.

Johnson, F. (ed.), 1949, *The Boylston Street Fishweir II,* Paper No. 4, R. S. Peabody Foundation for Archaeology, Phillips Academy, Andover, Massachusetts.

Kaye, C. A., and Barghoorn, E. S., 1964, Late Quaternary Sea-Level Changes and Crustal Rise at Boston, Massachusetts, with Notes on Autocompaction of Peat, *Geological Society of America Bulletin* 75:63–80.

Kazanskiy, A. B., 1985, A Hypothesis for the Sawlike Pattern of World Sea-Level Fluctuations, *Quaternary Research* 24:285–294.

Keene, H. W., 1971, Postglacial Submergence and Salt Marsh Evolution in New Hampshire, *Maritime Sediments* 7:64–68.

Kelley, J. T., Belknap, D. F., and Shipp, R. C., 1984, The Marine Geology of Casco Bay, Maine, *Geological Society of America Program with Abstracts* 17:28.

Kellogg, D. C., 1982, *Environmental Factors in Archaeological Site Location for the Boothbay Maine Region, with an Assessment of the Impact of Coastal Erosion on the Archaeological Record,* M.S. Thesis, Institute for Quaternary Studies, University of Maine at Orono, unpublished.

Kellogg, D. C., 1984, Site Survey and Settlement Pattern Analysis of the Muscongus/St. George Region of the Maine Coast, Report submitted to the Maine Historic Preservation Commission, Augusta, and Maine Sea Grant, Orono, Maine.

Kellogg, D. C., 1985a, 1984 Survey, Testing, and Shore Studies in Muscongus Bay, Maine, Open File Report, Maine Geological Survey, Augusta, Maine.

Kellogg, D. C., 1985b, Implications of Archaeological Deposits for Coastal Stability of the West Central Maine Coast, Open File Report, Maine Geological Survey, Augusta, Maine.

Kidson, C., 1982, Sea-Level Changes in the Holocene, *Quaternary Science Reviews* 1:121–151.

Killingley, J. E., 1983, Effects of Diagenetic Recrystallization on $^{18}O/^{16}O$ Values of Deep-Sea Sediments, *Nature* 301:594–597.

Knebel, H. J., 1986, Holocene Depositional History of a Large Glaciated Estuary, Penobscot Bay, Maine, *Marine Geology* 73:215–236.

Knebel, H. J., and Scanlon, K. M., 1985, Sedimentary Framework of Penobscot Bay, Maine, a Large Glaciated Estuary, *Marine Geology* 65:305–324.

Kraft, J. C., 1976, Geological Reconstructions of Ancient Coastal Environments in the Vicinity of the Island Field Archaeological Site, Kent County, Delaware, *Transactions of the Delaware Academy of Sciences* 7:31–66.

Kraft, J. C., 1985, Marine Environments: Paleogeographic Reconstructions in the Littoral Region, in: *Archaeological Sediments in Context* (J. K. Stein and W. R. Farrand, eds.), Center for the Study of Early Man, University of Maine, Orono, pp. 111–125.

Kraft, J. C., and John, C. J., 1978, Paleogeographic Analysis of Coastal Archaeological Settings in Delaware, *Archaeology of Eastern North America* 6:41–60.

Kraft, J. C., Rapp. J. R., and Aschenbrenner, S. E., 1975, Late Holocene Paleogeography of the Coastal Plain of the Gulf of Messenia, Greece, and Its Relationship to Archaeological Settings and Coastal Change, *Geological Society of America Bulletin* 86:1191–1208.

Kraft, J. C., Aschenbrenner, S. E., and Rapp, G., 1977, Paleogeographic Reconstructions of Coastal Aegean Archaeological Sites, *Science* 195:941–947.

Kraft, J. C., Belknap, D. F., and Kayan, I., 1983, Potentials of Discovery of Human Occupation Sites on the Continental Shelves and Nearshore Coastal Zone, in: *Quaternary Coastlines and Marine Archaeology* (P. M. Masters and N. C. Felmming, eds.), Academic Press, New York, pp. 87–120.

Kranck, K., 1972, Geomorphological Development and Post-Pleistocene Sea-Level Changes, Northumberland Straight, Maritime Provinces, *Canadian Journal of Earth Sciences* 9:835–844.

Lanesky, D. E., Logan, B. W., Brown, R. G., and Hine A. C., 1979, A New Approach to Portable Vibracoring Underwater and on Land, *Journal of Sedimentary Petrology* 49:654–657.

Letzsch, W. S., and Frey, R. W., 1980, Deposition and Erosion in a Holocene Salt Marsh, Sapelo Island, Georgia, *Journal of Sedimentary Petrology* 50:529–542.

Lisitzin, E., 1974, *Sea-Level Changes,* Elsevier Oceanographic Series No. 8, Elsevier, Amsterdam.

Luedtke, B. E., 1980, The Calf Island Site and the Late Prehistoric Period in Boston Harbor, *Man in the Northeast* 20:25–76.

Lyons, C. J., and Goldthwait, J. W., 1934, An Attempt to Cross-Date Trees in Drowned Forests, *Geographical Review* 24:606–614.

Macintyre, I. G., Blackwelder, B. W., Land, L. S., and Stuckenrath, R., 1975, North Carolina Shelf-Edge Sandstone: Age, Environment of Origin, and Relationship to Pre-existing Sea Level, *Geological Society of America Bulletin* 86:1073–1078.

Macintyre, I. G., Pilkey, O. H., and Stuckenrath, R., 1978, Relict Oysters on the United States Atlantic Continental Shelf: A Reconsideration of their Usefulness in Understanding Late Quaternary Sea-Level History, *Geological Society of America Bulletin* 89:277–282.

MacIntyre, W. G., 1971, Methods of Correlating Cultural Remains with the Stages of Coastal Development, in: *Introduction to Coastal Development* (J. A. Steers, ed.), Coastal Studies Institute, Louisiana State University, Baton Rouge, pp. 188–203.

Mangerud, J., 1972, Radiocarbon Dating of Marine Shells, Including a Discussion of the Apparent Age of Recent Shells from Norway, *Boreas* 1:143–172.

Mangerud, J., and Gulliksen, S., 1975, Apparent Radiocarbon Ages of Recent Marine Shells from Norway, Spitsbergen, and Arctic Canada, *Quaternary Research* 5:263–273.

Marcus, L. F., and Newman, W. S., 1983, Hominid Migrations and the Eustatic Sea-Level Paradigm: A Critique, in: *Quaternary Coastlines and Marine Archaeology* (P. M. Masters and N. C. Flemming, eds.), Academic Press, New York, pp. 51–85.

Masters, P. M., and Flemming, N. C., 1983, *Quaternary Coastlines and Marine Archaeology*, Academic Press, New York.

Mayewski, P. A., Denton, G. H., and Hughes, T. J., 1981, Late Wisconsin Ice Sheets in North America, in: *The Last Great Ice Sheets* (G. H. Denton and T. J. Hughes, eds.), Wiley-Interscience, New York, pp. 67–178.

McFarland, E., Jr., 1961, Radiocarbon Dating of Late Quaternary Deposits, South Louisiana, *Geological Society of America Bulletin* 72:129–158.

McGinnis, L. D., 1968, Glacial Crustal Bending, *Geological Society of America Bulletin* 79:769–776.

McManamon, F. P. (ed.), 1984, *Chapters in the Archaeology of Cape Cod, I,* 2 volumes, Cultural Resource Management Study No. 8, North Atlantic Regional Office, National Park Service, Boston, Massachusetts.

McMaster, R. L., and Garrison, L. E., 1967, A Submerged Holocene Shoreline near Block Island, Rhode Island, *Journal of Geology* 75:335–340.

Merrill, A. S., Emery, K. O., and Rubin, M., 1965, Ancient Oyster Shells on the Atlantic Continental Shelf, *Science* 147:398–400.

Milliman, J. D., and Emery, K. O., 1968, Sea Levels during the Past 35,000 Years, *Science* 162:1121–1123.

Mix, A. C., and Ruddiman, W. F., 1984, Oxygen-Isotope Analysis and Pleistocene Ice Volumes, *Quaternary Research* 21:1–20.

Moran, G. P., 1980, Problems in the Preservation of Archaeological Resources in the Coastal Zone of Rhode Island: A Management Perspective, in: *Proceedings of the Coastal Archaeology Session* (Commission on the Coastal Environment, International Geophysical Union, Newport, Rhode Island, 1979), pp. 1–7.

Morner, N. A., 1971, Eustatic Changes during the Last 20,000 Years and the Method of Separating the Isostatic and Eustatic Factors in an Uplifted Area, *Palaeogeography, Palaeoclimatology, and Palaeoecology* 9:153–181.

Morner, N. A., 1976, Eustacy and Geoid Changes, *Journal of Geology* 84:123–151.

Morner, N. A. (ed.), 1980, *Earth Rheology, Isostasy, and Eustacy,* Wiley-Interscience, New York.

Nelson, B. W., 1979, *Shoreline Changes and Physiography of Maine's Sandy Coastal Beaches,* M.S. Thesis, Department of Oceanography, University of Maine at Orono, Orono, Maine, unpublished.

Newman, W. S., 1974, A Comment on Snow's "Rising Sea Level and Cultural Ecology in Northern New England," *American Antiquity* 39:135–136.

Newman, W. S., and Marsh, S., 1968, Littoral of the Northeastern United States: Late Quaternary Warping, *Science* 160:1110–1112.

Newman, W. S., and Rusnak, G. A., 1965, Holocene Submergence of the Eastern Shore of Virginia, *Science* 148:1464–1466.

Newman, W. S., Clinqemani, L. J., Pardi, R. R., and Marcus, L. F., 1980, Holocene Delevelling of the United States' East Coast, in: *Earth Rheology, Isostasy, and Eustacy* (N. A. Morner, ed.), Wiley-Interscience, New York, pp. 449–463.

Oldale, R. N., 1985, Rapid Postglacial Shoreline Changes in the Western Gulf of Maine and the Paleoindian Environment, *American Antiquity* 50:145–150.

Oldale, R. N., 1986, Late-Glacial and Postglacial Sea-Level History of New England: A Review of Available Sea-Level Curves, *Archaeology of Eastern North America* 14:89–99.

Oldale, R. N., and O'Hara, C. J., 1980, New Radiocarbon Data from the Inner Shelf of Southern Massachusetts and a Local Sea-Level Rise Curve for the Past 12,000 Years, *Geology* 8:102–106.

Oldale, R. N., Wommack, C. E., and Whitney, A. B., 1983, Evidence for a Postglacial Low Relative Sea-Level Stand in the Drowned Delta of the Merrimack River, Western Gulf of Maine, *Quaternary Research* 19:325–336.

Paterson, W. S. B., 1972, Laurentide Ice Sheet: Estimated Volumes during the Wisconsin, *Reviews of Geophysics and Space Physics* 10:855–917.

Pearson, C. E., Kelley, R. A., Weinstein, R. A., and Gagliano, S. M., 1986, *Archaeological Investigations on the Outer Continental Shelf: A Study of the Sabine River Valley, Offshore Louisiana and Texas,* Minerals Management Service, U.S. Department of the Interior, Reston, Virginia.

Peltier, W. R., 1984, The Rheology of the Planetary Interior, *Journal of Rheology* 28:665–697.

Peltier, W. R., Farrel, W. E., and Clark, J. A., 1978, Glacial Isostacy and Relative Sea Level: A Global Finite Element Model, *Tectonophysics* 50:81–110.

Perlman, S. M., 1979, The Sandy Point Site: An Example of Research Designs in Contract Archaeology, *Man in the Northeast* 17:55–124.

Perlman, S. M., 1980, An Optimal Diet Model, Coastal Variability, and Hunter-Gatherer Behavior, in: *Advances in Archaeological Method and Theory,* Vol. 3 (M. B. Schiffer, ed.), Academic Press, New York, pp. 257–310.

Poag, C. W., and Ward, L. W., 1987, Cenezoic Unconformities and Depositional Supersequences of North Atlantic Continental Margins: Testing the Vail Model, *Geology* 15:159–162.

Pratt, R. M., and Schlee, J., 1969, Glaciation on the Continental Margin off New England, *Geological Society of America Bulletin* 80:2335–2342.

Quinlan, G., and Beaumont, T. C., 1981, A Comparison of Observed and Theoretical Postglacial Sea Level in Atlantic Canada, *Canadian Journal of Earth Sciences* 18:1146–1163.

Quinlan, G., and Beaumont, T. C., 1982, The Deglaciation of Atlantic Canada As Reconstructed from the Postglacial Relative Sea-Level Record, *Canadian Journal of Earth Sciences* 19:2232–2246.

Ralph, E. K., Michael, H. N., and Han, M. C., 1974, Radiocarbon Dates and Reality, *Archaeology of Eastern North America* 2:1–29.

Redfield, A. C., 1967, Postglacial Changes in Sea Level in the Western North Atlantic, *Science* 157:687–691.

Redfield, A. C., and Rubin, M., 1962, The Age of Salt Marsh Peat and Its Relation to Recent Changes in Sea Level at Barnstable, Massachusetts, *Proceedings of the National Academy of Sciences of the USA* 48(10):1728–1735.

Richardson, J. B., III, 1981, Modeling the Development of Sedentary Maritime Economies on the Coast of Peru: A Preliminary Statement, *Annals of the Carnegie Museum* (University of Pittsburgh) 50:139–149.

Roberts, M. (ed.), 1979, *Summary and Analysis of Cultural Resource Information on the Continental Shelf from the Bay of Fundy to Cape Hatteras,* Institute for Conservation Archaeology, Peabody Museum, Harvard University, Cambridge, Massachusetts.

Robinson, B. S., 1985a, An Intertidal Archaeological Survey of Scarborough Marsh, *Maine Archaeological Society Bulletin* 25:17–45.

Robinson, B. S., 1985b, The Nelson Island and Seabrook Marsh Sites: Late Archaic, Marine-Oriented People on the Central New England Coast, *Occasional Publications in Northeastern Archaeology* 9:1–107.

Ronen, A., 1983, Late Quaternary Sea Levels Inferred from Coastal Stratigraphy and Archaeology in Israel, in: *Quaternary Coastlines and Marine Archaeology* (P. M. Masters and N. C. Flemming, eds.), Academic Press, New York, pp. 121–134.

Ruddiman, W. F., and Duplessy, J. C., 1985, Conference on the Last Deglaciation: Timing and Mechanism, *Quaternary Research* 23:1–17.

Salwen, B., 1962, Sea Levels and Archaeology in the Long Island Sound Area, *American Antiquity* 28:46–55.

Salwen, B., 1975, Postglacial Environments and Cultural Change in the Hudson River Basin, *Man in the Northeast* 10:43–70.

Sanger, D., 1975, Cultural-Change As an Adaptive Process in the Maine–Maritime Region, *Arctic Anthropology* 12:60–75.

Sanger, D., 1980, Coastal Archaeology and Site Loss in Maine, in: *Proceedings of the Coastal Archaeology Session* (Commission on the Coastal Environment, International Geophysical Union, Newport, Rhode Island, 1979), pp. 8–11.

Sanger, D., 1982, Changing Views of Aboriginal Seasonality in the Gulf of Maine, *Canadian Journal of Archaeology* 12:195–203.

Sanger, D., 1985a, Cultural Ecology in Passamaquoddy Bay, New Brunswick, *Maine Archaeological Society Bulletin* 25:10–16.

Sanger, D., 1985b, Sea-Level Rise in Passamaquoddy Bay: Archaeology and Sediment Cores, in: *Investigations of Recent Crustal Movements in Maine during Fiscal Year 1983,* (W. B. Thompson and J. T. Kelley, eds.), Maine Geological Survey, Augusta, Maine, pp. 47–58.

Sanger, D., 1985c, Sea-Level Rise and Archaeology in the Damariscotta River, in: *Investigations of Recent Crustal Movements in Maine during Fiscal Year 1983* (W. B. Thompson and J. T. Kelley, eds.), Maine Geological Survey, Augusta, Maine, pp. 59–72.

Sanger, D., and Belknap, D. F., 1987, Human Responses to Changing Marine Environments in the Gulf of Maine, Paper presented at the Chacmool Conference, Department of Anthropology, University of Calgary, Canada.

Sanger, D., and D. C. Kellogg, In press, Prehistoric Archaeology and Evidence for Coastal Subsidence on the Coast of Maine, Maine Geological Survey, Augusta, Maine.

Sanger, D., and Sanger, M. J., 1986, Boom and Bust on the River: The Story of the Damariscotta Oyster Shell Heaps, *Archaeology of Eastern North America* 14(Fall):65–78.

Scarre, C., 1984, Archaeology and Sea Level in West-Central France, *World Archaeology* 16(1):98–107.

Schafer, J. P., and Hartshorn, J. H., 1965, The Quaternary of New England, in: *The Quaternary of the United States* (H. E. Wright and D. G. Frey, eds.), Princeton University, Princeton, New Jersey, pp. 113–127.

Schiffelbein, P., and Hills, S., 1984, Direct Assessment of Stable Isotope Variability in Planktonic Foraminifera Populations, *Palaeogeography, Palaeoclimatology, and Palaeoecology* 48:197–214.

Schnitker, D., 1974, Postglacial Emergence of the Gulf of Maine, *Geological Society of America Bulletin* 85:491–494.

Scholl, D. W., and Stuiver, M., 1967, Recent Submergence of Southern Florida: A Comparison with Adjacent Coasts and Other Eustatic Data, *Geological Society of America Bulletin* 78:437.

Scholl, D. W., Craighead, F. C., and Stuiver, M., 1969, Florida Submergence Curve Revised: Its Relation to Coastal Sedimentation, *Science* 163:562–564.

Scott, D. B., and Greenberg, D., 1983, Relative Sea-Level Rise and Tidal Development in the Fundy Tidal System, *Canadian Journal of Earth Sciences* 20:1554–1564.

Scott, D. B., Medioli, F. S., and Duffett, T. E., 1984, Holocene Rise of Relative Sea Level at Sable Island, Nova Scotia, *Geology* 12:173–176.

Shackleton, N. J., 1967, Oxygen Isotopes and Pleistocene Temperatures Reassessed, *Nature* 215:15–17.

Shackleton, N. J., and Opdyke, N. D., 1973, Oxygen Isotope and Paleomagnetic Stratigraphy of Equatorial Pacific Core V28-238: Oxygen Isotope Temperature and Ice Volumes on a 10^5 and 10^6 Year Scale, *Quaternary Research* 3:39–55.

Shalowitz, A. L., 1964, *Shore and Sea Boundaries,* Vol. 1, U.S. Department of Commerce, Coast and Geodetic Survey, Washington, D.C.

Shepard, F. P., 1964, Sea-Level Changes in the Past 6,000 Years: Possible Archaeological Significance, *Science* 143:575–576.

Shepard, F. P., and Suess, H. E., 1956, Rate of Postglacial Rise of Sea Level, *Science* 123:1082–1083.

Simonsen, B. J., 1978, *Attrition of Coastal Archaeological Resources in the Maritime Provinces of Canada,* Part I, Reports in Archaeology No. 2, Council of Maritime Premiers, Halifax, Nova Scotia.

Smith, D. A., Scott, D. B., and Medioli, F. S., 1984, Marsh Foraminifera in the Bay of Fundy: Modern Distributions and Application to Sea-Level Determinations, *Maritime Sediments and Atlantic Geology* 20(3):127–142.

Smith, D. C., 1979, Historical Evidence of Sea-Level Change along the Maine Coast, in: *Annual Report and Maps for Brittle Fracture and Crustal Warping Studies in the State of Maine during Fiscal Year 1979* (W. B. Thompson, ed.), Maine Geological Survey, Augusta, Maine, pp. 41–52.

Smith, D. C., and Bridges, A. E., 1981a, The Historical Dating of Salt Marsh Dikes, in: *New England Seismotectonic Study Activities in Maine during Fiscal Year 1980* (W. B. Thompson, ed.), Maine Geological Survey, Augusta, Maine, pp. 91–98.

Smith, D. C., and Bridges, A. E., 1981b, The Historical Dating of Salt Marsh Dikes, Part ll, in: *New England Seismotectonic Study Activities in Maine during Fiscal Year 1981* (W. B. Thompson, ed.), Maine Geological Survey, Augusta, Maine, pp. 75–78.

Smith, G. W., 1985, Chronology of Late Wisconsinan Deglaciation of Coastal Maine, *Geological Society of America Special Paper* 197:29–44.

Snow, D. R., 1972, Rising Sea Level and Prehistoric Cultural Ecology in Northern New England, *American Antiquity* 37:211–221.

Snow, D. R., 1974, Reply to Newman, *American Antiquity* 39:136–137.

Spiess, A. E., 1981, Coastal Erosion: Worst Current Threat to Maine Archaeology, *Contract Abstracts and CRM Archaeology* 1:4–10.

Spiess, A., Bourque, B. J., and Cox, S., 1983, Five Thousand Years of Hunting and Gathering in Penobscot Bay: Thoughts on the Presence of Cultural Complexity in the Northeast, in: *Cultural Complexity and Cultural Development on the Northeast and Northwest Coasts* (R. Nash, ed.), Department of Archaeology, Simon Fraser University, Burnaby, British Columbia, Canada, pp. 98–101.

Stephens, N., and Synge, F. M., 1966, Pleistocene Shorelines, in: *Essays in Geomorphology* (G. H. Dury, ed.), Heinmann, London, pp. 1–51.

Stright, M. J., 1986, Human Occupation of the Continental Shelf during the Late Pleistocene/Early Holocene: Methods of Site Location, *Geoarchaeology* 1:347–364.

Stuiver, M., and Borns, H. W., Jr., 1975, Late Quaternary Marine Invasion in Maine: Its Chronology and Associated Crustal Movements, *Geological Society of America Bulletin* 86:99–104.

Stuiver, M., and Daddario, H. J., 1963, Submergence of the New Jersey Coast, *Science* 142:951.

Stuiver, M., and Kra, R. (eds.), 1986, Calibration Issue, *Radiocarbon* 28(2B).

Stuiver, M., Pearson, G. W., and Braziunas, T., 1986, Radiocarbon Age Calibration of Marine Samples Back to 9,000 cal. yr. B.P., *Radiocarbon* 28(2B):980–1021.

Swift, D. J. P., and Borns, H. W., Jr., 1967, A Raised Fluviomarine Outwash Terrace, North Shore of the Minas Basin, Nova Scotia, *Journal of Geology* 75:693–709.

Tanaka, N., Monaghan, M. C., and Rye, D. M., 1986, Contribution of Metabolic Carbon to Mollusc and Barnacle Shell Carbonate, *Nature* 320:520–523.

Terasmae, J., 1984, Radiocarbon Dating: Some Problems and Potential Developments, in: *Quaternary Dating Methods* (W. C. Mahaney, ed.), Elsevier, Amsterdam, pp. 1–15.

Thompson, S. W., 1973, *Sea-Level Rise along the Maine Coast during the Last 3,000 Years,* M.S. Thesis, Department of Geology, University of Maine, Orono, Maine, unpublished.

Tooley, M. J., 1985, Sea Levels, *Progress in Physical Geography* 9(1):113–120.

Tuck, J. A., 1975, The Northeastern Maritime Continuum: 8,000 Years of Cultural Development in the Far Northeast, *Arctic Anthropology* 12:139–147.

Tyler, D. A., Ladd, J., and Borns, H. W., Jr., 1979, Interim Report on Modern Crustal Subsidence in Eastern Maine, Open File Report 79-1, Maine Geological Survey, Augusta, Maine.

Vail, P. R., Mitchum, R. M., Jr., and Thompson, S., III, 1977, Seismic Stratigraphy and Global Changes of Sea Level, Part 4: Global Cycles of Relative Changes of Sea Level, *American Association of Petroleum Geologists Memoir* 26:83–97.

Van Andel, T. H., and Lianos, N., 1984, High-Resolution Seismic Reflection Profiles for the Reconstruction of Postglacial Transgressive Sequences: An Example from Greece, *Quaternary Research* 22:31–45.

Van Andel, T. H., and Shackleton, J. C., 1982, Late Paleolithic and Mesolithic Coastlines of Greece and the Aegean, *Journal of Field Archaeology* 9:445–454.

Van Andel, T. H., Jacobsen, T. W., Jolly, J. B., and Lianos, N., 1980, The Quaternary History of the Coastal Zone near Franchthi Cave, Southern Argolid, Greece, *Journal of Field Archaeology* 7:389–402.

Vanicek, P., and Hamilton, A. C., 1972, Further Analysis of Vertical Crustal Movement Observations in the Lac St. Jean Area, Quebec, *Canadian Journal of Earth Sciences* 9:1139–1147.

Walcott, R. I., 1972a, Past Sea Levels, Eustacy, and Deformation of the Earth's Crust, *Quaternary Research* 2:1–14.

Walcott, R. I., 1972b, Late Quaternary Vertical Movements in Eastern North America: Quantitative Evidence of Glacio-Isostatic Rebound, *Reviews of Geophysics and Space Physics* 10:849–884.

Walcott, R. I., 1980, Rheological Models and Observational Data of Glacio-Isostatic Rebound, in: *Earth Rheology, Isostasy, and Eustacy* (N. A. Morner, ed.), Wiley-Interscience, New York, pp. 3–10.

Watts, A. B., 1982, Tectonic Subsidence, Flexure and Global Changes of Sea Level, *Nature* 297:469–474.

Weertman, J., 1978, Creep Laws for the Mantle of the Earth, *Philosophical Transactions of the Royal Society of London, Series A* 288:9–26.

Williams, D. F., Be, A. W. H., and Fairbanks, R. G., 1979, Seasonal Oxygen Isotopic Variations in Living Planktonic Foraminifera off Bermuda, *Science* 206:447–449.

Wolf, D., 1985, An Improved Estimate of Lithospheric Thickness Based on a Reinterpretation of Tilt Data from Pleistocene Lake Algonquin, *Canadian Journal of Earth Sciences* 22(5):768–773.

Yesner, D. R., 1980, Maritime Hunter-Gatherers: Ecology and Prehistory, *Current Anthropology* 21:727–750.

Yesner, D. R., 1983, On Explaining Changes in Prehistoric Coastal Economies: The View from Casco Bay, in: *The Evolution of Maritime Cultures on the Northeast and Northwest Coasts of America* (R. J. Nash, ed.), Department of Anthropology, Simon Fraser University, Burnaby, British Columbia, Canada, pp. 77–89.

Zeigler, J. M., Tuttle, S. D., Tasha, H. J., and Geise, G. S., 1965, The Age and Development of the Province Lands Hook, Outer Cape Cod, Massachusetts, *Limnology and Oceanography* 10:R298–R311.

Part II

People on the Landscape

The primary record of human behavior during the Holocene is derived from archaeological patterns set within the context of reconstructed landscapes. In order to observe the ecological processes implicit in those patterns, we need to humanize the natural history of Holocene landscape development. People are an integral part of a landscape that is constantly changing; they must therefore respond to different levels of environmental change at both individual and community levels. People are also an active agent of change within their own environment. Human communities respond not only to fluctuations in the resource base, for example, but to political and demographic pressures resulting from complex patterns of cultural/environmental interaction and feedback. In some cases, the landscape itself may actually be the product of cultural modification. Diversity in the way that people interact with their environment is investigated here through three studies on prehistoric land-use patterns and economic organization.

Human land-use practices have had a significant impact upon the vegetation of the northeastern United States, and the intentional modification of resources by controlled burning has been widely cited in North America and elsewhere. In their evaluation of natural and culturally induced fire histories in New England, Patterson and Sassaman evaluate the behavioral implications of set fires on resource distribution, settlement location, and economic organization. For example, burning can be used both to maintain critical resources and to increase the size of productive lands, while differential access to fire-controlled resources may provide a currency in intergroup social relations. This type of management strategy may also have lessened the effect of regional landscape trends by allowing people to control specific aspects of their habitat or range. Although several possibilities are proposed to explain the strong correlation noted between intensity of prehistoric site distribution and sedimentary charcoal in their data, Patterson and Sassaman stress that bridging assumptions between regional fire histories, demographic pressures, resource productivity, and settlement patterns must be carefully constructed and evaluated.

The introduction of horticulture in the Northeast had a more uncontrolled impact upon both aboriginal populations and the landscape. Mulholland postulates that some diversity in late prehistoric land-use patterns can be attributed to the different responses to environmental and social stress exhibited by foragers and forager–horticulturalists. As horticulture increased in importance during the late Holocene, dependent groups became more sedentary and both predictable resource zones and arable lands increased in importance. As a result, land use became more localized, leading to increased competition for—and defense of—those areas. One other impact of this shift in land use was that increased dependence on the limited arable lands may have resulted in degradation of local environments that was out of phase with regional trends.

Thorbahn and Cox examine a pattern of land use within a small coastal drainage that contrasts with the regional prehistoric record. Their data indicate that it was not estuarine resources at this location that were the primary criterion for settlement, as might be expected. They suggest, instead, that land-use behavior there (and perhaps elsewhere) was oriented to multiple rather than single environmental zones. One implication of this study is that greater diversity may exist in the relationship between coastal and interior land use. Unfortunately, the type of settlement orientation evidenced in the Pettaquamscutt basin may be obscured within larger, more densely occupied drainages.

Chapter 5

Indian Fires in the Prehistory of New England

WILLIAM A. PATTERSON III AND
KENNETH E. SASSAMAN

1. INTRODUCTION

Humans have inhabited the northeastern United States for approximately 12,000 years (Figure 1). During that period the vegetation of the region changed first from tundra to boreal spruce and fir forests. For the past 8,000 years or so, mixtures of pines, spruces, fir, hemlock, and a variety of hardwoods including beech, several oaks, maples, hickories, and birches have occupied the landscape. Changing climate and the differential migration of plant species were responsible for the vegetation changes that occurred over time periods spanning several hundred to a few thousand years (Davis 1983). Today we see a vegetation pattern that can be related to regional variations in climate, topography, soils, and human settlement. Alpine tundra is found at high elevations in the White Mountains, spruce and fir forests cover much of northern Maine, Vermont, and New Hampshire, and pitch pine and oaks occupy the xeric, sandy soils of southern coastal areas. Mesic forest species occur over a broad area of central New England (Westveld *et al.* 1956).

Historically, ecologists have focused their study of vegetation dynamics on long-term changes that are largely a function of climate. Because climate changes

WILLIAM A. PATTERSON III • Department of Forestry and Wildlife Management, University of Massachusetts, Amherst, Massachusetts 01003. **KENNETH E. SASSAMAN** • Department of Anthropology, University of Massachusetts, Amherst, Massachusetts 01003.

TIME	PERIOD		SOUTHERN NEW ENGLAND/LONG ISLAND	NORTHERN NEW ENGLAND
	Historic		Pawtucket Massachusett Pokanoket Narragansett Pequot-Mohegan Others	Western Abenaki Eastern Abenaki Passamaquoddy Maliseet
1600	WOODLAND	Late		Unnamed and little known Middle to Late Woodland Complexes
1500			Shantok, Niantic	
1400				
1300			Sebonac, Clason's Point	
			Bowman's Brook	
1000		Middle		
800				
600				
400			Clearview	
200				
AD			North Beach	
0		Early		
BC			Lagoon	Adena-Middlesex
1000			Orient, Coburn	
2000	ARCHAIC	Late	Atlantic	Atlantic
			Squibnocket	Moorehead
3000				
				Vergennes
4000		Middle	Merrimac	
5000			Stark	Stark
			Neville	Neville
6000		Early	poorly known complexes with bifurcate points	
8000			traces of Palmer, Kirk, Eva and others	
	PaleoIndian		traces of late PaleoIndian groups	
10,000	?		Eastern fluted-point tradition	

Figure 1. Cultural sequence and chronology for New England. Adapted from Funk 1983:306–307.

slowly, vegetation change related to it tends to be slow. This research emphasis on long-term changes in vegetation gave rise, during the early decades of this century, to the concept of climax vegetation, i.e., types that remain stable over broad areas for long periods of time. The gradual change in vegetation that occurs with time is known as succession, with a series of successional stages (or seres) leading ultimately to a climax vegetation type that maintains a stable species composition as long as climate is constant (Clements 1916).

Although the concept of climatically controlled climax vegetation types has been useful in comparing vegetation on a continental scale, recent research has shown that at both the regional and local levels vegetation is constantly changing in response to local perturbations (Brown 1960; Henry and Swan 1974; Spurr 1956; Stephens 1955). Disturbances such as wind storms, flooding, insect and disease outbreaks, grazing and browsing by animals, and fire can cause vegetation to change despite stable climate, topography, and soils. Of these disturbances, fire is of particular interest because it can affect very large areas (up to several thousand square kilometers) in a short period of time and because it can occur both naturally due to lightning and as a result of human activity. In the absence of human suppression activity, fires burn much of the earth's vegetation at varying intervals and intensity, depending upon ignition sources and local weather conditions and fuel accumulations (Kozlowski and Ahlgren 1974).

Lightning has caused wildland fires for millions of years, and humans have used fire for at least the last several hundred thousand years. Thus, fire has been an ecological factor long enough to affect both landscape composition and species evolution. Many plant and animal species show specific adaptations to living in environments that burn periodically. Humans are no exception, but, significantly, humans have the ability not only to exploit resources made available by wildland fires, but also to influence fire behavior and occurrence.

Ecologists and archaeologists have long recognized that fires had an important effect on the vegetation of North America prior to the Colonial period. Evidence from areas as widely separated as Alaska (Shackleton 1979), Minnesota (Craig 1972), and Maine (Anderson 1979) shows that fires burned since before the time when humans first emigrated to the continent at the end of the last ice age. It seems likely that the early inhabitants of North America were accustomed to living in environments that were periodically affected by fire.

2. EXTENT OF INDIAN BURNING

There is little agreement on the extent to which humans may have actually influenced fire occurrence in North America. Stoltman and Baerreis (1983) postulate that in the Northeast, Indians have used fire to clear land for agriculture for at least the last 1,000 years. Accounts by early explorers and missionaries imply that Indians used fire to drive game, communicate among groups, clear away underbrush, and prepare land for planting (Day 1953). Interviews with elders from Indian villages in northern Alberta led anthropologist Henry Lewis to conclude that Indians had a sophisticated knowledge of fire effects and use and that they burned "to manipulate and eventually to create local environments of their own design" (Lewis 1982a:65). Of interest to ecologists is the fact that, according to Lewis, Indian fires "differed significantly from natural fires in terms of seasonality, frequency, intensity and selectivity" (Lewis 1982a:45).

Bromley (1935:64) reviewed accounts of early colonists in southern New England, concluding that, "all are in accord . . . that the original forest was, in most places, extremely open and park like, due to the universal factor of fire, fostered by the original inhabitants to facilitate travel and hunting." He estimates that "there was probably a sufficient population (of Indians) to bring about an annual burning of most of the country sufficiently dry for a conflagration," and cites Thomas Morton, who wrote in 1632 that "the Salvages are accustomed to set fire of the country in all places where they come; and to burn it, twize a yeare."

Russell (1983) notes that many early accounts were second hand in nature and were often based upon hearsay. She speculates that some writers may have had economic motives for portraying the land as open and parklike as a result of frequent burning (i.e., many of the early colonists hoped to encourage others to emigrate from the Old World, where few forests remained after centuries of exploitation for agriculture and fuelwood). Russell cites Brown and Davis (1973:16), who state that, "it is at least a fair assumption that no habitual or systematic burning was carried out by Indians." She concludes that "there is no strong evidence that Indians purposely burned large areas of the forested northeastern United States frequently," although she acknowledges that "the presence of Indians did, however, undoubtedly increase the frequency of fires above the low numbers caused by lightning." Forman and Russell (1983:5) expanded Russell's conclusions to North America as a whole: "This example touches on probably the most widespread problem of human disturbance history in North American ecology, regular and widespread Indian burning (Day 1953), an unlikely hypothesis that regretfully has been accepted in the popular literature and consciousness (Russell 1983)." This statement drew immediate and strong criticism from Myers and Peroni (1983), and controversy over the extent and purpose of Indian burning continues.

3. THE NEED FOR ADDITIONAL STUDY

The question of Indian burning deserves continued study because of the potential impact that frequent fires could have had on the vegetation of New

England. Analyses to date have suffered, on one hand, from rather subjective evaluations of sketchy historical accounts, as Russell (1983) observes, and on the other, from too strict adherence to the Clementsian concept of succession toward climatic climax vegetation associations.

For the Northeast in particular, too little emphasis has been placed upon evaluating both the factors that influence fire occurrence and the evidence for human cultural development available from archaeological and anthropological studies. We disagree with Russell (1983:78), who states that "the importance of man in the pre-Colonial forests can be approached directly only through the use of historical records." This statement would be true only if human and vegetation development were static for a long period of time prior to European settlement. It may never be possible to examine "the importance of man" in the pre-Colonial forest directly, but we feel that by examining indirect evidence covering long periods of time it should be possible to develop a logical construct that will help us to understand the development of forest conditions as they existed at the time of settlement. Although such a construct will be continually revised as new evidence accumulates, it is our purpose to review information and to provide a basis for further development of our understanding of the use of fire by Indians in New England.

4. FACTORS INFLUENCING FIRE OCCURRENCE

Pyne (1984:208) defines fire regime as "a particular complex of fuel, manifested as a biota, and a particular pattern of fire occurrence and behavior, described generally as its fire history." A fire regime describes, for a particular area, the way in which the combustion process (fire) interacts with fuels, topography, and weather conditions (i.e., the fire environment) to produce a specific frequency and pattern of burning leading to predictable effects on the plants and animals occupying the area. Pyne notes that "everywhere, and from the earliest times, humans have altered the natural fire regimes they have entered" (Pyne 1984:223).

Today humans are certainly an important component of fire regimes of New England, causing more than 95% of all fires in the eastern United States (Pyne 1984); similarly, Indians may have significantly altered natural fire regimes prior to European settlement. Before evaluating the potential impact of Indians, we must, however, understand (1) the elements that comprised prehistoric fire environments, and (2) how these elements could have interacted to produce a fire regime. Although there is no way to evaluate directly early fire environments, indirect evidence (e.g., fossil plant assemblages) provides information about paleoclimate and vegetation. We can also study the characteristics of modern vegetation, topography, and climate to identify potential analogs for past environments. Recently completed work allows us to do these things for at least some areas of

New York and New England (Backman 1984; Patterson *et al.* 1983, 1984a,b; Rudnicky 1984).

Wildland fires occur when favorable weather, fuels sufficiently dry to burn, and an ignition source combine to initiate and sustain the combustion process. Fires will not burn if any of these three elements is missing. Weather parameters, fuel moisture content, and ignition sources each vary seasonally within the year and from year to year. On a given day, each may vary within an area the size of New England.

Weather is a function of climate, and the present climate of New England is humid continental, with summers that are warm in the south and cool in the north (Espinshade 1976). Rainfall is ample for forest growth and evenly distributed throughout the year, although short periods of intense drought occur irregularly from year to year. Warmer temperatures in the south increase evapotranspiration, and coarse-textured soils on the coastal plain support a typically xeric vegetation comprised of pitch pine, scrub oak, and several ericaceous shrub species (e.g., blueberry, huckleberry). Fuels dry rapidly, and these forests are among the most flammable in New England. To the north, and inland at higher elevations, humidity increases and cooler temperatures result in reduced evapotranspiration. Soils retain water more effectively and physiological drought conditions are likely only on shallow soils on steep slopes.

Overall, air temperature, wind speed, relative humidity, and the duration of precipitation are the most important factors influencing the moisture content of forest fuels and the likelihood that ignitions will occur. Once a fire ignites, wind speed, fuel accumulations, and topographic factors influence the rate at which it will spread. In New England, weather conditions are generally most conducive to burning in exposed coastal areas and on hill tops.

Forest fuels may be living or dead, with the latter generally the most flammable, although the leaves of some ericaceous shrubs such as huckleberry and mountain laurel contain volatile organic compounds that allow them to burn even when green. Dead grass, herbs and ferns, and the needles and leaves that comprise the forest floor are referred to as fine fuels. They dry rapidly when temperatures rise in the spring prior to leaf out. The amount of fine fuels and their moisture content are important factors determining whether or not ignitions will occur. Once a fire is ignited, larger fuels (chiefly dead, woody material either standing or fallen) burn and generate the heat that further dries material in advance of a flaming front. Although initially driven by winds, free-burning fires produce convection columns as warm air rises. Air is drawn in from the perimeter of the fire, and in this way wildland fires can become self-propogating. As dead material accumulates within forests over time, the potential for large, self-propogating crown fires increases. Frequent fires prevent the accumulation of fuel and promote the growth of grasses and herbs that become fine fuels during the dormant season. These fine fuels further increase the likelihood of ignitions.

Frequent fires tend to burn near the ground. Fires at longer intervals allow heavier fuels to accumulate, and these promote larger "catastrophe" fires (Heinselman 1981).

Fire scientists classify ignition sources as being either natural or human. Lightning is the most important source of natural ignitions, but lightning ignitions are less important in the East than elsewhere in the United States (Pyne 1984). Lightning occasionally causes fires in interior sections of northern and central Maine (Fobes 1944), but along the coast and in southern New England, lightning fires are rare. Of 209 fires reported since 1937 at Acadia National Park on the Maine coast, only five were ignited by lightning and none of these fires burned more than 0.25 acre (Patterson *et al.* 1983). There were no lightning fires reported at Cape Cod National Seashore during 1974–1983 (Patterson *et al.* 1984b), but 112 were caused by humans during that period. Apparently the abundant moisture associated with coastal thunderstorms extinguishes the few lightning strikes that occur along the coast. Today humans are the primary cause of New England's wildland fires, and fires appear to be most common where population densities are greatest. It seems likely that this pattern also occurred prior to the Colonial period.

5. NEW ENGLAND CULTURAL ECOLOGY

5.1. Introduction

It is clear that for North America as a whole Indians were, when first encountered by Europeans, an important ignition source for wildland fire (Barrett and Arno 1982; Lewis 1973, 1977, 1982a,b; Lutz 1959; Stewart 1956). The same can be said for tribes of New England (Cronon 1983; Day 1953; Pyne 1982), although the extent to which these fires were ecologically significant is open to question (Forman and Russell 1983; Russell 1983). Comparable documentation of prehistoric fires does not exist. This precludes a straightforward evaluation of the evolutionary context of fire utilization in New England. Archaeological evidence of hearths indicates that fire was used domestically since the time humans first entered the region about 12,000 years ago, but nothing more specific is known about other uses of fire, or where and when these may have developed. Thus, the following discussion first focuses on Native American burning at the time of European contact. We then build upon this information to explore the potential for prehistoric fire use.

5.2. Protohistoric/Historic Fire Use

When European colonization began in the early seventeenth century the Atlantic seaboard from Nova Scotia to North Carolina was inhabited by peoples collectively referred to as Eastern Algonquians. Territories occupied by Eastern Algonquian tribes generally corresponded to watersheds of major rivers and were bounded to the west by land of Iroquian-speaking peoples (Salwen 1978). The Eastern Algonquian shared many cultural traits, yet differences are evident between northern and southern New England groups with respect to subsistence economy, including the use of fire.

5.2.1. Southern New England

Early seventeenth century Indian populations in southern New England were divided into five major coastal groups and two series of smaller interior groups (Figure 2). The protohistoric subsistence economy of southern New En-

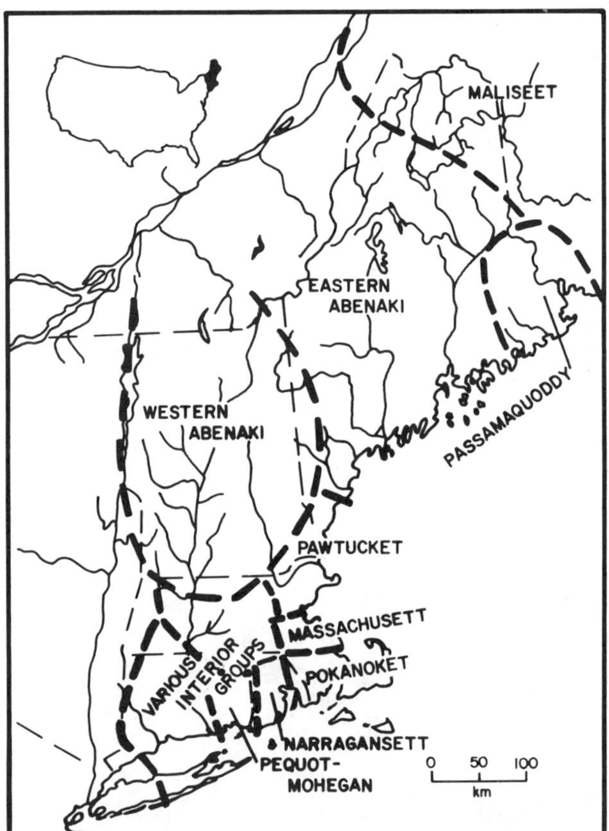

Figure 2. Approximate locations of New England Indian tribes at time of European contact.

gland was a diverse mix of hunting, fishing, shellfish collecting, plant gathering, and agriculture. The relative contributions of these pursuits to particular economies depended on natural distributions of resources, as well as on regional distributions of people, technology, and labor. Corn agriculture, for instance, became relatively more important in southern coastal areas, where the growing season was long, fertile land was plentiful, and populations tended to be concentrated (Snow 1978a). In these areas, fire became essential to the success of agriculture (e.g., Brasser 1971). Early accounts of coastal settlements describe vast open fields resulting from the Indian's slash-and-burn (i.e., swidden) technique of farming. At Narragansett Bay, Rhode Island, Verrazano noted "open plains twenty or thirty leagues in extent, entirely free from trees" (cited in Pyne 1982:48). Similar observations were made by Champlain at Cape Ann in 1606 (Champlain 1907). Letters by Higgeson and Graves in 1629 also describe numerous barren fields near Salem, Massachusetts (Byers 1946). Reports such as these were often biased toward economic or religious self-interests (Russell 1983), so we must exercise caution in interpreting their accuracy.

A more conservative interpretation based on vegetational history is that Indian burning resulted in a mosaic of forests and fields in varying stages of succession. This mosaic effect can be attributed to the shifting patterns of settlement that characterized slash-and-burn agriculturalists (Harris 1972). Equally important was the intensity to which fixed locations were used for planting, plant gathering, firewood collecting, and hunting (Cronon 1983). Once sites were cleared for cultivation they were usually reoccupied each growing season for a decade or more (Cronon 1983; Snow 1978a). Often these sites were reburned to replenish soil nutrients and to expand fields, or to maintain them as hunting and gathering locations by encouraging new browse.

Hunting territory surrounding coastal farming sites may also have been modified by fire (Brasser 1971). In his 1634 account of coastal Massachusetts, Wood observes,

> . . . for it being the custome of the Indians to burne the wood in November, when the grasse is withered, and leaves are dryed, it consumes all the underwood, and rubbish, which otherwise would over grow the Country, making it unpassable, and spoil their much affected hunting: so that by this means in those places where the Indians inhabit, there is scarce a brush or bramble, or any cumbersome underwood to bee seene in the more champion ground. (Wood 1634, quoted in Byers 1946:19)

Besides leaving the forest passable for hunting and gathering, broadcast burning would have had other advantages. One objective of Indian fires "was to produce fresh and sweet pasture, for the purpose of alluring the deer to the spots on which they had been kindled" (Dwight 1823:50). White-tailed deer seem to have been the most economically important terrestrial species hunted in southern New England (Salwen 1978; Snow 1978a). By improving the quality and quantity of browse for deer with fire, hunters effectively increased the carrying capacity of the environment for this species. These improvements would also have increased the growth rates and reproductive success of game (Mellars 1976). Moreover, the efficiency of hunting was enhanced through fire. Improved visibility within burned hunting grounds allowed hunters to be more selective in their kills. By concentrating deer in burned areas, Indians reduced the uncertainty of hunting and created better opportunities for cooperative hunting ventures (Mellars 1976).

Indian-set fires were apparently also common to some portions of interior southern New England. In Massachusetts, histories of North Brookfield (Temple 1887), Palmer (Temple 1889), and Hadley (Judd 1863) (all cited by Byers 1946:22) contain references to parklike forests and open meadows, although the latter could have been marshy areas in some instances. Though common, fires in the interior were probably not as widespread as those on the coast. Almost all accounts of the interior describe cleared or open land in or near the valleys of major rivers. Native populations favored these locations for settlement and subsistence activities. Annual fish runs attracted many groups to the banks of the Connecticut and Housatonic Rivers and their larger tributaries. At the time of contact, maize agriculture was also being practiced in the bottomland of interior drainages (Thomas 1976). Agricultural sites in these locations appear to have been moved with greater frequency than coastal ones, but abandoned plots were often kept clear with fire for hunting and gathering activities. Upland forests along major interior drainages were also thinned, though not cleared, by Indian fires (Byers 1946). Interriverine uplands, on the other hand, were only temporarily used and fired less often, if at all.

5.2.2. Northern New England

Historical accounts of aboriginal burning in northern New England are more difficult to find. The lack of information and evidence for burning in the north is partly a function of relatively few first-hand accounts and less intensive burning practices. At the time of European colonization, northern New England was inhabited by four major groups: Western Abenaki, Eastern Abenaki, Passamaquoddy, and Maliseet (Figure 2).

A major contrast between northern and southern New England subsistence economies was the near absence of agriculture in the north. Referring to tribes north of the Kennebec River, Verrazano reported "no sign of cultivation" in 1524 (Wroth 1970, quoted in Cronon 1983:38). Champlain also noted a distinction between communities of the Saco River and those to the northeast (Snow 1978a). Maize cultivation was, in fact, a late arrival for the Eastern Abenaki in southern Maine, and it never gained the economic importance it did farther south. Instead, northern populations drew most of their sustenance from the seashore and major rivers. According to ethnohistorical documents, the spring months were spent exploiting fish runs in the rivers, whereas summer brought groups to the coast to collect a diversity of plant and animal resources. In midfall, populations dispersed inland along river tributaries to hunt and trap through the winter.

Prior to contact, the settlement patterns of northern populations were quite different. Faunal remains from prehistoric sites on the coast of Maine document late winter–early spring occupations (Bourque 1973). Bourque (1973) hypothesized that a shift in settlement practices took place in the early historic period to accommodate trade with European explorers. Additional analyses of faunal collections at sites in the Gulf of Maine suggest that seasonal relocation was not practiced among all groups, and that a diverse range of adaptive strategies, including some year-round occupation, can be identified for the late prehistoric period (Sanger 1982).

At the time of contact, burning vegetation for purposes of hunting did not convey the same advantages for northern coastal groups as it did for groups in the south. During those seasons when the benefits of broadcast burning could be realized, fish, shellfish, and other aquatic resources were abundant and accessible. Canoes provided the chief mode of transportation in the north, so the travel advantages of open forests were inconsequential. Most importantly, as Cronon (1983:41–42) points out, because the northern New England Indians did not practice agriculture, they were less dependent on fixed locations (i.e., less sedentary) and thus less apt to alter their environment out of economic stress.

Nevertheless, the late historic record of northern New England does include a few descriptions of anthropogenic fires. For instance, Pyne cites one account describing the diminished pine forests near Lake Champlain: an explanation given for the shrinking forests was "the numerous fires that happen every year in the woods, through the carelessness of the Indians, who frequently make great fires when they are hunting, which spread over the fir woods when everything is dry" (Kalm 1770–1771, quoted in Pyne 1982:44). This account may indicate that burning was conducted by groups lacking access to coastal resources, who were consequently more dependent on the procurement of terrestrial species. Alternatively, post-Contact burning activity in the interior, as well as on the coast, was almost certainly intensified in conjunction with Indian participation in the fur trade.

6. ECOLOGICAL AND CULTURAL FACTORS AFFECTING FIRE USE

Our discussion of Indian burning at the time of contact indicates that (1) fire practices were not universal among New England tribes, and (2) beyond differences in fire regimes, variation in anthropogenic fires depended on the nature of human subsistence activities and the way those activities were distributed on the landscape. The evidence reviewed thus far shows that frequent burning would have been most advantageous to agricultural tribes of southern New England. Following the argument of Cronon (1983), we can further state that the nature of agriculturalists' fires was a function of relatively stationary land use. Once established, agriculture promoted higher population density and growth, placing a greater demand on limited resource space. Fire became important for two reasons. First, it afforded greater returns on investments in hunting and gathering by increasing the quality and quantity of resources, by reducing the risk of uncertainty, and by promoting cooperation in resource procurement. Second, fire became important to agricultural activity itself as a means of creating arable land and maintaining its fertility.

We can hypothesize that among agriculturalists, the economic importance of burning varied directly with population density. The ethnohistorical record indicates that agricultural tribes were concentrated in coastal and riverine areas of southern New England, so we may expect that our most abundant sedimentary records of fire would show strongest evidence for burning in this subregion (discussed below). Although Indians may have became especially dependent on fire in this subregion, this does not mean that fire had little relevance to adaptations elsewhere. If the question to debate is whether or not Indian fires were sufficiently widespread to have an impact on the ecology of New England (cf. Russell 1983), then the deciding issue is the way settlement and subsistence governed the ecological impacts of burning. The fire practices of agriculturalists were *intensive* and left a lasting mark on the landscape of New England largely because they were geographically redundant.

The burning activities of mobile hunters and gatherers may have been more *extensive,* but were not as redundant as those of agriculturalists. Over much of interior and northern New England, fires set by hunter–gatherers would have only complemented infrequent lightning-set fires, and these areas may have experienced only a few fires per millennium. The ability of hunter–gatherers to relocate camps and exploit diverse resources acted to prevent overexploitation of the land. Burning vegetation as a means of improving resource productivity was probably not a significant strategy for groups able to maintain relatively unrestricted settlement mobility. On the other hand, burning to facilitate travel, control pests, flush game, and clear settlements and burning by accident were probably common, albeit irregular, events in certain sections of New England for thousands of years.

An exception to the relatively low ecological impact of hunter–gatherer burning is resource specialization. For instance, hunters and gatherers on the Great Plains of North America specialized in the exploitation of bison for thousands of years. Plains hunters often took advantage of the bison's gregariousness by mounting cooperative drives and surround kills. Fire was an important tool for driving the game, and its benefits in manipulating herd size and distribution were long recognized. Enthusiastic use of broadcast burning for these purposes was a contributing factor in the expansion and maintenance of prairie vegetation (Pyne 1982; Stewart 1953). White-tailed deer, which are largely solitary animals, were the chief game in New England, and fire would have been most useful for improving habitat. Modern wildlife managers recognize that deer range need be burned only occasionally to sustain adequate levels of cover and browse.

Resource specialization among protohistoric Northeastern Indians was encouraged through participation in the swelling trade of Europe (Wolf 1982). Indian fire practices were undoubtedly affected by these economic changes. The Eastern Abenaki, for example, were eager participants in the French and English fur trade, and during the first half of the seventeenth century, they reorganized settlement and subsistence practices to accommodate full-time trapping (Snow 1978b). Similarly, coastal Algonquians on Long Island became specialized in the production of wampum for trade. On the basis of this change, some Long Island villages became completely sedentary and maize agriculture was intensified to offset the growing inaccessibility of seasonal wild food resources (Ceci 1979).

The foregoing discussion provides a substantive and theoretical basis for prehistoric fire utilization in New England. Several important factors have been identified. These include (1) population density, (2) the geographic structure of subsistence (i.e., land use), and (3) resource specialization. Processes affecting these variables are intrinsically linked, and prolonged debates have been conducted over the causal relationships between land use, subsistence, and population. In this paper we will use these three variables, which were important to the fire ecology of historic Indians, as a basis for reviewing current knowledge of prehistoric adaptation in New England to build inferences for burning during the pre-Contact era.

7. ARCHAEOLOGICAL BASIS FOR INFERRED PREHISTORIC FIRE IN NEW ENGLAND

7.1. Southern New England

Demographic patterns and the nature of land use observed in seventeenth century southern New England began to take shape around 3,000 years ago. Snow (1980) refers to this episode as the "Early Horticulture" (or Woodland) period, based upon expectations for parallel developments in the cultivation of native plants documented in the Midwest. Changes in settlement and subsistence in southern New England during the third millennium are consistent with patterns existing among contact period agriculturalists, although direct proof of early horticulture is lacking. The beginnings of horticulture in southern New England were apparently embedded in a shift to increased utilization of lowland habitats. Site locational data document a rise in coastal occupation accompanied by a regional drop in the number of sites during the third millennium (Mulholland 1984). Mulholland suggests that a decrease in the amount of oak forests beginning around 4000 B.P. made inland and upland habitats less attractive to human foragers. At the same time, stabilization of tidal waters enhanced the productivity and attractiveness of coastal zones.

Trends toward increased utilization of coastal habitats continued over the subsequent two millennia. For instance, on the basis of site counts, we can infer that Cape Cod supported relatively dense populations throughout this time (Mulholland 1984). Similar densities could be expected for Long Island, although comparable site data are not readily available. The development of agriculture in these areas during the last millennium was probably predicated on an established horticultural economy. The maintenance of garden plots (as opposed to fields during the later agricultural period) may have involved regular burning (Stoltman and Baerreis 1983). Given the general similarities in population and land use between the late prehistoric and contact periods, it is likely that coastal groups of the third and second millennia relied on burning for hunting and gathering.

During the mid-Holocene, southern New England provided a benevolent and productive temperate forest habitat for the expansion of hunter–gatherer populations [the Foraging Ecosystem Type of Stoltman and Baerreis (1983)]. Warming temperatures during this time encouraged the growth of mast-producing species. Site distributions tend to correspond with areas where pollen assemblages are characterized by high oak percentages, perhaps because mast resources were important to the hunter–gatherer economy (Dincauze and Mulholland 1977; Mulholland 1984).

By the fifth millennium, site frequencies reached their highest levels in southern New England. Unlike subsequent patterns, the fifth millennium distribution demonstrates a wider use of available habitats. Large clusters of sites are located in estuarine, coastal, riverine, and inland lowland locations (Mulholland 1984). Intraregional variation in patterns of subsistence and technology reflect broad trends toward regional specialization and demographic circumscription. Concentrations of people in these habitats placed greater demands on local environments and thus promoted the need to manipulate resources with fire. The mid-Holocene fire environment of New England may also have been characterized by more lightning fires (due to generally warmer and drier conditions that would be more likely to produce convective thunderstorms). Populations may have developed strategies for reaping the benefits of these natural fires.

The early prehistory of New England was probably characterized by less intensive land use and low population density. The original colonizers of New England entered the region when late glacial tundra had given way to spruce parkland between 11,000 and 12,000 years ago. One model suggests that early groups adapted to the low density and diversity of early postglacial forests by utilizing a wide variety of resources and by being highly mobile (Dincauze and Curran 1983). Lowland areas and river valleys seem to have been slightly favored for settlement, particularly during the later part of the period, when deciduous species began to migrate into the region. Despite preferences for certain landforms, land use was generally nonredundant. Accordingly, the need for burning to enhance productivity may have been low. If they did burn, early postglacial populations may have done so for reasons that were quite different from those of populations encountered by the first European explorers. By way of analogy, Indians of northern Alberta regularly set fire to grasslands within the boreal forest habitats that are important to elk, bison, deer, and moose. By burning in early spring, the Indians encourage grasses to sprout 2–3 weeks earlier than they might otherwise (Lewis 1982b). Such a strategy applied to the early postglacial period would be advantageous to populations trying to glean resources from environments characterized by short growing seasons.

Models of late-glacial and early postglacial subsistence/settlement that are based on assumptions of low population density and low intensity land use are being reevaluated. Nicholas (1982, 1983, 1987) argues that the initial drainage of glacial lakes in New England created landscape mosaics with resource potentials greater than either the early riverine or full lake stages (cf. Curran and Dincauze 1977). Human exploitation of these basins during the early Holocene is predicted to be more intensive and regular than in nonbasin areas. As the productivity of basin habitats diminished through time, strategies to maintain existing exploitation patterns, including the use of intentional burning, may have become important (Nicholas 1987).

7.2. Northern New England

North of Massachusetts the changes in vegetation and climate described above were delayed by several centuries. The potential constraints this may have put on the early occupation of northern New England are important to an understanding of the variability throughout the region. Dincauze and Mulholland suggest that intensive exploitation of northern New England was limited until 8000 B.P. By this time the oak-forest ecotone [represented by 20% oak isopoll of Bernabo and Webb (1977)] had penetrated southeastern New Hampshire and extended up the Maine coast as far as Penobscot Bay (Dincauze and Mulholland 1977). Areas north and west of the oak-forest ecotone were probably less attractive to southern New England foragers. Certain habitats, especially the Champlain Lowlands, probably did not suffer these environmental limitations, as they seem to have supported occupation continuously throughout the Holocene (Thomas 1980). Thus, despite local variations, regional population density in northern New England remained low until about the fifth millennium, when populations expanded and diversified in much the same way as those of southern New England.

During the third millennium, distinctive coastal and inland cultural patterns emerged in northern New England. This development basically established the pattern of cultural variability that was evident at the time of contact. Over the course of the last two millennia, site concentrations formed at Casco Bay, Boothbay, Penobscot Bay, Frenchman Bay, Machias Bay, and the Quoddy Group (Passamaquoddy and Cobscook Bays). Among these, relatively higher site densities are found at Penobscot and Frenchman Bays (Sanger 1979). Sites of the interior cluster along large streams and lake shores; few sites are located away from permanent water (Sanger 1979). Given this distribution, the potential for pre-Contact fire in northern New England was greatest on the coast around bays and estuaries, and least along inland waterways. It seems unlikely that fire reached the level of economic importance that it did farther south, however.

8. PALEOECOLOGICAL EVIDENCE

8.1. Introduction

We can observe directly and record the effects of modern (and to a lesser extent historic) fires, but not so those of prehistoric times. Climate is constantly changing, but it is difficult to envision changes in weather patterns during the past two to three millennia that could have significantly altered the frequency or intensity of fires in New England. Changes comparable to the midpostglacial climatic optimum would be required to alter burning patterns at the regional level.

The structure and composition of New England's vegetation have changed since pre-Colonial time, and we might reasonably expect that the pattern and distribution of fuels have changed also. Similarly, modern human population numbers far exceed those of prehistoric times, and it is likely that with more humans inhabiting the landscape, more ignitions have occurred and fires have been more pervasive. Fahey and Reiners (1981) found that during the twentieth century, fire incidence (i.e., number of fires per land area) has varied directly with population density in Maine and New Hampshire. Even with modern fire prevention and suppression techniques, as many as 10,000 wild fires occur in Massachusetts each year.

Given the differences in fuels and human population levels outlined above, the frequency and effects of modern fires probably differ from those of the pre-Colonial period. According to Neiring and Goodwin (1962), fire adaptations in oak, chestnut, hickory, and pitch pine suggest that these species, which were common in southern New England when Europeans arrived, formed plant communities that were subjected to at least occasional burning. We cannot, however, study modern fire regimes and extrapolate directly to the past. We must turn to paleoecological studies for evidence of fire occurrence prior to European colonization.

8.2. Methods of Constructing Fire Histories

A number of methods are available for investigating past fires. Analyses of fire records maintained by federal and state agencies provide information on the number, size, and cause of fires. These records are only available for the twentieth century, a period when fire occurrence has been influenced by rapidly expanding populations, increased industrial activity, and increasingly efficient fire prevention, detection, and suppression programs. Study of these records can tell us little about the number and extent of fires prior to the Colonial period, but modern records do provide clues about fire causes. The incidence of lightning fires prior to settlement, for example, was probably similar to the low levels observed today.

Tree ring studies and fire scar analyses have been helpful in reconstructing prehistoric fire occurrence in some areas (e.g., Frissell 1973 and Heinselman 1973 in Minnesota, Arno 1980 in the northern Rocky Mountains). These methods can not be used in New England, however, because few stands dating to pre-Colonial times remain. We have used fire scars to reconstruct fire occurrence during historic times (Patterson *et al.* 1983), but not for the prehistoric period.

Written records and accounts yield information about fire occurrence at the time of contact. Russell's (1983) reservations notwithstanding, we believe that the early accounts from New England clearly indicate that the Indians understood the importance of fire to the environment in which they lived, and that, at least in southern coastal and riverine areas, Indians intentionally used fire to enhance resource availability. The argument for intentional burning by New England tribes is strengthened by later, more authoritative accounts of Indian burning elsewhere on the continent. We feel that Forman and Russell's (1983) opinion about regular and widespread Indian burning being "The most widespread problem of human disturbance history" and "an unlikely hypothesis" is overstated, especially when applied to North America as a whole. We agree with Myers and Peroni (1983:218), who state that the "use of historical accounts alone would only have provided information for a few locations over a relatively brief period of less than four centuries."

The hypothesis that the forests of the Northeast had, at the time of European settlement, been modified by centuries of prehistoric burning is testable by pollen analysis, charcoal counts, and archaeological investigations (Davis 1983). We have reviewed the archaeological evidence for cultural adaptations that would encourage the use of fire. This evidence suggests that fire *might* have been employed for resource augmentation. It *does not* prove that fire was actually present in the pre-Colonial landscape. To do this, we must turn to sedimentary records of fossil pollen and charcoal.

Wildland fires are characterized by incomplete combustion; uneven aeration and varying moisture contents of fuels ensure that large quantities of charcoal result from most fires. Some charcoal is carried into the air and away from the site of a fire by the strong winds that typically accompany large fires. Most charcoal remains on or near the burned area, however. Some of this charcoal is carried by surface runoff to streams and then to lakes or ponds, where it settles through the water column. Charcoal from successive fires is buried as sediments accumulate over time, and in this way a history of fire in a basin's watershed is preserved. Fires alter the species composition and age structure of vegetation, and changes associated with fires are reflected in variations in the types and amount of pollen reaching a basin. Thus, sediments also preserve a history of postfire vegetation development. Pollen and charcoal analyses have been used to detect the effects of prehistoric fires on vegetation of several areas (e.g., Cwynar 1978; Davis 1967; Fredskild 1967; Iversen 1941; Swain 1973; Tolonen 1978; Tsukada 1966; Tsukada and Deevey 1967; Waddington 1968). In New England, Anderson (1979), Backman (1984), Winkler (1982), W. A. Patterson (unpublished), and A. M. Swain (unpublished) provide stratigraphic evidence for pre-Colonial fires. Data are available for 11 sites (Table 1). Most involve records of no more than the past 1,000 years, but two (Upper South Branch Pond in Maine and Duck Pond on Cape Cod) span the Holocene. Sites are available for both inland and coastal as well as northern and southern areas (Figure 3).

Charcoal abundance is often measured as the ratio of the sum of the areas (measured in square micrometers) of fragments tallied on microscope slides prepared for pollen analysis to fossil pollen found on the slides (Swain 1973). Typical charcoal area to pollen (C:P) values for New England lake sediments range from <100 to 3000 or higher. One New England site (The Bowl) lies within the boundaries of the Great Bar Harbor Fire that burned nearly one-fourth of Mount Desert Island, Maine, during October 1947. The charcoal profile for this pond shows C:P values of 600–800 in sediments that date to the late 1940s (Figure 4). The C:P values for sediments of similar age in Sargent Mountain Pond, which is near The Bowl but outside the boundaries of the 1947 fire, are <200 (Backman 1984). The results from these two sites demonstrate that sedimentary charcoal can be used to detect the local occurrence of fire (Patterson *et al.* 1984a).

Average C:P values for sediments of pre- and post-Colonial age (Table 2) show wide variation within the region. With the exception of Duck Pond on Cape Cod, charcoal is more abundant in recent sediments than in those deposited prior to European settlement. Differences are greatest at the inland sites where pre-Colonial values are very low. At these locations, fires were apparently rare before Europeans arrived. Along the coast, and especially in southern New England,

Table 1. Site Information for Eleven New England Basins with Sedimentary Charcoal Analysis

Site	Location	Physiographic region	Modern vegetation type[a]	Time span (B.P.)	Analyst[b]	Reference
Larkum Pond	Otis, MA	Central-inland	Northern hardwoods–hemlock–white pine	0–800	A. Backman (P), A. Hine (C)	Backman 1984
Basin Pond	Kenebec Co., ME	North-inland	Transition hardwoods–white pine–hemlock	0–1600	K. Gajewski (P), M. Winkler (C)	Gajewski 1983, Backman 1984
Conroy Lake	Aroostook Co., ME	North-inland	Spruce–fir–northern hardwoods	0–2000	K. Gajewski (P), M. Winkler (C)	Gajewski 1983, Backman 1984
Upper South Branch Pond	Piscataquis Co., ME	North-inland	Spruce-fir-northern hardwoods	0–11,000	R. Anderson (P&C)	Anderson 1979
The Bowl	Bar Harbor, ME	North-coastal	Spruce–fir–northern hardwoods	0–500	W. Patterson (P), A. Hine (C)	Backman 1984
Sargent Mt. Pond	Bar Harbor, ME	North-coastal	Spruce–fir–northern hardwoods	0–280	A. Backman (C&P)	Backman 1984
Long Pond	Isle au Haut, ME	North-coastal	Spruce–fir–northern hardwoods	0–430	A. Backman (C&P)	Backman 1984
Duck Pond	South Wellfleet, MA	Central-coastal	Pitch pine–oak	0–5000	A. Backman (P), A. Hine (C)	Backman 1984
				0–12,000	M. Winkler (C&P)	Winkler 1982
Charge Pond	Wareham, MA	Central-coastal	Pitch pine–oak	0–800	A. Backman (C&P)	Backman and Patterson 1984
Widgeon Pond	Wareham, MA	Central-coastal	Pitch pine–oak	0–1000	A. Backman (C&P)	Backman and Patterson 1984
Deep Pond	Wading River, NY	South-coastal	Pitch pine–oak	0–2100	A. Backman (C&P)	Backman 1984

[a] From Westveld *et al.*, 1956.
[b] P = pollen; C = charcoal.

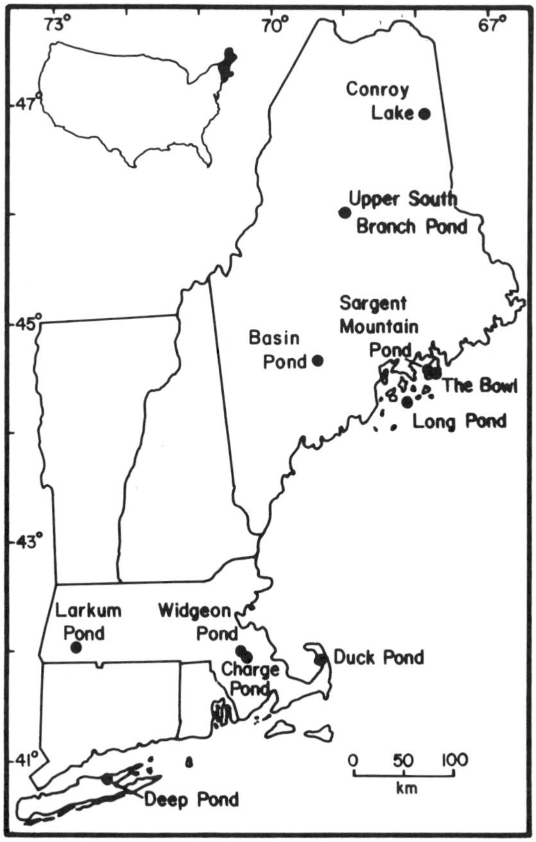

Figure 3. Locations of sedimentary fire-history studies in the Northeast.

differences are not so great, with average pre-Colonial values for the southern sites comparable to those observed at The Bowl for sediments dating to the 1947 fire. Pre-Colonial fires must have been common within the drainage basins of the four southern sites. Occasional pre-Colonial fires apparently burned near three coastal Maine sites. By contrast, there is no evidence for fire in the 500 years prior to local settlement at Larkum Pond in the Berkshire Mountains of western Massachusetts (Backman 1984).

The results of these sedimentary studies suggest that prior to European settlement, fires were most common in areas where, on the basis of archaeological site distributions, Indian populations were greatest. We lack charcoal data for the inland riverine environments of central New England, but the data for coastal areas seem convincing. We know of no archaeological sites within the watersheds

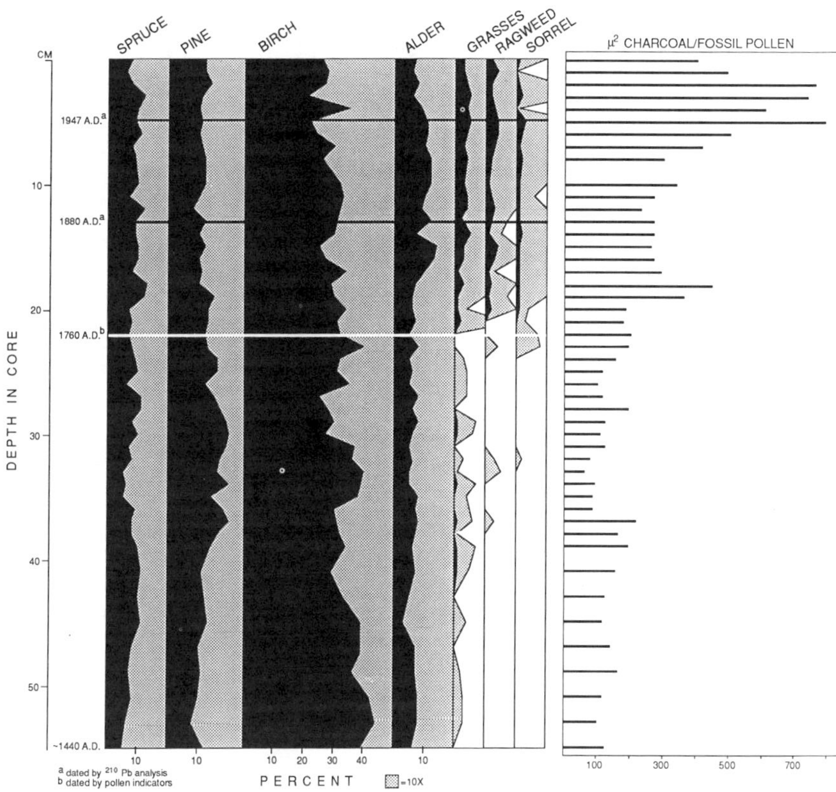

Figure 4. Pollen and charcoal profiles for The Bowl, Mount Desert Island, Maine. Evidence for the 1947 Bar Harbor Fire is seen in the peak in charcoal between 3 and 6 cm. Dating is by ^{210}Pb analysis (*a*) or pollen indicators (*b*) (see Clark and Patterson 1984).

of our study sites[1] and the pollen profiles show no evidence for local cultivation [e.g., the presence of corn (*Zea mays*) pollen or the pollen of other cultigens]. Thus, sedimentary charcoal in these basins probably reflects more generalized burning as opposed to the burning of small, highly localized garden plots. McAndrews' (1976) studies of Crawford Lake in Ontario suggest that local

[1]A review of Mulholland's (1984) data files and the site files of the Massachusetts Historical Commission, Boston, indicated no recorded sites within the watersheds of sample locations in Massachusetts. Likewise, no sites are known from sample locations on the coast of Maine (David Sanger, personal communication, 1985). Site records from sample areas of interior Maine and Long Island were not consulted. We should note that a lack of *recorded* sites does not preclude the possibility of aboriginal settlement in these areas. The distribution of recorded sites is as much, if not more, an artifact of archaeological sampling as it is of prehistoric land use (Wobst 1983).

Table 2. Average Charcoal Abundance in Pre- and Post-Colonial Sediments of New England

		Charcoal: Pollen value		
Site	Location	Average pre-Colonial	Average post-Colonial	Average post–pre-Colonial
Larkum Pond	MA (central-inland)	27.5	190.5	6.9
Basin Pond	ME (north-inland)	80.9	385.5	4.8
Conroy Lake	ME (north-inland)	38.7	291.1	7.5
The Bowl	ME (north-coastal)	123.4	390.9	3.2
Sargent Mt. Pond	ME (north-coastal)	131.3	161.2	1.2
Long Pond	ME (north-coastal)	151.9	320.9	2.1
Deep Pond[a]	NY (south-coastal)	650.1	1040.2	1.6
Duck Pond	MA (central-coastal)	250.7	160.7	0.6
Charge Pond	MA (central-coastal)	713.9	2895.0	4.1
Widgeon Pond	MA (central-coastal)	580.7	968.2	1.7

[a]Values are ~50% low due to pyrite.

cultivation and swidden agricultural practices would be detected in the stratigraphic record if they were present.

Trends that we see for the past several centuries are evident also in the two sites for which we have longer records. The central Maine forests near Upper South Branch Pond have burned only rarely during the past 10,000–11,000 years (Anderson 1979). Major fires apparently burned at the 1,000- to 2,000-year intervals predicted by Lorimer's (1977) analysis of natural fire rotations. At Duck Pond on Cape Cod, however, abundant charcoal throughout the stratigraphic column suggests that fire has played an important role in maintaining pine–oak forests throughout the Holocene (Winkler 1982).

9. DISCUSSION

Fire occurrence in the eastern United States is a function of broad regional variations in climate and vegetation. To the south and west of New England, drier climate and/or seasonal periods of prolonged drought promote the development of complexes of vegetation and fuel that burn more readily than many of those of the Northeast. Within this broad regional framework, ignition sources, be they natural or human, influence the seasonality and frequency of fires. As an example, the average annual precipitation of 45–65 in. in south Florida equals or exceeds the amount of precipitation received throughout much of New England. The seasonal distribution of rainfall is quite uniform in the Northeast, however, whereas Florida is characterized by dry winters and wet summers (Wade *et al.* 1980). Florida is among the most fire-prone regions of North America (Pyne 1984), with human ignitions important during the winter and lightning ignitions important during spring and early summer (Taylor 1981). By contrast, fires are less common in New England, and some areas of interior, mountainous terrain have the lowest occurrences of fire in the East (Bormann and Likens 1979). Human ignitions are usually restricted to a short period between late March and early May and following short periods of late-summer drought that are the exception rather than the rule.

Lightning accounts for a small fraction of the total ignitions in the Northeast (Pyne 1984), and it is generally true that human factors become more important in wildland fire ignitions toward the east and north in North America. This pattern, which is so evident today, probably duplicates patterns that prevailed during prehistoric times. It is clear that Indians were an important source of ignitions in other portions of North America prior to European settlement (Lewis 1982a), and it seems likely that they were important in New England as well. Restricted opportunities for burning may have limited the areal extent and frequency of fires, however, and thus it is all the more important that we understand the demography of prehistoric Indian populations if we are to understand the role of Indians in pre-Colonial fire regimes of the Northeast.

Our comparison of sedimentary charcoal counts to archaeological site distributions indicates that fires were most common where Indian populations were greatest and their land-use practices most intensive. Lewis (1982b) discusses a number of reasons why Indians used fire to enhance resource availability. It is tempting to conclude that demographic pressure to intensify resource production caused Indians to use fire to modify prehistoric environments (Cohen 1977). We hesitate to correlate sedimentary charcoal concentrations with intensity of land-use, however, because charcoal influx to sediments is influenced by many factors besides the simple occurrence of fire within the watershed of a lake or pond (Patterson *et al.* 1987). Other things being equal, however, intense fires or repeated low-intensity fires should result in greater accumulations of sedimentary charcoal than infrequent, low-intensity fires. Swain (1980) has shown this to be true for northeastern North America as a whole, and it is likely that it is also true within a region the size of New England and for individual basins subjected to varying fire occurrence through time. In New England, the abundant charcoal in sediments from ponds in areas burned by south-coastal Indians probably reflects intensive land-use practices. Although burning documented by abundant charcoal in sediments from Cape Cod and Long Island may represent fires set for nonagricultural purposes, these fires appear to be geographically and temporally congruent with dense occupations of late-prehistoric agriculturalists.

Ethnohistorical accounts of fire use in New England suggest that generalized burning among hunter–gatherers may have been as intensive as that of agriculturalists. Extrapolation to prehistoric times is difficult, however, because ethnohistorical documentation describes patterns of land use among hunter–gatherers, as well as agriculturalists, that were being affected by European trade opportunities, introduced disease, and ensuing intergroup conflict. Intentional burning may have been common to prehistoric hunter–gatherers throughout New England, but charcoal records are probably inadequate to distinguish between the various types of fires that burned.

Problems associated with interpreting sedimentary charcoal records underscore the importance of developing more rigorous theory and method for examining prehistoric fire practices in New England. Additional research would benefit from closer attention to studies of prehistoric burning in Europe, where the environmental impacts of prehistoric populations has long been a subject of interdisciplinary research. Recent studies in Great Britain have begun to explore relationships between Mesolithic site locations, vegetation, soils, and fire. Simmons *et al.* (1981) suggest that hunter–gatherer fires were concentrated in areas of natural forest edge, particularly along the upland tree line and in coastal zones. Forest edge in these areas was economically attractive to Mesolithic groups and was especially prone to burning, and thus easy to manage with fire. Similar conclusions are drawn by Mellars and Reinhardt (1978) from comparisons of site locations and soil types. They document a preponderance of Mesolithic sites in southern England on well-drained, coarse sandy soils. They argue that these locations may have been selected for settlement because of their susceptibility to fire. Evidence for repeated Mesolithic burning at these sites includes vegetational change (Dimbleby 1962; Mellars and Reinhardt 1978), soil deterioration (Keef *et al.* 1965; Rankine and Dimbleby 1960, 1961), and increased erosion (Mellars and Reinhardt 1978:264).

Ecological and archaeological research in New England should focus on the identification of landscapes especially prone to fire. These locations can be compared to existing archaeological site records, and thus provide the basis for directed archaeological surveys. The success of regional approaches will depend, however, on our ability to recover specific evidence for burning at multiple sites. British researchers have employed a wide range of techniques and data sets to detect local fire use. In North America, wood charcoal from stratified sites was used to reconstruct changes in forest composition and structure, and together with pollen cores, is providing direct evidence for human prehistoric burning practices in Tennessee (Chapman *et al.* 1982).

10. CONCLUSION

Our work has implications beyond the role of Indians in prehistoric fire regimes of the Northeast. The methods brought to bear on prehistoric fire use will be limited only by the types of questions posed. Models that emphasize environment-dependent relationships for hunter–gatherers obscure evidence for environmental modification. The construction of models that emphasize the impacts of hunter–gatherer land-use activities allows us to look at old data in new ways and to develop methods for analyzing additional evidence. More importantly, a change in perspective toward interactive human–land relationships should provide new insight into the causes and consequences of population growth, settlement change, and food production.

Acknowledgments

We thank Dena Dincauze, George Nicholas, Fred Dunford, Michael Nassaney, Arthur Keene, and David Foster for comments on earlier drafts of the manuscript. Mitch Mulholland, Eric Johnson, Catherine Carlson, Omer Stewart, and David Sanger provided unpublished reports and data that were helpful in our analysis. The study was supported in part by the McIntire-Stennis Cooperative Forestry Research Program (Grant MS-33 to W.A.P.) and was completed while the senior author was a Charles Bullard Fellow at Harvard University.

11. REFERENCES

Anderson, R. S., 1979, *A Holocene Record of Vegetation and Fire at Upper South Branch Pond in Northern Maine,* M.Sc. Thesis, University of Maine, Orono, Maine.

Arno, S. F., 1980, Forest Fire History in the Northern Rockies, *Journal of Forestry* 78:460–465.

Backman, A. E., 1984, *1000-Year Record of Fire–Vegetation Interactions in the Northeastern United States: A Comparison between Coastal and Inland Regions,* M.Sc. Thesis, University of Massachusetts, Amherst, Massachusetts.

Backman, A. E., and Patterson, W. A., III, 1984, Vegetation and Fire History of Myles Standish State Forest Interpreted from Sedimentary Pollen and Charcoal Analyses, University of Massachusetts Cooperative Forestry Research Unit, Report no. 2, 21 p.

Barrett, S. W., and Arno, S. F., 1982, Indian Fires As an Ecological Influence in the Northern Rockies, *Journal of Forestry* 80:647–651.

Bernabo, J. C., and Webb, T., III, 1977, Changing Patterns in the Holocene Pollen Record of Northeastern North America, *Quaternary Research* 8:64–96.

Bormann, F. H., and Likens, G. E., 1979, Catastrophic Disturbances and the Steady State in Northern Hardwood Forests, *American Scientist* 67:660–669.

Bourque, B., 1973, Aboriginal Settlement and Subsistence on the Maine Coast, *Man in the Northeast* 6:3–20.

Brasser, T. J. C., 1971, The Coastal Algonkians: People of the First Frontier, in: *North American Indians in Historical Perspective* (E. B. Leacock and N. O. Lorie, eds.), Random House, New York, pp. 64–91.

Bromley, S. W., 1935, The Original Forest Type of Southern New England, *Ecological Monographs* 5:61–89.

Brown, A. A., and Davis, K. P., 1973, *Forest Fire Control and Use,* 2nd ed., McGraw-Hill, New York.

Brown, J. H., Jr., 1960, The Role of Fire in Altering the Species Composition of Forests in Rhode Island, *Ecology* 41:310–316.

Byers, D., 1946, The Environment of the Northeast, in: *Man in Northeastern North America* (F. Johnson, ed.), Robert S. Peabody Foundation for Archaeology, Andover, Massachusetts, pp. 3–32.

Ceci, L., 1979, Maize Cultivation in Coastal New York: The Archaeological, Agrinomical and Documentary Evidence, *North American Archaeologist* 1:45–74.

Champlain, S. de, 1907, *Voyages of Samuel de Champlain 1604–1618. Original Narratives of Early American History,* Vol. 3 (W. L. Grant, ed.), Barnes and Noble, New York.

Chapman, J., Delcourt, P. A., Cridlebaugh, P. A., Shea, A. B., and Delcourt, H. R., 1982, Man–Land Interaction: 10,000 Years of American Indian Impact on Native Ecosystems in the Lower Little Tennessee River Valley, Eastern Tennessee, *Southeastern Archaeology* 1:115–121.

Clark, J. S., and Patterson, W. A., III, 1984, Pollen, Pb-210, and Opaque Sperules: An Integrated Approach to Dating and Sedimentation in the Intertidal Environment, *Journal of Sedimentary Petrology* 54:1253–1265.

Clements, F. E., 1916, *Plant Succession: An Analysis of the Development of Vegetation,* Publication No. 242, Carnegie Institute, Washington, D.C.

Cohen, M. N., 1977, *The Food Crisis in Prehistory: Overpopulation and the Origins of Agriculture,* Yale University Press, New Haven, Connecticut.

Craig, A. J., 1972, Pollen Influx to Laminated Sediments: A Pollen Diagram from Northeastern Minnesota, *Ecology* 53:46–57.

Cronon, W., 1983, *Changes in the Land: Indians, Colonists and the Ecology of New England,* Hill and Wang, New York.

Curran, M. L., and Dincauze, D. F., 1977, Paleoindians and Paleolakes: New Data from the Connecticut Drainage, *Annals of the New York Academy of Science* 288:333–348.

Cwynar, L. C., 1978, Recent History of Fire and Vegetation from Laminated Sediments of Green Lake, Algonquin Park, Ontario, *Canadian Journal of Botany* 56:10–21.

Davis, M. B., 1983, Holocene Vegetational History of the Eastern United States, in: *Late Quaternary Environments of the United States* (H. E. Wright, ed.), University of Minnesota Press, Minneapolis, pp. 166–181.

Davis, R. B., 1967, Pollen Studies of Near-Surface Sediments in Maine Lakes, in: *Quaternary Paleoecology* (E. J. Cushing and H. E. Wright, Jr., eds.), Yale University Press, New Haven, Connecticut, pp. 144–173.

Day, G., 1953, The Indian as an Ecological Factor in the Northeastern Forests, *Ecology* 34:329–346.

Dimbleby, G. W., 1962, *The Development of British Heathlands and Their Soils,* Oxford Forestry Memoirs.

Dincauze, D. F., and Curran, M. L., 1983, Paleoindians as Generalists: An Ecological Model, Paper presented at the 48th Annual Meeting of the Society for American Archaeology, Pittsburgh, Pennsylvania.

Dincauze, D. F., and Mulholland, M. T., 1977, Early and Middle Archaic Site Distributions and Habitats in Southern New England, *Annals of the New York Academy of Science* 288:439–456.

Dwight, T., 1823, *Travels in New England and New York,* W. Baynes, London.

Espinshade, F. B. (ed.), 1976, *Goode's World Atlas,* 14th ed., rev., Rand McNally, Chicago, Illinois.

Fahey, T. J., and Reiners, W. A., 1981, Fire in the Forests of Maine and New Hampshire, *Bulletin of the Torrey Botanical Club* 108:362–373.

Fobes, C. B., 1944, Lightning Fires in the Forests of Northern Maine, *Journal of Forestry* 42:291–293.

Forman, R. T. T., and Russell, E. W. B., 1983, Evaluation of Historical Data in Ecology, *Bulletin of the Ecological Society of America* 64:5–7.

Fredskild, B., 1967, Paleobotanical Investigations at Sermermiut, Jakobshavn, West Greeland, *Meddeleser om Grønland* 178.

Frissell, S. S., Jr., 1973, The Importance of Fire as a Natural Ecological Factor in Itasca State Park, Minnesota, *Quaternary Research* 3:397–407.

Funk, R. E., 1983, The Northeastern United States, in: *Ancient Northern Americans* (J. D. Jennings, ed.), Freeman, New York, pp. 303–371.

Gajewski, K., 1983, *On the Interpretation of Climatic Change from the Fossil Record: Climatic Change in Central and Eastern United States over the Past 2000 Years Estimated from Pollen Data,* Ph.D. Dissertation, University of Wisconsin, Madison, Wisconsin.

Harris, D. R., 1972, Swidden Systems and Settlement, in: *Man, Settlement and Urbanism* (P. J. Ucko, R. Tringham, and G. W. Dimbleby, eds.), Duckworth, London, pp. 245–262.

Heinselman, M. L., 1973, Fire in the Virgin Forests of the Boundary Waters Canoe Area, Minnesota, *Quaternary Research* 3:329–382.

Heinselman, M. L., 1981, Fire Intensity and Frequency as Factors in the Distribution and Structure of Northern Ecosystems, in: *Fire Regimes and Ecosystem Properties* (H. A. Mooney, ed.), U.S.D.A. Forest Service General Technical Report WO-26, pp. 7–57.

Henry, J. D., and Swan, J. M. A., 1974, Reconstructing Forest History from Live and Dead Plant Material—An Approach to the Study of Forest Succession in Southwestern New Hampshire, *Ecology* 55:772–783.

Iversen, J., 1941, Land Occupation in Denmark's Stone Age, *Danmarks Geologiske Undersøgelse. Series II,* No. 66.

Keef, P. A. M., Wymer, J. J., and Dimbleby, G. W., 1965, A Mesolithic Site on Iping Common, Sussex, England, *Proceedings of the Prehistoric Society* 31:85–92.

Kozlowski, T. T., and Ahlgren, C. E. (eds.), 1974, *Fire and Ecosystems,* Academic Press, New York.

Lewis, H. T., 1973, Patterns of Indian Burning in California: Ecology and Ethnohistory, *Anthropological Papers,* No. 1, Ballena Press, Ramona, California.

Lewis, H. T., 1977, Maskuta: The Ecology of Indian Fires in Northern Alberta, *Western Canadian Journal of Anthropology* 7:15–52.

Lewis, H. T., 1982a, Fire Technology and Resource Management in Aboriginal North America and Australia, in: *Resource Managers: North American and Australian Hunter–Gatherers* (N. M. Williams and E. S. Hunn, eds.), AAAS Selected Symposia 67, Westview Press, Boulder, Colorado, pp. 45–67.

Lewis, H. T., 1982b, A Time for Burning, University of Alberta Boreal Institute of North America, Occasional Paper 17.

Lorimer, C. G., 1977, The Presettlement Forest and Natural Disturbance Cycle of Northeastern Maine, *Ecology* 58:139–148.

Lutz, H. J., 1959, Aboriginal Man and White Man As Historical Causes of Fires in the Boreal Forest, with Particular Reference to Alaska, *Yale University School of Forestry Bulletin,* No. 65.

McAndrews, J. H., 1976, Fossil History of Man's Impact on the Canadian Flora: An Example from Southern Ontario, *Canadian Botanical Association Bulletin* 9:1–6.

Mellars, P. A., 1976, Fire Ecology, Animal Populations and Man: A Study of Some Ecological Relationships in Prehistory, *Proceedings of the Prehistoric Society* 42:15–46.

Mellars, P. A., and Reinhardt, S. C., 1978, Patterns of Mesolithic Land-Use in Southern England: A Geological Perspective, in: *The Early Post-Glacial Settlement of Northern Europe* (P. A. Mellars, ed.), University of Pittsburgh Press, Pittsburgh, Pennsylvania, pp. 243–294.

Mulholland, M. T., 1984, *Patterns of Change in Prehistoric Southern New England: A Regional Approach,* Ph.D. Dissertation, University of Massachusetts, Amherst.

Myers, R. L., and Peroni, P. A., 1983, Approaches to Determining Aboriginal Fire Use and Its Impact on Vegetation, *Bulletin of the Ecological Society of America* 64:217–218.

Neiring, W. A., and Goodwin, R. H., 1962, Ecological Studies in the Connecticut Arboretum Natural Area: Introduction and a Survey of Vegetation Types, *Ecology* 43:41–54.

Nicholas, G. P., 1982, Former Glacial Lake Basins and Early Postglacial Settlement, *AMQUA Abstracts* (American Quaternary Association) 7:148.

Nicholas, G. P., 1983, A Model for the Early Postglacial Settlement of the Central Merrimack River Basin, New Hampshire, *Man in the Northeast* 25:43–63.

Nicholas, G. P., 1987, Rethinking the Early Archaic, *Archaeology of Eastern North America,* 15:99–124.

Patterson, W. A., III, Saunders, K. E., and Horton, L. J., 1983, Fire Regimes of the Coastal Maine Forests of Acadia National Park, U.S.D.I. National Park Service Office of Scientific Programs, Report OSS 83-3, Boston, Massachusetts.

Patterson, W. A., III, Backman, A. E., and Davis, R. B., 1984a, Sedimentary Charcoal as an Indicator of Forest Fires in Acadia National Park, Maine, *Sixth International Palynological Congress Abstracts,* p. 126.

Patterson, W. A., III, Saunders, K. E., and Horton, L. J., 1984b, Fire Regimes of Cape Cod National Seashore, U.S.D.I. National Park Service Office of Scientific Programs, Report OSS 83-1, Boston, Massachusetts.

Patterson, W. A., III, Edwards, K. J., and Maguire, D., 1987, Microscopic Charcoal as a Fossil Indicator of Fire, *Quaternary Science Reviews* 6:3–23.

Pyne, S. J., 1982, *Fire in America: A Cultural History of Wildland and Rural Fire,* Princeton University Press, Princeton, New Jersey.

Pyne, S. J., 1984, *Introduction to Wildland Fire,* Wiley, New York.

Rankine, W. F., and Dimbleby, G. W., 1960, Further Investigations at a Mesolithic Site at Oakhanger, Selbourne, Hants, *Proceedings of the Prehistoric Society* 26:246–262.

Rankine, W. F., and Dimbleby, G. W., 1961, Further Excavations at Oakhanger, Selbourne, Hants, *Wealdon Mesolithic Research Bulletin.*

Rudnicky, J. L., 1984, Vegetation and Fuel Types on Fire Island National Seashore and the William Floyd Estate, M.Sc. project, University of Massachusetts, Amherst, unpublished.

Russell, E. W. B., 1983, Indian-Set Fires in the Forests of the Northeastern United States, *Ecology* 64:78–88.

Salwen, B., 1978, Indians of Southern New England and Long Island: Early Period, in: *Handbook of North American Indians: Northeast,* Vol. 15 (B. G. Trigger, ed.), Smithsonian Institute, Washington, D.C., pp. 160–175.

Sanger, D., 1979, The Ceramic Period in Maine, in: *Discovering Maine's Archaeological Heritage* (D. Sanger, ed.), Maine Historic Preservation Commission, Augusta, Maine, pp. 99–115.

Sanger, D., 1982, Changing Views of Aboriginal Seasonality and Settlement in the Gulf of Maine, *Canadian Journal of Anthropology* 2:195–202.

Shackleton, J., 1979, *Paleoenvironmental Histories from Whitefish and Imuruk Lakes, Seward Peninsula, Alaska,* M.Sc. Thesis, Ohio State University, Columbus, Ohio.

Simmons, I. G., Dimbleby, G. W., and Grigson, C., 1981, The Mesolithic, in: *The Environment in British Prehistory* (I. G. Simmons and M. J. Tooley, eds.), Cornell University Press, Ithaca, New York, pp. 82–124.

Snow, D. R., 1978a, Late Prehistory of the East Coast, in: *Handbook of North American Indians: Northeast,* Vol. 15 (B. G. Trigger, ed.), Smithsonian Institute, Washington, D.C., pp. 137–147.

Snow, D. R., 1978b, Eastern Abenaki, in: *Handbook of North American Indians: Northeast,* Vol. 15 (B. G. Trigger, ed.), Smithsonian Institute, Washington, D.C., pp. 137–147.

Snow, D. R., 1980, *The Archaeology of New England,* Academic Press, New York.

Spurr, S. H., 1956, Natural Restocking of Forests Following the 1938 Hurricane in Central New England, *Ecology* 37:443–451.

Stephens, E. P., 1955, *The Historical-Developmental Method of Determining Forest Trends,* Ph.D Dissertation, Harvard University, Cambridge, Massachusetts.

Stewart, O. C., 1953, Why the Great Plains are Treeless, *Colorado Quarterly* 2:40–49.

Stewart, O. C., 1956, Fire as the First Great Force Employed by Man, in: *Man's Role in Changing the Face of the Earth* (W. L. Thomas, ed.), University of Chicago Press, Chicago, Illinois, pp. 115–133.

Stoltman, J. B., and Baerreis, D. A., 1983, *The Evolution of Human Ecosystems in the Eastern United States* (H. E. Wright, Jr., ed.), University of Minnesota Press, Minneapolis, pp. 252–268.

Swain, A. M., 1973, A History of Fire and Vegetation in Northeastern Minnesota as Recorded in Lake Sediments, *Quaternary Research* 3:383–396.

Swain, A. M., 1980, Environmental Changes between 600–100 B.P. in Mid-Latitudes of North America: Pollen and Charcoal Evidence from Annually Laminated Lake Sediments, *5th International Palynological Conference Abstracts.*

Taylor, D. L., 1981, Fire History and Fire Records for Everglades National Park, U.S.D.I., National Park Service Everglades National Park Report T-619.

Thomas, P. A., 1976, Contrastive Subsistence Strategies and Land-Use: 20 Factors for Understanding Indian–White Relations in New England, *Ethnohistory* 23:1–18.

Thomas, P. A., 1980, Comments on Recent Trends in Vermont Archaeology, *Man in the Northeast* 19:3–14.

Tolonen, K., 1978, Effects of Prehistoric Man on Finnish Lakes, *Polskie Archiwum Hydrobiologii* 25:419–421.

Tsukada, M., 1966, Late Postglacial Absolute Pollen Diagrams in Lake Nojiri, *The Botanical Magazine* (Tokyo) 79:179–184.

Tsukada, M., and Deevey, E. S., Jr., 1967, Pollen Analysis from Four Lakes in the Southern Maya Area of Guatemala and El Salvador, in: *Quaternary Paleoecology* (E. J. Cushing and H. E. Wright, Jr., eds.), Yale University Press, New Haven, Connecticut, pp. 303–331.

Waddington, J. C. B., 1968, *The Postglacial Vegetation History of the Big Woods Area of Minnesota,* M.Sc. Thesis, University of Minnesota, Minneapolis, Minnesota.

Wade, D., Ewel, J., and Hofstetter, R., 1980, Fire in South Florida Ecosystems, U.S.D.A. Forest Service General Technical Report SE-17.

Westveld, M., *et al.*, 1956, Natural Forest Vegetation Zones of New England, *Journal of Forestry* 56:332–338.

Winkler, M., 1982, *Late-Glacial and Post-Glacial Vegetation History of Cape Cod and the Paleolimnology of Duck Pond, South Wellfleet, Massachusetts,* M.Sc. Thesis, University of Wisconsin, Madison, Wisconsin.

Wobst, H. M., 1983, We Can't See the Forest for the Trees: Sampling and the Shapes of Archaeological Distributions, in: *Archaeological Hammers and Theories* (J. A. Moore and A. S. Keene, eds.), Academic Press, New York, pp. 37–85.

Wolf, E., 1982, *Europe and the People without History,* University of California Press, Berkeley, California.

Chapter 6

Territoriality and Horticulture

A Perspective for Prehistoric Southern New England

MITCHELL T. MULHOLLAND

1. INTRODUCTION

Several archaeologists in southern New England have observed that during the Archaic period of gathering and hunting in the Northeast, between approximately 10,000 and 3,000 years ago, sites proliferated in the interior and were located in a variety of environmental zones (Dincauze 1974; Thorbahn *et al.* 1980). These studies further suggest, on the basis of subregional site distributions, that prehistoric people abandoned the interior at the beginning of the Woodland period, 3,000 years ago, and shifted the focus of their subsistence to the coastal zone. This paper evaluates the validity of these propositions and attempts to answer the question: why did prehistoric land use shift in focus from a diversity of habitats throughout southern New England to the major river valleys and coastal areas, once horticulture became an important element in the subsistence mix? Archaeological site distributions from southern New England for the past 12,000 years are used to evaluate geographic changes in land use, while documentary evidence written by European observers of Native Americans is used to compare differences in territoriality and land use between gatherer–hunters and foraging horticulturalists.

MITCHELL T. MULHOLLAND • Department of Anthropology, University of Massachusetts, Amherst, Massachusetts 01003.

2. BACKGROUND

Archaeological evidence from southern New England indicates that for most of the 12,000 years of prehistoric occupation, a primarily foraging subsistence pattern predominated, which was enhanced by the introduction of horticulture only quite recently. Culture-historical models for the Northeast invoke a continuous occupation of the region from the late Pleistocene period through to the present (e.g., Dincauze 1974; Funk 1976; Salwen 1975; Snow 1980). For 9,000 years, a foraging subsistence pattern was followed in which a diversity of resources was available for exploitation. Site frequencies increased during this period, and expanded into a variety of environmental zones, including the upland interior, large and small interior river valleys, and the coast and coastal plain, thus reflecting the use of a broad-spectrum resource base (Dincauze 1974). At some point during the past three millennia, horticulture was introduced and combined with foraging, which resulted in changes in land-use patterns. This occurred primarily because horticulture is dependent upon the arable, alluvial soils of the river valleys. As subsistence focused increasingly on these riverine zones, mobility and the opportunity to expand territories into distant areas was reduced. By the first millennium, foraging horticulturalists exhibited a settlement distribution that was distinct within the prehistoric record. While inland/upland areas were still in use, most sites were located in the coastal zone and on the finite arable lands of the major river valleys.

2.1. Subsistence and Settlement Patterns in Southern New England

The sequence of prehistoric adaptation in southern New England begins with a 2,000-year period of park tundra Paleoindian foraging adaptation that ultimately focused on the developing deciduous forests about 10,000 B.P. The Archaic period (10,000–3000 B.P.) has been characterized as a foraging adaptation to wild resources in a temperate forest (Ritchie 1969a:31). In the Northeast, this period is characterized by a diversity of artifact styles, as well as by the highest frequency of sites in the regional prehistoric record. Archaic sites are found in a wide range of habitats (Dincauze 1974:48) in a dispersed pattern through the interior (Thorbahn *et al.* 1980). Regional site records show that while site frequencies increased on the coast and in the uplands, the interior coastal plain and river valleys were still well-represented (Mulholland 1984:547). In southern New England, a riverine/coastal foraging subsistence strategy developed that was oriented to seasonally available temperate resources, the most important of which were probably mast-oriented foods, such as the acorn/white-tailed deer association and, after 8000 B.P., anadromous fish, resources from freshwater lakes, and ocean fish from the coastal zone. Areas with high wetland densities and lakes were also important sources of freshwater aquatic species and were occupied heavily (Mulholland 1984; Nicholas, Chapter 10). Sites continued to increase during the Archaic, reaching a peak some 4500 B.P. (Figure 1), partly in response to environmental conditions (Dincauze 1974; Mulholland 1979, 1984). Pollen records for the region (Davis 1981; Davis *et al.* 1980; Whitehead 1980) suggest increases in the frequency of mast species (e.g., oak and hickory) in southern New England at approximately 4500 B.P. This change corresponded with a dramatic decline in hemlock in the region's forests from 5200 to 3900 B.P. (Webb 1982:570), which is attributed by Davis (1981) to an as-yet-unidentified insect predator and/or disease, rather than to climate. Hemlock is today a very competitive species in the region, the loss of which would have enhanced the competitive ability of nutritionally important mast species. This combination of events (added to generally warm/dry conditions) would have enhanced the attraction to human populations of a variety of environmental zones that had been less frequently or intensively utilized in the past (Mulholland 1979, 1984).

Transgressing sea levels that followed the recession of the Laurentide ice sheet have undoubtedly obscured much evidence of coastal exploitation prior to

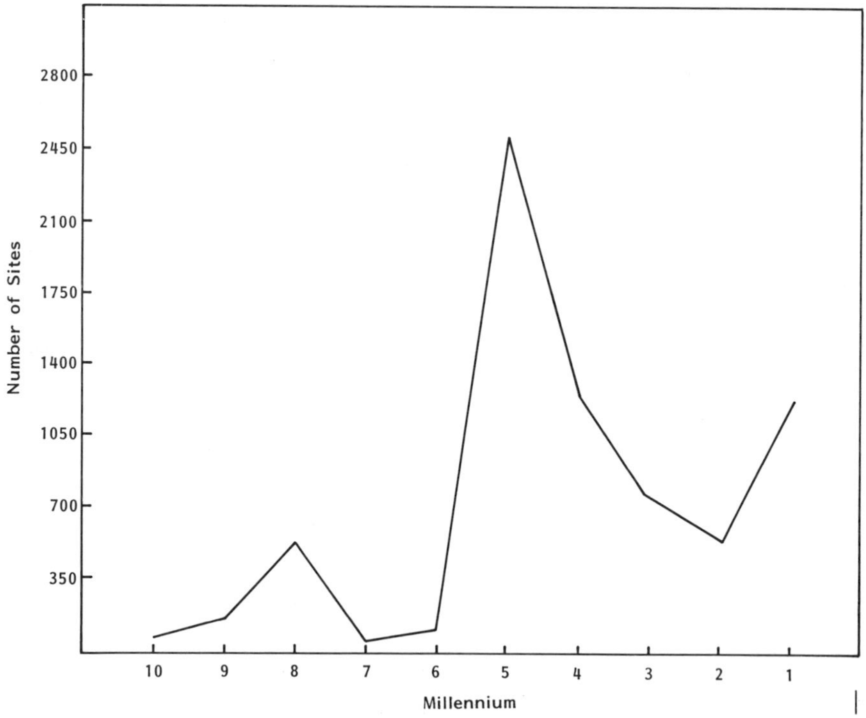

Figure 1. Prehistoric site frequency in southern New England by millennium before present (Mulholland 1984).

the fourth and fifth millennia in southern New England (Braun 1972; Kellogg, Chapter 4). There are few reasons why this rich habitat was not as attractive then as it was following the fourth millennium once sea-level transgression had slowed, permitting the coastal estuaries to develop. Site distributions evident in regional site records show that site frequencies increased along the coast and in the interior and declined somewhat on the coastal plain and river valleys (Mulholland 1984:547).

The next chronological/cultural period defined in the regional literature is the Woodland period, which began at approximately 3000 B.P. and ends with European contact. The period name, which was first coined by McKern (1939), carries with it an assumed adaptation to a horticultural subsistence base, with manufacture of ceramics the major technological milestone. The exact timing of the introduction of horticulture into the region is controversial and current estimates range in time from 3000 B.P. (Snow 1980) to the first millennium (e.g., Dincauze 1974; Ritchie 1969a; Salwen 1975). Radiocarbon dates for maize from archaeological sites in southern New England support the more recent date (Ritchie 1969b; Salwen 1975). Settlement patterns during this period traditionally have been considered to have involved a gradually increasing level of sedentism, particularly in the lowlands and along the coast, but are, on the whole, considered to be much the same as during the preceding Late Archaic (i.e., 6000–3000 B.P.). On the basis of a decrease in the number of sites, regional site records are suggestive of a dramatic population decline following 3000 B.P. (Figure 1), a decline that is accompanied by a decline in the frequency of oak in the forest and, presumably, by a decline in mast-related fauna (Mulholland 1984:332). Site frequencies for the region drop from 1,263 sites in the fourth millennium to 711 sites in the third millennium (Mulholland 1984:445). This time is also characterized by the start of a decline in regional temperatures (Davis *et al.* 1980).

A proportional increase in the number of sites in the coastal zone is indicated during Early Woodland times approximately 3000 B.P. (Figures 2 and 3) (Mulholland 1984:506–509). There is also an increase in the number of shell middens throughout coastal southern New England (Figure 4) (Mulholland 1984:487). The data concerning shell middens must be used with caution because of the effect rising sea levels had on many such sites. However, in areas such as the coast of Connecticut, where the horizontal impact of sea-level rise is less than in areas such as southeastern Massachusetts, the rise in the frequency of shell middens is similar (Mulholland 1984).

Although the traditionally suggested shift from terrestrial resources to the use of coastal resources has been interpreted as an evolutionary inevitability that resulted from technological development (e.g., Ritchie 1969a; Snow 1972), coastal resources were undoubtedly an important part of the subsistence mix throughout much of Northeastern prehistory. Site distributional data for the

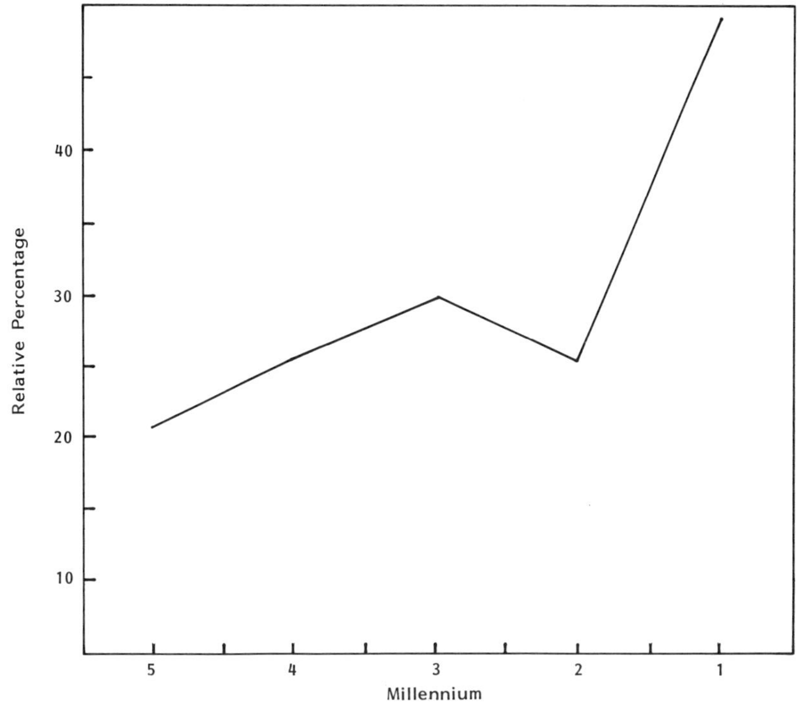

Figure 2. Relative percentage of components within 10 km of the modern coastline, Greater Boston area, studied by Dincauze (1974). Data from Mulholland (1984).

second millennium before present show a slight decrease in the number of sites (from 711 to 551) and a continued usage of coastal areas (Mulholland 1984). On the other hand, increases in site frequency for the first millennium, during which maize horticulture is believed by some to have been introduced, support the hypotheses of population increase suggested by Dincauze (1974:53) and Thorbahn *et al.* (1980). The data suggest a dramatic increase in population in the first millennium comparable in many ways to Late Archaic times, with the exception that the upland pattern is not quite as extensive.

This paper examines the ways that territoriality and coastal/riverine land use are expressed in the subsistence strategies of foragers and foraging horticulturalists. Insights are derived from both the archaeological and historical record and from regional site distributions. An important premise is that as horticultural groups became more reliant upon arable land, which was geographically finite, the availability of that land would have decreased in the wake of an increase in population density. Another major premise is that as subsistence strategies

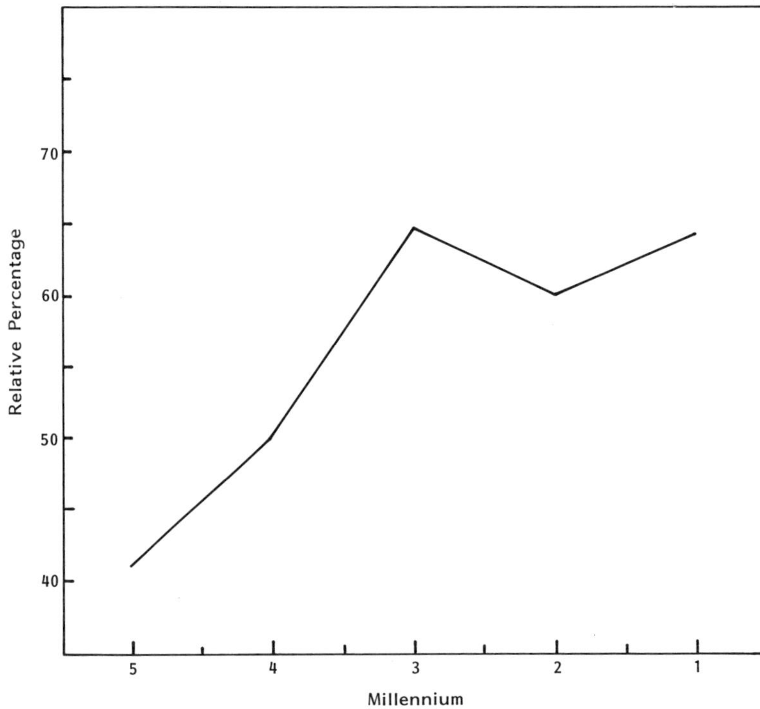

Figure 3. Relative percentage of components within 10 km of the modern coastline, southeastern Massachusetts, study area of Thorbahn *et al.* (1980). Data from Mulholland (1984).

changed from foraging alone to foraging from a horticultural economic base, groups became more dependent on finite geographic areas in the river valleys and in the coastal zone. This change in resource focus ultimately resulted in a reduction in the availability of arable lands as land became overexploited and population increased.

2.2. Sources of Evidence

The relationships among land use, subsistence strategies, territoriality, and environmental degradation are examined here in light of evidence in the historical and archaeological record. It is difficult to observe archaeologically many of the cultural subsystems that are affected by territorial behavior, however. On the basis of material evidence alone, only broad statements can be made concerning prehistoric foraging behavior. Greater resolution is therefore provided by the use

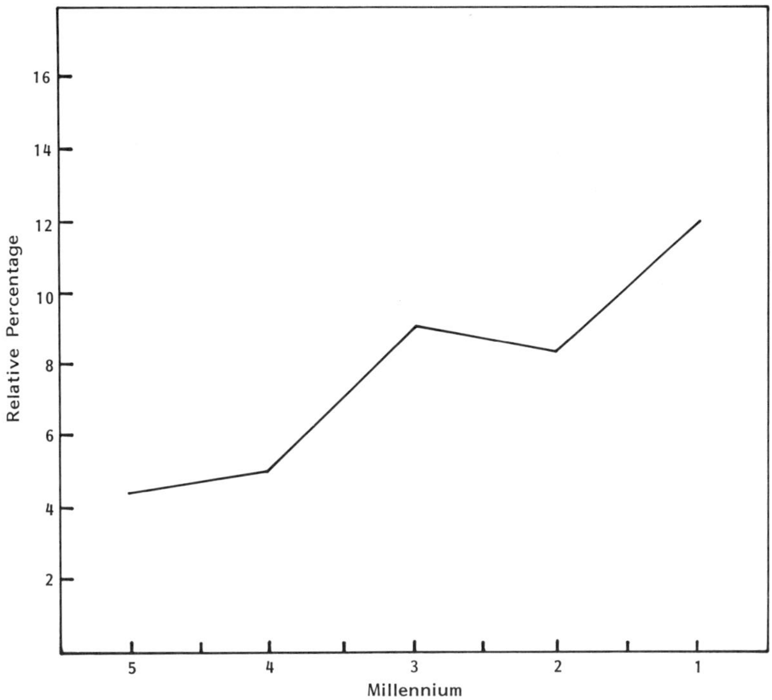

Figure 4. Percentages of sites with shell middens in southern New England by millennium before present (Mulholland 1984).

of historical records derived during the early seventeenth century, which represents the recent end of the period of foraging and horticulture.

Settlement patterns and site distributional models in southern New England frequently have been constructed on the basis of local archaeological observations and then extrapolated to the region as a whole (cf. Ritchie 1969a). Unfortunately, many of these early culture-historical studies focused on a very few, but highly visible sites, using them as though they typified the total range of prehistoric behavior for the period being considered. This approach has the disadvantage that inferred behaviors may be responses to only local factors, and that such behaviors tend to amplify disproportionately those behaviors that made the site visible in the first place. Although observations at the local level are important, because of the fine resolution of the data upon which they are based, their utility is limited when comparisons with other local areas and to the region as a whole are not made. As an alternative, site distributions that are evident in a data base derived from the major site repositories for the region (e.g., site survey files) are used to evaluate traditional assumptions and to explain changes in settlement patterns as groups relied on a different resource base.

3. PREHISTORIC SUBSISTENCE PATTERNS

3.1. Historical Considerations

Although plant cultivation began relatively recently in southern New England, following 9,000 years of foraging, *both* foraging and foraging–horticulture were practiced in historic times. In the sixteenth century, for example, Samuel de Champlain and many other early European observers of aboriginal behavior in North America noted that both mobile foraging groups and sedentary cultivators could be recognized among coastal populations (Champlain 1907:4), depending upon latitude.

An important working assumption in the use of historical documents in the Northeast is that aboriginal groups east and north of the Kennebec River in Maine may be used as an analog to test predictions of prehistoric foraging group behavior. The Kennebec River represents approximately the known northern limit of maize cultivation (Thomas 1978:18). If such an analog is considered reasonable, then documentary reports of historical foragers can provide insight into prehistoric patterns during the Archaic through Middle Woodland times; similarly, reports concerning sedentary cultivators that subsisted south of this boundary may be considered analogous to Late Woodland patterns in southern New England. It is recognized that the European fur trade dramatically affected aboriginal behavior by modifying settlement patterns, subsistence patterns, size and location of settlements, and the like. In addition, the various epidemics that decimated the aboriginal population from 1610 to 1635 destroyed pre-Contact period behavior patterns (Brasser 1978). Therefore, the early historical accounts are used here. Additionally, archaeological site distributional data provide a comparison with historical data and shed light on behaviors that occurred prior to the documentary record.

During the middle and late Holocene in southern New England, riverine and coastal areas are expected to have had the highest resource diversity, density, and predictability (Mulholland 1984; Nicholas, Chapter 10). It is these areas, therefore, that foragers are expected to have exploited most extensively for seasonally available, aquatic, coastal, and mast-oriented resources. Interior wetlands and lakes, however, were also important, especially in southeastern Massachusetts (cf. Thorbahn and Cox, Chapter 7), northwestern Connecticut (Nicholas and Handsman 1984, 1986), and western Massachusetts (Shaw *et al.* 1987). With some exceptions, interior site densities and the frequency of large occupation sites tend to be lower than in the lowland river valleys and along the coast. This is especially true of the higher elevations (Mulholland 1984). Dodge (1957:68) has demonstrated that groups of foragers and foraging horticulturalists from Long Island Sound to the Saint Lawrence River were represented in the literature as a "series of tribes centered on the large river valleys each varying only slightly from its neighbors." Similarly, Snow (1968:1147) has shown that the site distribution of the Wabanaki group, which included the Micmac, Malecite, Passamaquoddy, Penobscot, and Abenaki, coincided with the dendritic stream pattern of major streams and their tributaries. River basin territoriality was also a focal argument in Snow's *Prehistory of New England* (1980:2–3) and has been adopted by other researchers (cf. Kenyon and McDowell 1983), although Carlson (1986) suggests that this is a post-Contact period pattern resulting from the fur trade.

Historical evidence of riverine-based territoriality of foraging groups is given in reports of Captain Gilbert's excursion up the Kennebec River in Maine in 1607. Gilbert encountered the sachem Sabenoa, who he observed "macks himselffe unto us to be the Lord of the ryver of Sagadehock [Kennebec]" (Anonymous 1932:415), and Basabe, "who is the cheeffe commander of the ryver of Penobscot" (Anonymous 1932:405). The difficulty with relying heavily on historical accounts is that they tend to emphasize the coastal areas, and it is not easy to separate patterns that are vestiges of the pre-Contact period from those patterns that are reported (cf. Carlson 1986).

3.2. Foragers

Foraging groups who maintain a system of seasonal mobility move in response to resource yields of high predictability and availability (Binford 1982; Curran 1987). While the subsistence mix generally includes only a few predictable, high-energy foods, alternative edible resources are usually abundant, depending upon the richness and diversity of the habitat. Barring major environmental fluctuations, foraging groups in southern New England functioned under relatively stable conditions during the middle and late Holocene.

In such a highly productive environment, there are few constraints on population size. As population increased over time and the drain on exploited resources increased, the quality and width of the currently exploited environmental niche would be expected to decrease. Because foraging groups are mobile, the breadth of the food niche could be expanded to include previously less attractive zones and resources (Fretwell and Lucas 1969). Site distributions of known Late Archaic sites in southern New England are suggestive of just such an expansion (Mulholland 1979, 1984), indicating a utilization of upland zones that were less exploited in the past.

The defense of foraging territories is only practical when critical resources are sufficiently abundant and predictable (Dyson-Hudson and Smith 1977). This is supported by the earlier demonstration by Schoener (1971) that defense is too costly unless resource density is high. Territorial defense is a high-energy-use activity and can occur only when population is large enough to absorb the cost of defense. Wilson's (1975) "Sufficient Resource Hypothesis" further suggests that those species that must control large multipurpose territories generally defend an adequate food supply (see also Verner 1977:771). By slightly modifying Wilson's model, it can be postulated that mobile foragers defend a multiresource territory that requires exclusive control of a large geographic area. Foraging territories should therefore be expected to encompass those resources needed for survival of the group, thereby keeping defense costs at a minimum. This large "sufficient resource" territory should encompass many biotic zones, but would be most intensively utilized in the coastal and riverine zones, where maximum clustering of nutritionally important resources would occur. The upland interior, especially in areas with high lake and wetland densities, was also occupied, but territories would be larger because nutritionally important resources are of lower density, and the distance between areas of predictable subsistence resources is frequently greater than in the lowlands.

Thus, foraging groups are generally seasonally mobile and would defend large geographic territories that include sufficient resources for group survival. In the face of population increase, they could, as an alternative strategy, expand the niche to include utilization of zones that were previously less attractive. Such a process of expansion should minimize the need for territorial maintenance through overt aggressiveness.

3.3. Foraging Horticulturalists

The precise timing of the introduction of horticulture to the region is poorly understood. Snow (1980:262), for example, argues that by 2700 B.P., potential domesticates were in use by Native Americans. More research is clearly needed to substantiate this in southern New England. By approximately 1000 B.P., foraging groups in southern New England had expanded the diet breadth and subsistence strategy to include cultivated foods. Evidence for known maize cultivation prior to that time is lacking throughout the Northeast. The earliest maize in southern New England has been dated at approximately A.D. 1200 (790 +/− 80 B.P., Y-1653) (Ritchie 1969a,b), and in New York State at A.D. 1070 (880 B.P.) (Salwen 1975:56; Ritchie 1969b). Other dates for maize in the region also fall within the first millennium and include 445 +/− 90 B.P. (QC-974B) and 460 +/− 100 B.P. (QC-940A) from South Windsor, Connecticut (Kevin McBride, personal communication, 1984, and 445 +/− 100 B.P. (QC-949) from the Indian Crossing site in Chicopee, Massachusetts (Hasenstab *et al.* 1986). Historical records suggest that horticulture did not replace foraging as a subsistence strategy in the Northeast, but contributed to the subsistence mix as an additional, but important storable resource (cf. Thomas 1978; Winthrop 1908; Wood 1865).

Perhaps the most important effect of increasing reliance on this new resource was that group settlements became more permanent. Salwen (1975:57) has argued that this new "concentrated settlement system" could have resulted in "overexploitation and long term degradation of habitat around valleys." Occupation areas that once clustered in the location of predictable aquatic resources also would have focused increasingly on arable lands probably in the vicinity of the fishing stations. This would have resulted in intensified land use in riverine areas for longer periods of time. The horticultural cycle requires attention for several months of the year, whereas the most important fish runs occur in the spring. Short-distance movement from horticultural areas would occur only as soil fertility became exhausted (Salwen 1975). Denton reported in 1670 that the Indians of Manhattan and Jamaica, New York "remove two or three times a year, having their principal quarters where they plant their corn"; next in importance were their hunting grounds and their fishing quarters (Denton 1937:7). Aboriginal groups on Long Island were observed in 1626 to remain in one location from December until October, leaving only to hunt deer (de Rassieres 1909:103). Whereas this observation was made after the epidemics of the sixteenth and seventeenth centuries, inhabitants of this area appear to have been less affected than people in eastern New England (Salwen 1978), and the settlement pattern may be similar to that of pre-European contamination. Unfortunately, it is difficult to measure the effect of European contact on these patterns.

The increasingly concentrated use of a finite geographic area as necessitated by a reliance on horticulture would have resulted in significant changes in aboriginal cultural processes. Initially, maize was a nutritious and highly-storable addition to the subsistence mix, and, as in a purely foraging subsistence pattern, many alternative resources were available during periods of food stress. Dincauze has suggested that in the early years the practice of shifting cultivation may have actually improved the attractiveness of the riverine areas, because the fallowing fields created a resource-rich and diverse habitat (Dena F. Dincauze, personal communication, 1984). Historical observations of aboriginal swidden horticulture in the Northeast suggest that fallow periods could last from 5 to 20 years, depending on soil types (van der Donck 1841; Thomas 1976; Winslow 1885; Wood 1865). This relatively stable environmental situation would have produced few constraints on population expansion, but population increase among land-restricted horticulturalists could ultimately have led to an intensification of land use, which would have resulted in an increased competition for arable land. Site distributions in the region support this increase. From the second to the first millennium B.P., sites in southern New England increased from 551 to 1,274 (Figure 1), a site frequency which is half the frequency of the Late Archaic period (2,497 sites), the most highly populated period in prehistory (Mulholland 1984:547). This frequency is probably lower than it could have been, because the prehistoric first millennium extends only from approximately

1000 to 400 B.P., when it was terminated by European contact. One way in which this condition could have been satisfied is through a reduction in population density. This could obviously occur as a result of starvation or reduced fertility caused by dietary stress, or by warfare.

The same processes that justify territorial defense efforts among foraging groups are relevant in a combined foraging horticultural system. Defense costs are reduced as population expansion increases the number of potential defenders. Defense needs are particularly apparent in a horticultural system, because a major portion of the group's annual caloric uptake is produced from small geographic areas that are in high demand. Once population is large enough, competition from outside the group could have disastrous consequences. Given the development of an excessively large population approaching the carrying capacity of available arable areas, land would be so intensively exploited that the usual attractiveness of the riverine zone to prey species would be significantly reduced. If this occurred, populations of foraging horticulturalists would be highly susceptible to starvation. Their situation would be severely complicated by any environmental perturbation that could negatively affect crop yields, or by the cyclical fluctuations in the size of animal populations. Unlike the more mobile hunter–gatherers, foraging horticulturalists would have less of an opportunity to expand into unutilized zones, because of their reliance on cultigens and the restriction of horticultural activities to alluvial areas.

Susceptibility to food stress could be reduced by periodic reliance on perennially available coastal resources. Following the middle Holocene, the coastal zone was an attractive biotic zone for human groups. Combining terrestrial and aquatic resources, the coastal zone undoubtedly contained the highest diversity and density of human subsistence resources throughout the region. For cultivators, the most productive area within the coastal zone would be the freshwater floodplain nearest the estuaries. The mouths of estuaries are highly productive food-producing areas for marine resources, such as fish, shellfish, and water fowl (Carlson 1986; Dincauze 1974). Coastal and pelagic species could be added to the diet as desired, and would have been available as starvation foods during periods of dietary stress. Thus, for both inland and coastal foraging horticulturalists with large populations, coastal foods became important critical resources for group survival. Inland groups would have had to rely on the coastal group for access to those resources during periods of stress. There are frequent accounts in the historical record of inland groups making trips to the coastal zone for collecting and drying shellfish and other marine species (Brereton 1906; Champlain 1907; Mourt 1963). Thus, territoriality among foraging horticulturalists would encompass both river basins and the coastal zones, with the most strategic and important defense positions being the coastal fringe.

Thomas (1976:13) has suggested that segmentary tribes were the units controlling arable land and hunting territories. If groups occupying the river basin territory were related through kinship (e.g., Brasser 1971), intraterritorial aggressive behavior would be expected to be minimized. However, once populations are large enough, a foraging horticultural system is replete with interterritorial warfare. Reduction per capita of available horticultural land would have been felt within each territory, and there is the tendency for a territorial group experiencing dietary pressure to attempt to obtain the necessary resources from better endowed groups, through aggressive means.

3.4. Territoriality among Foraging Horticulturalists

Historical discussions of observed political power of aboriginal sachems in the Northeast maintain that coersive power was limited, and that power was exercised through persuasive influence and generosity (Denton 1937; Gookin 1970; Salwen 1978). Generosity, however, implies that the more powerful group had something important that other groups needed. Among the northern foraging groups, defended territories included sufficient resources for group survival, and the same would have been true for foraging horticulturalists. Because they were situated on the richest and most productive land within the entire drainage, the major territorial concerns of coastal groups were threefold: (1) to maintain control over their own prime land so that they had adequate resources for survival, (2) to restrict access to coastal shellfish resources by outsiders in order to minimize overexploitation of those critical resources, and (3) to minimize the threat of warfare from without.

Several different strategies would be appropriate for maintaining a controlling position in the coastal zone. First, the coastal group must practice interference competition by dominating the coastal zone, thus controlling access to coastal resources. By controlling a disproportionate share of available resources, potential competitors are kept in a suboptimal and dependent position. Second, the coastal group must maintain a system of reciprocity within the drainage or territory in order to maximize internal cooperation among the various segmentary units. This could be accomplished through reciprocal kinship obligations and through a redistributive tribute system. The controlling group would not only monitor access to those resources as needed, but would exact tribute of preservable commodities for periods of stress. Finally, the coastal group must maintain a system of reciprocity with groups outside of the drainage territory. Resulting alliances would minimize the threat of warfare from other groups. Extraterritorial alliances should also tend to strengthen the political authority and legitimacy of the controlling coastal group over its own subordinates.

It is predicted then that the groups of foraging horticulturalists in southern New England would have had a territorial system that corresponded geographically to the dendritic drainage system of a coastally oriented pattern, in which political order focused on the mouths of tributaries off the major trunk streams, with the highest order political units on the arable land nearest the coast. The coastal group would maintain its strategic hegemony by controlling a disproportionate share of starvation foods, thereby keeping inland groups in a dependent role. Reciprocity is advantageous for the coastal group because it would improve access to anadromous fish and maize in the interior. Warfare and aggressive interference would be minimized through reciprocity among internal territorial segments and extraterritorial groups.

4. MODELING LATE PREHISTORIC LAND USE IN SOUTHERN NEW ENGLAND

4.1. Introduction

On the basis of the preceding discussion, a model can be constructed that explains the relationship among subsistence strategies, territoriality, land use, and environmental degradation. A primary assumption is that contrasts in territorial behavior between groups that subsist purely through hunting–gathering–foraging and those that add horticulture to the subsistence mix hinge on the dependence on arable lands, which are finite. Either group must control at least adequate resources for group survival, and both foragers and foraging horticulturalists are attracted to the riverine zone by seasonally predictable aquatic and terrestrial resources. Horticulturalists, however, are more intensively tied to the same geographic portion of the riverine zone for major portions of the year because of their commitment to horticultural activities. A second assumption is that coastal resources are a perennially available food that can be used during periods when terrestrial resources fail.

The following are the major components that comprise the foraging horticulturalist territorial model that was discussed above:

1. In a riverine horticultural subsistence pattern, arable alluvial land is a finite resource that is degradeable in the face of human population increase.
2. The reduction in the amount of arable land per capita that is available to a growing horticultural population can result in increased competition for that land, which in turn necessitates more structural political control.
3. As degradation and reduced availability of arable land in a horticultural system become severe, the susceptibility to food stress increases.
4. Among populations faced with periodic or long-term dietary stress, perennially available coastal resources become critical to the survival of both inland and coastal groups, and help minimize the overexploitation

of inland resources. It is therefore important for groups in the interior to maintain a friendly relationship with groups that have access to the coast.

5. Horticulturalists exploiting the floodplain of the major trunk stream nearest the coastal estuaries occupy the most important perennial energy source in the drainage system, and would therefore be most vulnerable to overt aggressive attacks from outside the territory.
6. An effective strategy for maintaining the superiority of coastal groups is to control a disproportionate share of resources in order to keep internal segmentary groups in a dependent suboptimal situation, thereby maximizing cooperation.
7. The controlling coastal groups can minimize the threat of internal dissidence by maintaining a reciprocal tribute system, whereby resources paid in tribute are maintained as redistributive stores for stress periods.
8. The controlling coastal group must also minimize the threat of external aggression and internal discordance by maintaining alliances with other powerful extraterritorial groups.

The model suggests that horticultural groups rely on arable land, which is a finite, geographically localized resource. As population density increases, this finite resource becomes overexploited and degraded, increasing the need for interaction with other groups that have access to seasonally or perennially available foods. Coastal groups, in an effort to control those resources, should maintain their position by controlling a disproportionate share of resources and maintaining a reciprocal tribute system.

4.2. Sources of Data and Test Implications

Direct evidence of prehistoric human behaviors such as territoriality is not available, given the incomplete nature of the region's archaeological site data base. Such behaviors, however, are reflected in other data sets. Detailed evaluation and description of artifacts and technological attributes may help to distinguish geographic patterning of styles that would be indicative of some types of territorial distinctions. Unfortunately, these are seldom recorded in the site records for the region. On the bright side, many of the artifacts recorded from early site records are presently housed in museums and collections throughout the region and are still available for further research. Evidence of settlement patterning is more widely available, however, in both the historical record and the region's site distributional data base. Both suggest an intensified use of alluvial areas for the third through the first millennium, which increases nearest the coast. The number of shell middens in archaeological context increased somewhat during the Woodland period (Figure 4), partly as a result of slowing

Holocene sea-level transgression, although this trend continued during the last three millennia after sea-level rise had slowed.

4.2.1. Ethnohistoric Implications

There are few reports in the historical record concerned with decreasing availability of horticultural lands. Most of the early reporters were in the employ of companies intent on colonizing the Northeast and probably minimized reports of negative impacts of colonization on aboriginal populations. Several aboriginal groups in the Hudson River Valley are recorded as having had to move village locations every 10–20 years; Salwen (1975:57) interprets this as indicative of decreased soil fertility or degradation of other critical resources. Thomas (1976:4) estimates that with native technology, 20% of New England's landscape was suitable for horticultural purposes. He estimates that a minimum aboriginal population of 90,000 in southern New England would have been exploiting 220,000–297,000 acres, and argues that competion for arable lands in the face of population increase may have been a common situation even before European contact (Thomas 1978:23).

The importance to aboriginal horticulturalists of the coastal floodplains and areas near the estuaries is well-documented in the historical record (Champlain 1907; de Laet 1909; Mourt 1963). Coastal adaptation, however, has been overemphasized in the literature because most of the early observations were made in coastal areas. Few reports were made inland. While this bias is acknowledged, the apparent pattern of coastally oriented land-use was a reality during pre-Contact times. The few accounts of early inland peoples suggest that most were allied with coastal groups. De Laet (1909), for example, noted between 1625 and 1640 that aboriginal peoples in the upper Housatonic River of present-day Connecticut were part of the group of people observed at the river *mouth*. De Laet also observed that there were few people living at the mouth of the Connecticut River, which, despite having the largest trunk stream in southern New England, has a small estuary. The nearest observed group of any appreciable size was a horticultural people called the Sequins, located roughly 30 km north of the river mouth. This area is the southernmost location of significant amounts of arable land in the river drainage. Perhaps because of its distance from the coast, and the lack of a large estuary at the mouth of the Connecticut River, the Sequins and their northern neighbors, the Nawaas, were allied with the Pequots, who controlled the nearest drainage system and major estuary to the east (Figure 5).

Historical observations of major political groups in the southern portion of southern New England are illustrated in Figure 5. De Laet's (1909) observations include the "Wapenocks" [Wampaoags] in the northern reaches of Narragansett Bay and the Narragansetts, who occupied the western side of the bay (see also Williams 1904). The next group to the west were the "Pequotoos" [Pequots], who

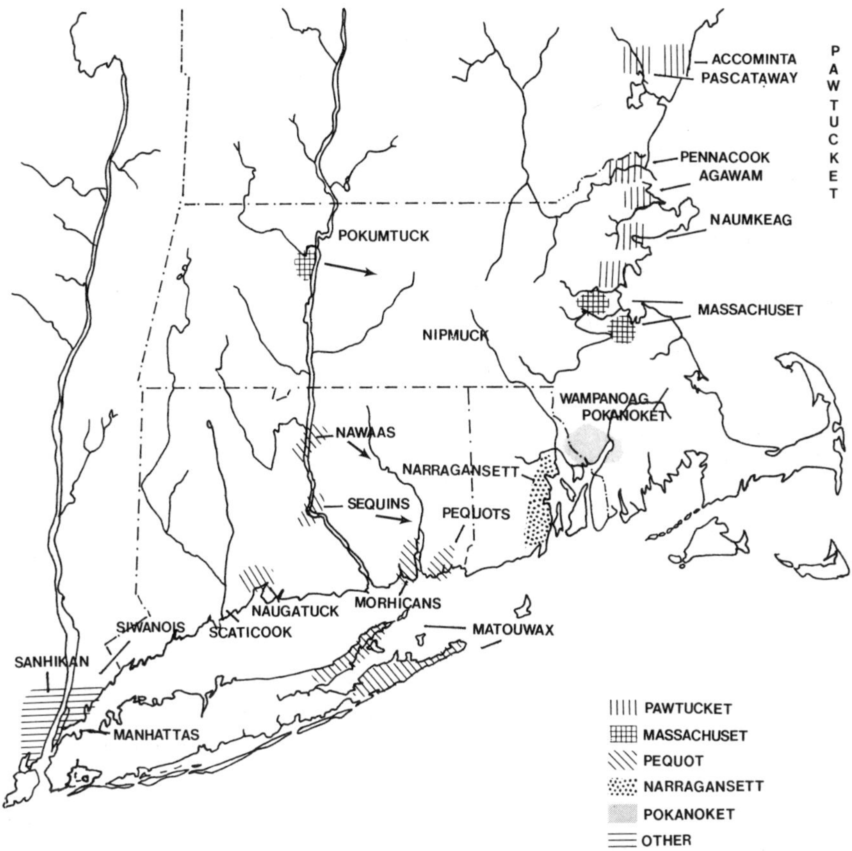

Figure 5. Groups of foraging horticulturalists in seventeenth century southern New England. Data derived from de Laet (1909), Gookin (1970), Mourt (1963), and Williams (1904).

inhabited the "River of Siccanomos," which may be either the Mystic River or the Thames River. Farther west were the Mohicans, situated on the "Fresian" River, which may have been either the Thames or the Niantic (de Laet 1909:39–42). Gookin (1970:10) observed that the Pequots dominated a group of "Nipmucks" to their north, and that, in the 1630s, the Pequot sachem Sassacus was recorded as the political leader of 26 petty sachems, who also controlled all of the territory between the Quinnipiac River and the present-day Rhode Island border (Salwen 1978; Winthrop 1908:22). Other major groups, such as the Massachusett in the southern part of Massachusetts Bay, maintained tribute relationships with bordering "Nipmucks" to the northwest and with groups as far west as the Pocumtuck area in the middle Connecticut River Valley (Figure 5). To the north of the

Massachuset, the Pawtucket nation extended from Massachusetts Bay north to the vicinity of York, Maine (Gookin 1970:10; Salwen 1978), with all major political groups located in the horticultural areas nearest the estuaries. All of the groups mentioned here are reported as the major political units in later accounts, and all had control of coastal and estuarine resources. Tribute alliances of inland groups with coastal people may be indicative of the attempt to maximize access to coastal resources by inland peoples.

The degree to which arable lands and coastal resources were in demand by the European contact period is evident in the many accounts in the historical record of warfare among aboriginal groups. The development of warfare in the Woodland period is problematic, but is suggested in the archaeological record. For example, material evidence of pallisaded forts is primarily from the European contact period. These forts extend well into the inland areas of the drainage systems (Thomas 1978). Mourt's *Relation* concerning the death of Nanapeshamet at the hands of the northern foraging Tarentines describes his fortified house surrounded by a 12 to 15 m wall of poles some 9–12 m high "stuck in the ground as thick as they could be set one by another" (Mourt 1963:78–79). The numerous attacks of the northern foragers against the foraging horticulturalists may be indicative of periodic reliance of mobile hunters on distant coastal resources, perhaps as a result of environmental degradation in northern territories. De Laet (1909:43) observed in 1625 that Indian forts near present-day Middletown, Connecticut, were used for protection in wartime, and Champlain (1907:4) recorded the existence of people in Maine who had been foraging horticulturalists displaced from the coastal fringe by warfare and forced to move inland.

4.2.2. Prehistoric Implications

Prehistoric data from the first millennium B.P. add another dimension to our understanding of horticulturalist land use. Site distributional data have been researched recently for more than 8,500 prehistoric components from the southern New England region (Mulholland 1984). These data have been compiled from various regional site data repositories, including the historical preservation offices of Connecticut, Massachusetts, and Rhode Island; from academic records, such as the site files at the University of Massachusetts and the University of Connecticut; from the regional literature; from archaeological societies; and from contributions of the numerous cultural resource management and archaeological research organizations across the region. The data base has been derived from a variety of sources, all of which have their own inherent bias. The value of such a data base is that localized observations may be compared with those at the regional level. Fortunately, these site records contain general locational, typological, and environmental information for each site. A geographic information system was used to add environmental variables to the data base that was not included in the original site records (Mulholland 1984). For the vicinity of each site, environmental variables were added to the data base and included density of wetlands; surficial area of wetlands, alluvium, lakes (and lake frequency), glacial outwash and till; zones of oak pollen; distance from the coast; and modern climatic variables, such as winter and summer temperatures, frost-free days, etc. This approach permits the evaluation of change at a regional, rather than a local scale.

Sites that contained artifacts with stylistic attributes that are temporally diagnostic (6,994 out of a sample of 8,533 sites) were used to compute the number of sites for each millennium (Figure 1). Mean dates for each style were computed from more than 300 radiocarbon dates for the region that were available in 1984 (Mulholland 1984). When more than one style for a millennium occurred at a site, only one was counted. Table 1 shows the styles assigned to each millennium.

In order to evaluate changes in the intensity of land use in alluvial areas, sites for each millennium were tabulated according to the number of alluvial square kilometers within the U.S. Geodetic Survey quadrangle in which they reside. U.S. Geological Survey quadrangle resolution was used because of the inadequate reporting of explicit site locations in regional site records. In the discussion that follows, the coastal zone is that area from the shoreline to approximately 6–14 km inland from the coast.

A number of the more pertinent implications of the model are described below, along with a summary of related archaeological data. An intensification of horticulture in the region should be accompanied by an intensified use of the floodplains, the most arable lands in the region. Coastal areas would also increase in importance because of their perennially available resources, particularly shellfish. An increase in the frequency of sites and a high or increased relative percentage of sites in the riverine and coastal zones would suggest such an intensification. If the occupants of the region were practicing horticulture intensively, then site distributions should also focus on the areas where climate is most conducive to cultivation.

4.2.2.1. Implication 1. Increases in Riverine and Coastal Land Use. On the basis of site locations throughout southern New England, there was a slight reduction (8.8%) in the importance of riverine areas from the third to the second millennium (Figure 6). During this time, sites declined in frequency in the interior and increased somewhat on the coast (Dincauze 1974; Mulholland 1984; Thorbahn *et al.* 1980). The decline may be partly related to a cooler climatic episode, which continued through the third millennium (Davis *et al.* 1980) and resulted in a decline of mast-related species in the interior that were important in human subsistence. Site distributional maps also support a slight decrease in the

Table 1. Artifact Styles by Millennium

Millennium	Time frame (B.P.)	Mean (B.P.)	Style
10	10,000–9000	9436	Fluted point
9	9000–8000	8270	Plano
		8239	Bifurcate-based
		8239	Hardaway–Dalton
8	8000–7000	7533	Neville
		7475	Stark
7	7000–6000	6393	Merrimack
		6060	Kirk-like
6	6000–5000	5391	Otter Creek
5	5000–4000	4525	Vosburg
		4398	Sylvan
		4366	Brewerton (combined)
		4086	Squibnocket stemmed
		4086	Squibnocket triangle
		4086	Beekman
		4067	Small stem (generalized)
		4067	Lamoka
4	4000–3000	3970	Wading River
		3584	Normanskill
		3323	Susquehanna
		3083	Orient
3	3000–2000	2948	Turkey-tail
		2791	Steatite artifacts
		2580	Vinette I ceramics
		2578	Meadowood
		2305	Rossville
		2000	Adena
2	2000–1000	1503	Steubenville
		1503	Fox Creek (combined)
		1475	Greene
		1036	Shell-tempered ceramics
1	1000–180	958	Ceramics
		933	(undifferentiated)
		933	Madison
		713	Levanna
			Jack's Reef

extent of occupation of the river valleys from the third to the second millennium (Figures 7 and 8 [see key, p. 157]). Forty-seven percent of all sites during the second millennium were located in the coastal zone, a 12.6% increase from the end of the fourth millennium. The increase in the importance of the alluvial floodplains during the first millennium is suggested by an 18.6% rise (Figure 6) in the relative percentage of sites that were located in areas of extensive alluvium

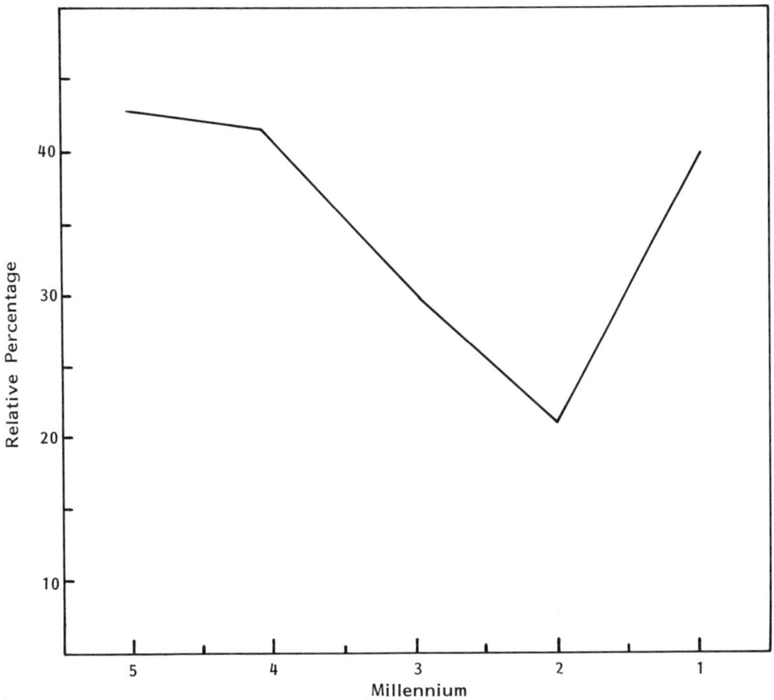

Figure 6. Relative percentage of archaeological components in areas of extensive alluvium in riverine areas in southern New England (Mulholland 1984).

Symbols					
LLLLLLLLL	....	+++++++++	000000000	θθθθθθθθθ	■■■■■■■■■
LLLLLLLLL	....	+++++++++	000000000	θθθθθθθθθ	■■■■■■■■■
LLLLLLLLL	....1....	++++2++++	000030000	θθθθ4θθθθ	■■■■5■■■■
LLLLLLLLL	....	+++++++++	000000000	θθθθθθθθθ	■■■■■■■■■
LLLLLLLLL	....	+++++++++	000000000	θθθθθθθθθ	■■■■■■■■■

Number of sites per level						
First millennium	0	1–5	6–9	10–13	14–17	18–21
Second millennium	0	1–3	4–6	7–9	10–12	13–15
Third millennium	0	1–3	4–7	8–10	11–13	14–17

Figure Key. The above key depicts shading symbols for Figures 7 (p. 158), 8 (p. 159), and 11 (p. 161). Percentage of total number of sites applying to each shading level is 20 percent.

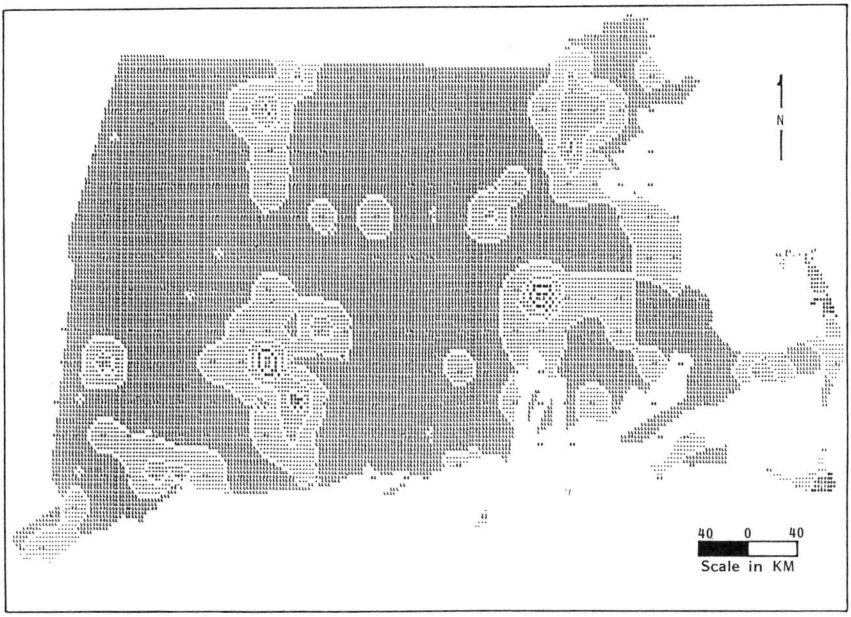

Figure 7. Distribution of prehistoric sites in southern New England for the third millennium before present (Mulholland 1984).

(Mulholland 1984:5). This pattern suggests that by the first millennium there was increased occupation in the riverine as well as coastal zones.

4.2.2.2. Implication 2. Increased Occupation of the Coastal Zone. The increased importance of the coastal area during the Woodland period is evident in the regional data base, which shows the relative percentage of sites in each prehistoric millennium on Cape Cod, Nantucket, and Martha's Vineyard (Figure 9). Similar increases are also evident in the Greater Boston area surveyed by Dincauze (1974) (Figure 2) and in a selective area of southeastern Massachusetts studied by Thorbahn *et al.* (1980) (Figure 3). In Connecticut (Figure 10), the picture is quite different because of the existence of extensive alluvium in the interior and smaller estuaries in comparison with those to the east. After the third millennium, the importance of interior areas of Connecticut increases dramatically. This reflects very high site densities in the Connecticut River Valley near Middletown, where the floodplain widens, and which was the most extensive area of alluvium nearest the coast. This high site density continues all the way to the Vermont/Massachusetts border (Figure 11 [see key, p. 157]).

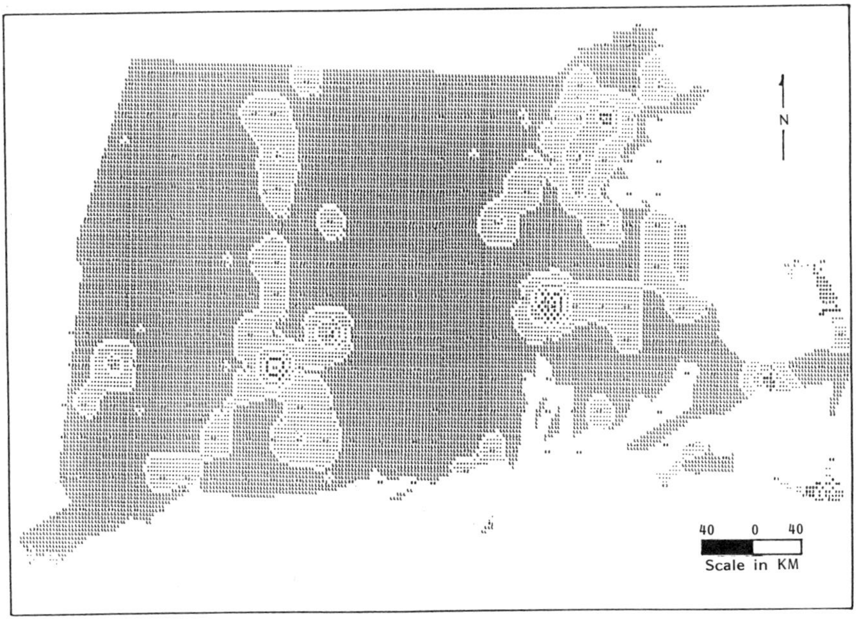

Figure 8. Distribution of prehistoric sites in southern New England for the second millennium before present (Mulholland 1984).

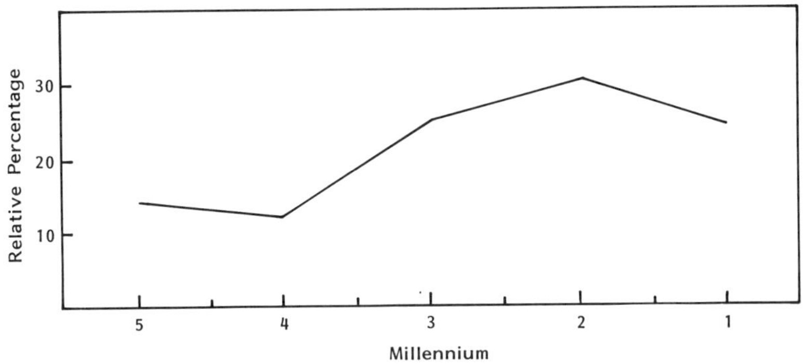

Figure 9. Relative percentage of prehistoric components on Cape Cod and the Massachusetts islands (Mulholland 1984).

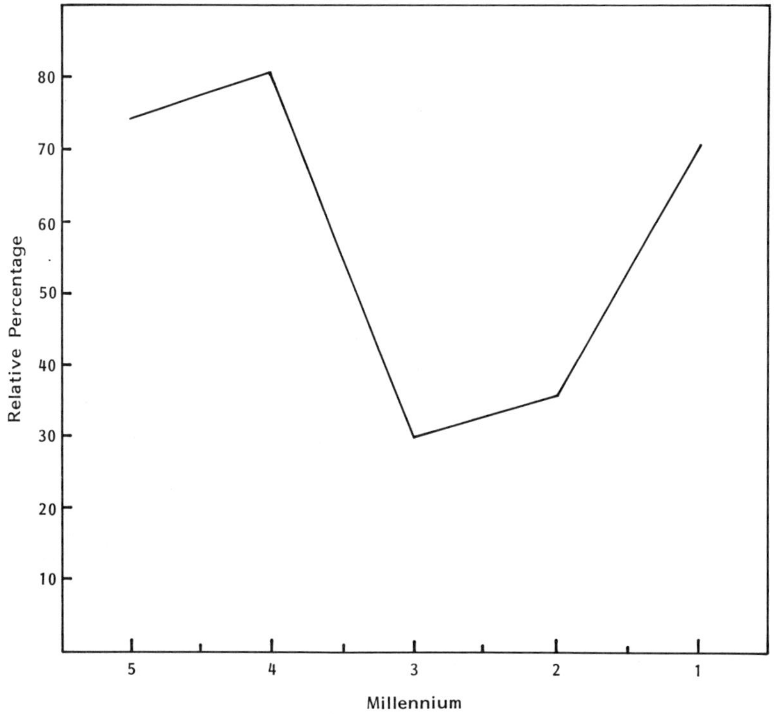

Figure 10. Relative percentage of archaeological sites within 10 km of the modern coastline in Connecticut (Mulholland 1984).

4.2.2.3. Implication 3. Increase in the Importance of Shellfish. The importance of shellfish also increased throughout the Woodland period. This trend is suggested by the increasing relative percentage of prehistoric millennial components that are situated on known shell middens (Figure 4). While the increase is undoubtedly valid, data from other sources are needed to strengthen the association, because of the effect of sea-level rise on surviving coastal shell middens. Sea-level rise, however, was not extensive for the past three millennia.

4.2.2.4. Implication 4. Shellfish Remains Found in the Interior. Archaeological evidence of shellfish use from the interior is scant, but substantiates a relationship between inland and coastal groups. There are numerous occurrences of the remains of marine shellfish in interior sites, especially in the Connecticut River Valley. Oyster, quahog, and whelk, for example, are represented at the Late Woodland period Indian Crossing site in the middle Connecticut River Valley of Massachusetts (Hasenstab *et al.* 1986), some 95 km from the coast. Unfortunate-

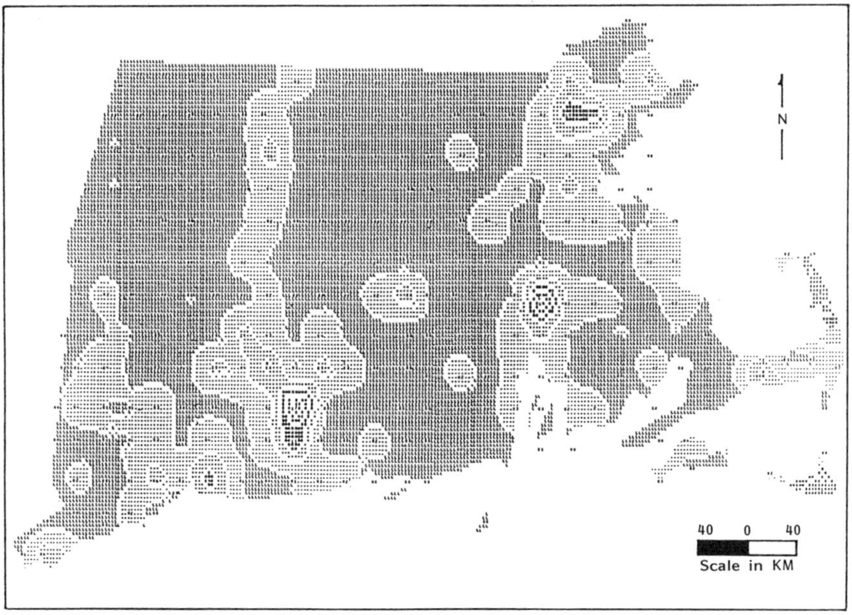

Figure 11. Distribution of prehistoric sites in southern New England for the first millennium before present (Mulholland 1984).

ly, the volume of shell is equivocal because of poor preservation in acidic New England soils. In addition, if dried shellfish were being obtained from the coast and transported to sites in the interior, few shell remains would be expected.

4.2.2.5. Implication 5. Focus of Sites in Areas Conducive to Horticulture. Finally, the predicted increase in horticultural activity is partially reflected by the increased frequency of sites in areas with more than 180–210 frost-free days/year (based on modern climatic records; Figure 12) (Mulholland 1984:486), a zone that is quite favorable for prehistoric maize cultivation (cf. Thomas 1976:8). From the fourth millennium, a time when the interior was heavily occupied (Dincauze 1974; Mulholland 1984; Thorbahn *et al.* 1980), to the first millennium, the period in which horticulture increased in importance, the percentage of sites located in the warmest frost-free day zone increased by 13.3% (Figure 13).

5. CONCLUSIONS

The analysis of the distribution of 8,500 prehistoric components throughout southern New England supports a considerable increase in the occupation of the

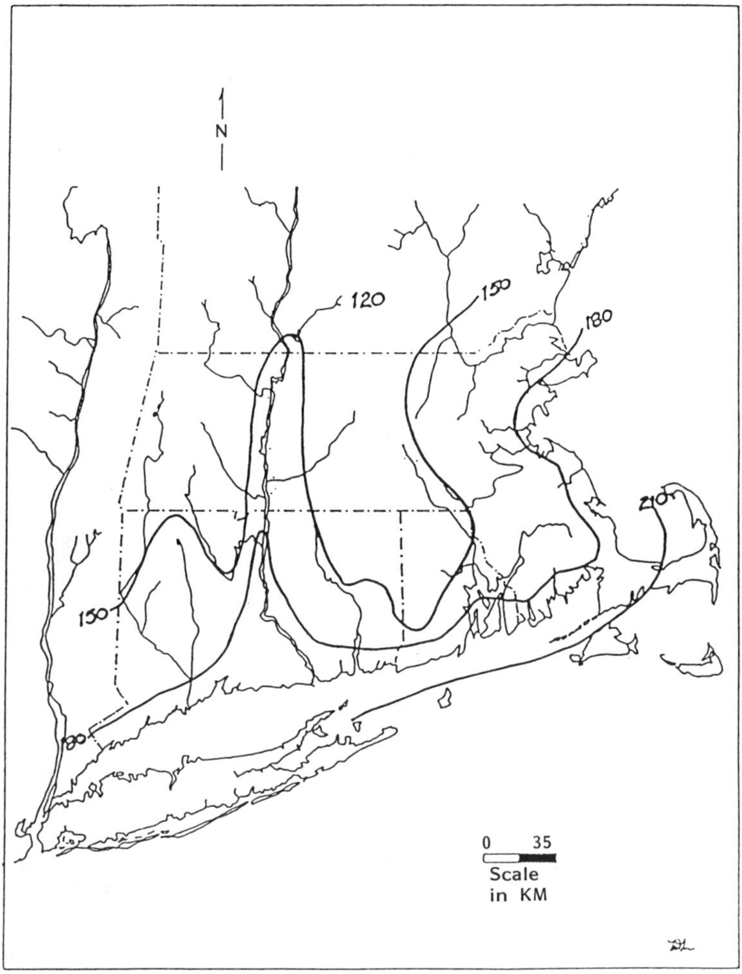

Figure 12. Zones of frost-free days in southern New England (Mulholland 1984, adapted from U.S. Department of Commerce data compiled from 1931 to 1965).

riverine and coastal zones by foraging horticulturalists by the first millennium before present, with a corresponding decrease in the frequency of sites in the interior. Archaeological data from the region support a model for territorial behavior that was manifested differently among foraging groups than among combined foraging horticulturalists. Both groups were subject to food stress over time, but responses to increases in land use differed between the two because of their distinctive settlement patterns.

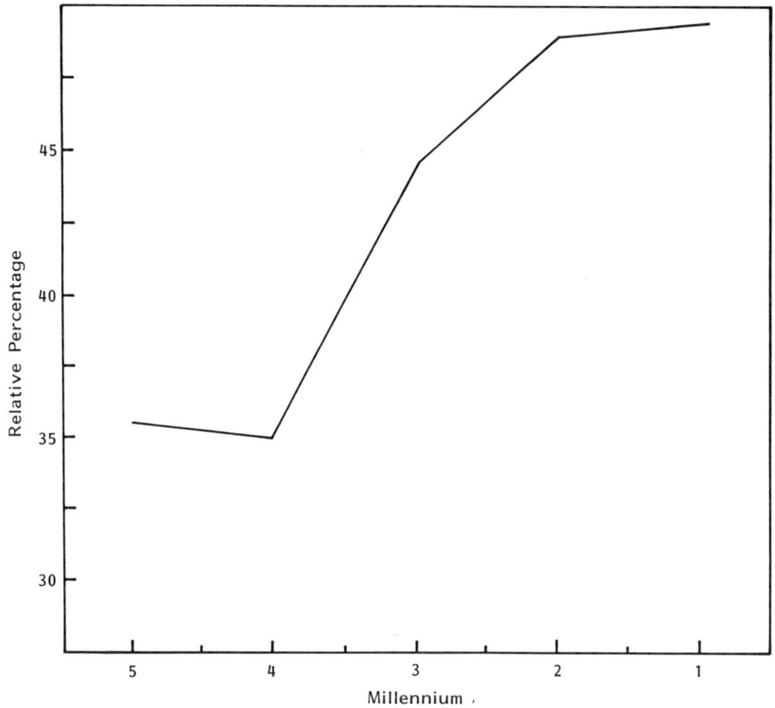

Figure 13. Relative percentage of prehistoric sites in southern New England situated in areas with a frost-free-day season of 180 or more days (Mulholland 1984, adapted from U.S. Department of Commerce data compiled from 1931 to 1965).

Because of the high mobility of foragers, defended territories could be expanded to include sufficient resources for survival, and people were capable of exploiting previously less attractive, unoccupied zones as population increased. Among foraging horticulturalists, settlement patterns were semipermanent and geographically fixed to arable land and aquatic resources for major portions of the year. The option of expansion was a less viable response to population increase, and these groups were more susceptible to food stress as available arable lands were reduced. As a result, competition for and defense of arable land became amplified. It may therefore have been necessary for foraging horticulturalists to maintain aggressively a disproportionate share of perennially available predictable resources to provide sufficient food for survival, and to keep potential competitors in a subordinate situation. This controlling position would have occurred in the high-resource-diversity, coastal/estuarine areas, and would have been maintained through systems of reciprocity from within and outside the territory. One consequence of this settlement/subsistence pattern would be the eventual degradation of arable land, the overexploitation of local resources, and a resultant shift in land-use behavior. The intensity of land use directed at specific geographic areas would have produced changes in local environmental conditions that were not necessarily in phase with natural environmental change.

Acknowledgments

The author would like to express appreciation to Dena Dincauze for her stimulus and encouragement during the research summarized here; and to Catherine Carlson and George Nicholas for reviewing the draft and providing critical and editorial comments. None of these individuals should be held accountable for the contents of the paper.

6. REFERENCES

Anonymous, 1932, A Relation of a Voyage to Sagadahoc 1607–1608, in: *Early English and French Voyages Chiefly from Hakluyt 1534–1608* (H. S. Burrage, ed.), Scribners, New York, pp. 395–419.

Binford, L., 1982, The Archaeology of Place, *Journal of Anthropological Research* 1:5–31.

Brasser, T. J. C., 1971, The Coastal Algonkians: People of the First Frontiers, in: *North Americans in Historical Perspective* (E. B. Leacock and N. Lurie, eds.), Random House, New York, pp. 64–91.

Brasser, T. J. C., 1978, Early Indian–European Contacts, in: *Handbook of North American Indians, Northeast,* Vol. 15 (B. Trigger, ed.), Smithsonian Institute, Washington, D.C., pp. 78–88.

Braun, D., 1972, Explanatory Models for the Evolution of Coastal Adaptation in Prehistoric Eastern New England, *American Antiquity* 39:582–596.

Brereton, J., 1906, Briefe and True Relation of the Discoverie of the North Part of Virginia Made this present Yeere 1602, in: *Earlv English and French Voyages Chiefly from Hakluyt 1534–1608* (H. S. Burrage, ed.), Scribners, New York.

Carlson, C., 1986, *Maritime Catchment Areas: An Analysis of Prehistoric Fishing Strategies in the Boothbay Region of Maine,* M.Sc. Thesis, Institute for Quaternary Studies, University of Maine, Orono, Maine.

Champlain, S. de, 1907, *The Voyages and Explorations of Samuel de Champlain,* Vol. 2 (W. L. Grant, ed.), Scribners, New York.

Curran, M., 1987, *The Spatial Organization of Paleoindian Populations in the Late Pleistocene of the Northeast,* Ph.D. Dissertation, Department of Anthropology, University of Massachusetts, Amherst, Massachusetts, unpublished.

Davis, M. B., 1981, Outbreaks of Forest Pathogens in Quaternary History, in: *Proceedings of Fourth International Conference of Palynology,* Lucknow, India, (1976–1977), pp. 216–227.

Davis, M. B., Spear, R. W., and Shane, L. C., 1980, Holocene Climate of New England, *Quaternary Research* 14:240–250.

De Laet, J., 1909, From the "New World" by John de Laet 1625, 1630, 1633, 1640, in: *Narratives of New Netherland 1609–1664* (J. Franklin Jameson, ed.), Scribners, New York, pp. 36–60.

Denton, D., 1937, *A Brief Description of New York,* Columbia University Press, New York.

De Rassieres, I., 1909, Letter of Isaak de Rassieres to Samuel Bloomaert 1628, in: *Narratives of New Netherland 1609–1644* (J. Franklin Jameson, ed.), Scribners, New York, pp. 97–115.

Dincauze, D. F., 1974, An Introduction to Archaeology in the Greater Boston Area, *Archaeology of Eastern North America* 2(1):39–67.

Dodge, E. S., 1957, Ethnology of Northern New England and the Maritime Provinces, *Bulletin of the Massachusetts Archaeological Society* 18:68–71.

Dyson-Hudson, R., and Smith, E. A., 1977, Human Territoriality: An Ecological Reassessment, *American Anthropologist* 80:21–41.

Funk, R., 1976, Recent Contributions to Hudson Valley Prehistory, *New York Museum and Science Service Memoir,* No. 22, Albany, New York.

Fretwell, S., and Lucas, L., 1969, On Territorial Behavior and Other Factors Influencing Habitat Distribution in Birds, I. Theoretical Development, *Acta Biotheoretica* 19:16–36.

Gookin, D., 1970, *Historical Collections of the Indians in New England—1674* (Jeffery Fiske, ed.), Towaid, Boston.

Hasenstab, R. H., Mulholland, M. T., and Ulrich, T., 1986, Indian Crossing—A First Millennium Horticultural Site on the Chicopee River, Massachusetts, unpublished manuscript.

Kenyon, V., and McDowell, P., 1983, Environmental Setting of Merrimack River Valley Prehistoric Sites, *Man in the Northeast* 25:7–23.

McKern, P., 1939, The Midwestern Taxonomic Method As an Aid to Archaeological Culture Study, *American Antiquity* 16:310–313.

Mourt, G., 1963, *A Journal of the Pilgrims at Plymouth. Mourt's Relation* (D. Heath, ed.), Corinth Books, New York.

Mulholland, M. T., 1979, Forest Succession and Population Change in a Temperate Forest Environment, in: *Ecological Anthropology of the Middle Connecticut River Valley* (R. Paynter, ed.), Research Reports No. 18, Department of Anthropology, University of Massachusetts, Amherst, Massachusetts, pp. 45–56.

Mulholland, M. T., 1984, *Patterns of Change in Prehistoric Southern New England: A Regional Approach,* Ph.D. Dissertation, Department of Anthropology, University of Massachusetts, Amherst, Massachusetts, unpublished.

Nicholas, G. P., and Handsman, R. G., 1984, Early Postglacial Landscape Development and Archaeology at Robbins Swamp, Northwestern Connecticut, *AMQUA Abstracts* (American Quaternary Association) 8:93.

Nicholas, G. P., and Handsman, R. G., 1986, Early Postglacial Archaeological Land-Use Patterns at Robbins Swamp, Northwestern Connecticut, *AMQUA Abstracts* (American Quaternary Association) 9:158.

Ritchie, W. A., 1969a, *The Archaeology of New York State,* Natural History Press, Garden City, New York.

Ritchie, W. A., 1969b, *The Archaeology of Martha's Vineyard,* Natural History Press, New York.

Salwen, B., 1975, Post Glacial Environments and Cultural Change in the Hudson River Basin, *Man in the Northeast* 10:43–70.

Salwen, B., 1978, Indians of Southern New England and Long Island: Early period, in: *Handbook of the North American Indians, Northeast,* Vol. 15 (B. Trigger, ed.), Smithsonian Institute, Washington, D.C., pp. 78–88.

Schoener, T. W., 1971, Theory of Feeding Strategies, *Annual Review of Ecology and Systematics* 2:369–404.

Shaw, L., Savulis, E., Mulholland, M. T., and Nicholas, G. P., 1987, Archaeological Locational Survey in the Central Berkshires, Pittsfield, Massachusetts, University of Massachusetts Archaeological Services Report Series, Amherst, Massachusetts.

Snow, D. R., 1968, Wabanaki "Family Hunting Territories," *American Anthropologist* 70:1143–1151.

Snow, D. R., 1972, Rising Sea-Level and Prehistoric Cultural Ecology in Northern New England, *American Antiquity* 37:211–221.

Snow, D. R., 1980, *The Prehistory of New England,* Academic Press, New York.

Thomas, P. A., 1976, Contrastive Subsistence Strategies and Land Use As Factors for Understanding Indian–White Relations in New England, *Ethnohistory* 23:1–18.

Thomas, P. A., 1978, Indian Subsistence and Settlement Patterns in Non-Coastal Regions: Early Historic Massachusetts, in: *Conservation Archaeology in the Northeast: Toward a Research Orientation* (A. E. Spiess, ed.), Peabody Museum Bulletin, No. 3, Harvard University, Cambridge, Massachusetts.

Thorbahn, P., Loparto, L., Cox, D., and Simon, B., 1980, Prehistoric Settlement Processes in Southern New England: A Unified Approach to Cultural Resource Management and Archaeological Research, Massachusetts Historical Commission, Department of the Secretary of State, Boston, Massachusetts.

Van der Donck, A., 1841, Descriptions of New Netherlands, Translated by J. Johnson, *New York Historical Society Collections* 1:135–242.

Verner, J., 1977, The Adaptive Significance of Territoriality, *American Naturalist* 3:769–775.

Webb, T., III, 1982, Temporal Resolution in Holocene Pollen Data, in: *Third North American Paleontological Convention, Proceedings* 3:569–572.

Whitehead, D. B., 1980, Late Glacial and Postglacial Vegetational History of the Berkshires, Western Massachusetts, *Quaternary Research* 4:9–24.

Williams, R., 1904, A Key to the Language of America—1643, in: *The Lands of Rhode Island As They Were Known to Caunounicus and Miantunnomu when Roger Williams Came in 1636* (S. Rider, ed.), Providence, Rhode Island.

Wilson, E. O., 1975, *Sociobiology,* Harvard University Press, Cambridge, Massachusetts.

Winslow, E., [1855] 1885, Governor Winslow's Account of the Natives of New England, Annexed to his Narratives of the Plantations A.D. 1624, in: *New England's Memorial, Nathaniel Morton,* Congregational Board of Publishers, Boston, Massachusetts.

Winthrop, J., 1908, *Winthrop's Journal, "History of New England: 1630–1649",* (J. K. Hosmer, ed.), Scribners, New York.

Wood, W., 1865, *New England's Prospect—1634,* Publications of the Prince Society, Vol. I, Boston.

Chapter 7

The Effect of Estuary Formation on Prehistoric Settlement in Southern Rhode Island

PETER F. THORBAHN AND DEBORAH C. COX

1. INTRODUCTION

There is a vigorous debate among archaeologists working in New England and eastern New York as to when and why prehistoric people began to focus on coastal and estuarine resources (cf. Dincauze 1973; Braun 1974; Brennan 1976; Snow 1980). However, there is no doubt that by 2,000 years ago, patterns of coastal land use left prominent features in the archaeological record of southern New England, with shell middens being the most visible class of remains. Studies of regional side distributions indicate a marked increase in the number of sites located along or near the coast of southern New England during the last 3,000 years of the prehistoric period (Barber 1979; Thorbahn *et al.* 1980). The estuaries of coastal rivers appear to have been a particularly attractive environmental zone for locating relatively large, semipermanent settlements (Dincauze 1973, 1974, 1980; Barber 1979).

Recent investigations along a coastal river in southern Rhode Island have produced a series of unexpected results in light of prehistoric land-use patterns reported from other rivers in southern New England. Initial survey, site testing,

---

PETER F. THORBAHN • Hammerway Research, West Barnstable, Massachusetts 02668. **DEBORAH C. COX** • Public Archaeology Laboratory, Pawtucket, Rhode Island 02860. Peter Thorbahn, former Executive Director of the Public Archaeology Laboratory, died unexpectedly in August 1987. He will be greatly missed by friends and colleagues.

and site excavations within the Pettaquamscutt River basin, Narragansett, Rhode Island (Figure 1), have yielded very little evidence for exploitation of coastal or estuarine resources, despite the fact that Narragansett Bay is within 3 km of any part of the Pettaquamscutt River. Only one shell midden and a few low-density shell deposits were found along the entire 12-km length of the river.

In addition, the temporal sequence of settlements along this river is unusual.

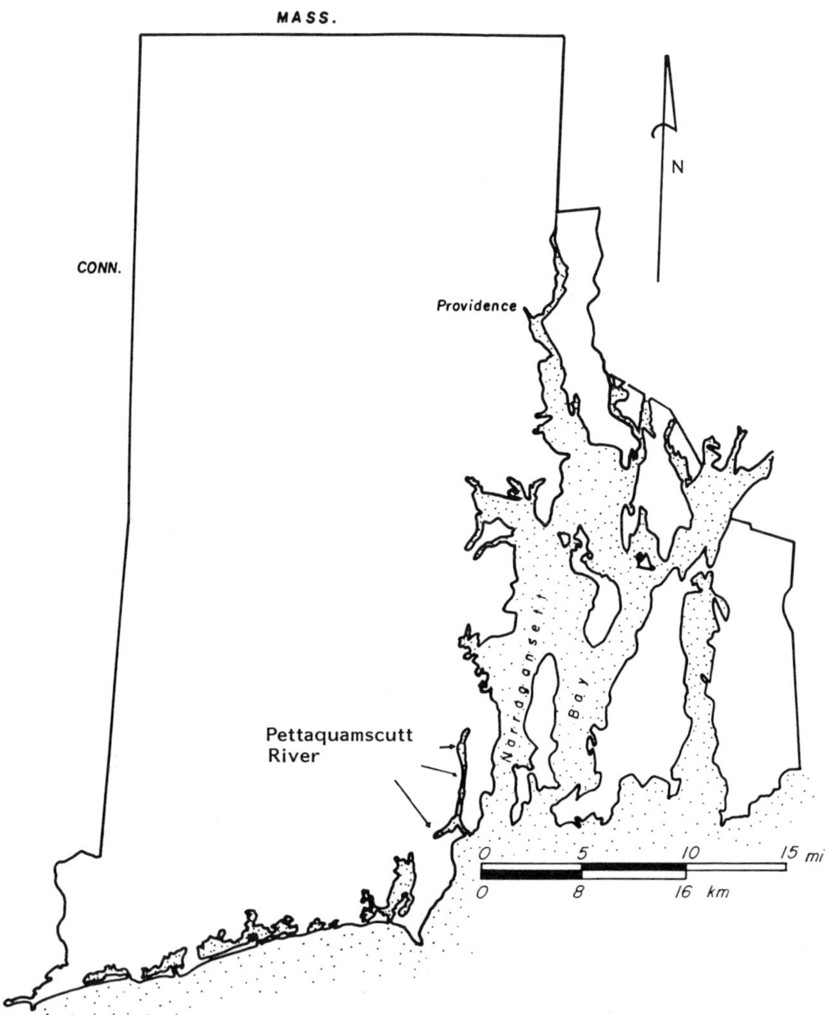

Figure 1. Location map of Pettaquamscutt River, Southeastern Rhode Island.

In river valleys such as the Charles (Dincauze 1974) and the Merrimack (Barber 1979), the earliest and longest occupied sites tend to be found on the lower parts of the river, nearest the oldest part of the estuaries. Along the Pettaquamscutt River, however, there are no sites dating to before 4000 B.P. After that date, the age of occupation is apparently scattered randomly up and down the river.

This paper focuses on the settlement pattern and environmental data collected during the first stages of archaeological investigations of the Pettaquamscutt drainage basin (Cox and Thorbahn 1982). The environmental aspects of this study consisted of recording salt marsh stratigraphy (Goldsmith 1980) and analyzing fossil pollen spectra from two sediment cores taken near prehistoric sites (Nelson 1980). In addition, data on primary and secondary production were gathered from the ecological literature for southern New England. The environmental data derived from these sources are compared with archaeological survey data for the entire river valley, which are supplemented with the results of several site excavations. The conclusion of this study is that the location of prehistoric settlements in the Pettaquamscutt River valley was conditioned by the mix of resources in several ecological zones rather than from a single zone, such as the estuary. We suggest that prehistoric land-use systems in southern New England were very dynamic, and that the process of choosing settlement location was highly complex.

2. ENVIRONMENTAL SETTING

The Pettaquamscutt River (Figure 2) lies west of the mouth of Narragansett Bay and west of Jamestown, an island that defines the boundary of the west passage from the lower Bay northward towards Providence. The sides of the river valley are marked by long, north/south-trending hills rising to a height of 55 m above sea level on the west side and 34 m on the east. The main body of the river does not empty directly into the ocean, but into the Narrows, an outlet channel that is at a right angle to the main north/south channel and the ocean. The distance from Carr Pond to Narragansett Bay is just over 12 km.

Narragansett Bay was once part of a system of river valleys that extended to the edge of the Continental Shelf and emptied into the Atlantic Ocean through the Block Island Channel. The north/south orientation of the Bay contrasts to that of other large bays (e.g., Buzzards Bay, Vineyard Sound, Long Island Sound) that lie parallel to the Continental Shelf. Glaciation in the area extended to what is now the southern limit of Rhode Island Sound. The advance and retreat of the ice sheets during the Pleistocene modified the topography of the area, depositing clay, silt, and sand, and was responsible for fluctuating sea levels (see Kellogg, Chapter 4). Since the final glacial retreat about 14,000 B.P., sea level rose and drowned the river valleys to form the Narragansett Bay as we know it today.

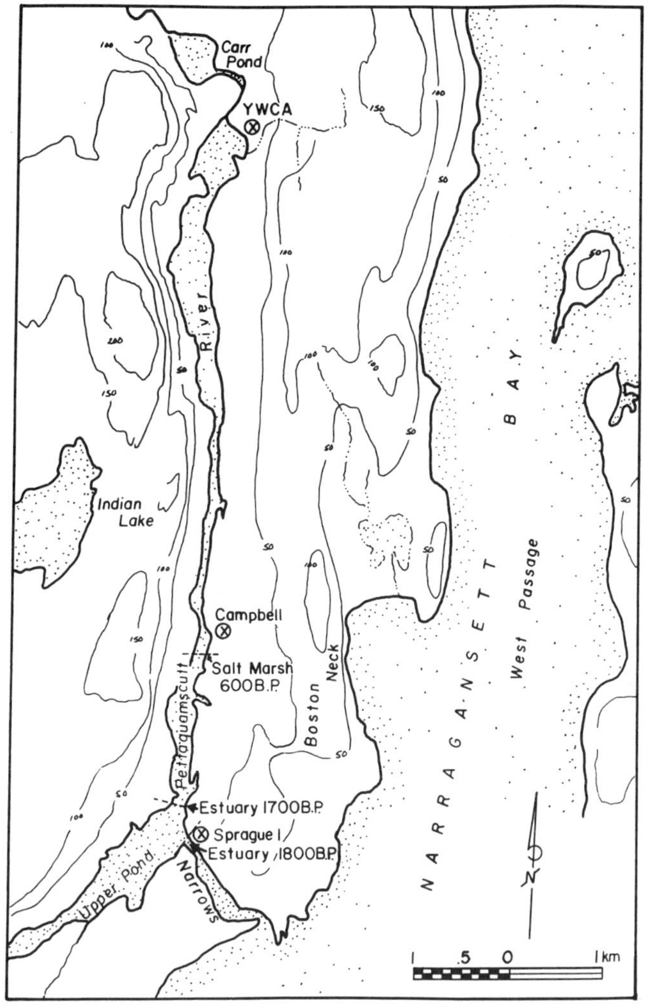

Figure 2. Pettaquamscutt River drainage, prehistoric sites, and marine inundation.

Estuarine environments, such as those at the mouth of the Blackstone or Taunton Rivers, have been continuously and significantly altered by the rising sea level since the Late Wisconsin. The case of the Pettaquamscutt River is somewhat different.

Radiocarbon dates obtained from sediment cores from the estuary zone of the river (Orr and Gaines 1973; Gaines 1975) indicate that organic deposits began forming in the Pettaquamscutt River basin at least by 11,308 ± 404 B.P.

This date correlates with other dates in New England for the disappearance of glacier ice, as complied by Emery (1969:26) and others. However, Gaines' date for the marine inundation of the Upper Pond area is only about 1700 ± 170 B.P. Radiocarbon dates at the base of marine deposits at the mouth of the Narrows mark the earliest flow of tidal energy into the Pettaquamscutt River at 1880 ± 80 B.P. (Nelson 1980). This is much later than the dates for the establishment of estuaries in other rivers in southern New England (O'Donnell and Leatherman 1980), and could be due to lack of a deep channel to bring saltwater upstream, to a now-submerged bar off the mouth of the Narrows, or to other, unknown factors.

3. ENVIRONMENTAL CHANGE IN THE LAST TWO THOUSAND YEARS

The first period for which we have reliable data is the second millennium before the present. Organic remains at the base of a sediment core taken near the Narrows have been radiocarbon dated to 1880 ± 80 B.P. This was one of two cores taken for pollen analysis by Sheldon Nelson of Brown University's Geology Department. Table 1 contains a summary of Nelson's interpretation of the pollen spectra from the core taken near the Spague I site and from the second core from near the Campbell site (see Figure 2). Both cores represent about 2,000 years of sedimentation (assuming uniform rates of deposition within cores and different

Table 1. Reconstructed Upland and Riverine Vegetation in the Pettaquamscutt River Basin[a]

Pollen zone	Dates (B.P.)	Upland vegetation	Riverine vegetation
A	0–600	Modern oak–pine forest with open fields (ragweed)	*Spartina* salt marsh established near Campbell site at 570 B.P.
B	600–1200	Upland forest more extensive, hickory and birch more common	Fresh water wetlands less extensive; probably drier conditions
C	1200–2500	Oak–pine forest with beech and tupelo, open-field expansion shown by rise of composites and ragweed	Red maple, winterberry, and alder wetlands more extensive; *Spartina* salt marsh established in lower river near Sprague site at 1880 B.P.

[a]From Nelson 1980.

rates between cores). Three distinct pollen zones are apparent for upland and riverine plant communities. The pollen from upland trees and ground cover is probably typical of the regional forest within a radius of 10–20 km of the sample loci. The riverine vegetation reflected in the fossil pollen is more likely to have been growing in the immediate vicinity of the streamside kettle holes from which the cores were taken.

Sometime around 2000 B.P., the *Spartina* salt grass marsh became established in the Narrows, which may have been an upriver extension of a drowned estuary that existed offshore in earlier times. The salt marsh in the Narrows bordered a freshwater wetland of sedges, red maple, and alder. This type of wooded swamp community was probably found in many areas of the floodplain of the river. On the higher terraces and uplands around the river, the forest was dominated by red oaks and white pines, with tupelo in lower, moister ares, and beech on higher, well-drained soils. Hickories and birch were also members of this forest. Thus, the upland forest matched the lower and midslope associations of the modern, southern New England oak forest (Jorgensen 1978). A typical ecological cross-section of the Pettaquamscutt River valley for ca. 1000 B.P. is shown in Figure 3.

A very interesting feature of the pollen record for around 1200 B.P. is a rise in the percentage of *Compositae* (the sunflower family) and *Ambrosia* spp. (ragweed and similar plants), which was due to the expansion of open fields at the expense of the closed canopy oak forest. This same pollen signal at around 1000 B.P. has been observed in other cores taken in the Narragansett Bay drainage basin (Anderson 1976; Bradshaw *et al.* 1980). This is at least 600 years before the arrival of European colonists in the region. In any case, the extensive forest clearing in the Colonial period is usually represented in the pollen record by a higher percentage of ragweed with less of a *Compositae* component.

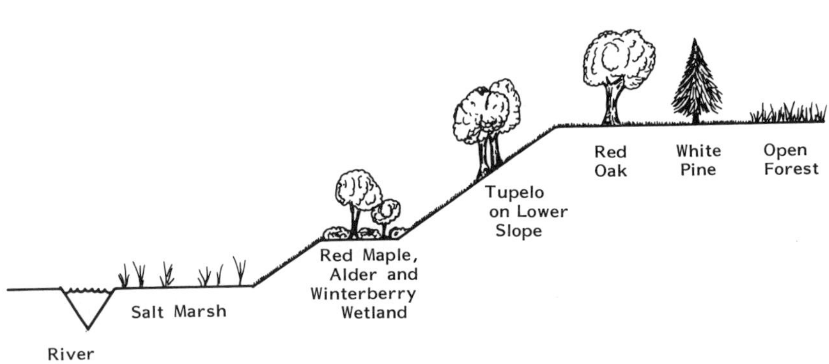

Figure 3. Schematic cross-section of plant communities in the Pettaquamscutt River valley, ca. 1000 B.P. (Nelson 1980).

Was this early expansion of open fields a direct result of prehistoric Native American activities? There is substantial evidence that aboriginal land-clearing activities were regularly practiced in southern New England (Cronon 1983; see also Mulholland, Chapter 6). Evidence for open-field land use in the Pattasquamcutt drainage is provided by the discovery of an aboriginal slate spade at the Browning site that was recovered during the survey of the middle section of the river. Although the Browning site has not been reliably dated, the slate spade is an intriguing discovery. Ethnohistoric data indicate that Native American horticulture was the slash-and-burn type, which would have opened the forest canopy and encouraged the growth of open-field plants when the fields were abandoned (see Patterson and Sassaman, Chapter 5).

Between 1200 and 600 B.P., the composition of the vegetation was changing and the freshwater wetlands were shrinking, according to shifts in the percentage of arboreal and sedge pollen. The pollen of open-field species declined substantially. Climatic conditions may have been significantly drier than those of the previous 1,300 years. Not only were wetland species less frequent in the pollen record from 1200 to 600 B.P., but forest species such as hickory and birch were common. These are typically midslope community trees, in contrast to the tupelos of moister, lower slopes in the earlier period (Jorgenson 1978).

If drier conditions were actually affecting the pollen record in the Pettaquamscutt drainage, then local climatic change should correlate with warmer, drier conditions around the North Atlantic and in the midwestern United States at this time (Bryson and Murray 1977; Gribben and Lamb 1978). Whether the environmental change in southern Rhode Island was part of a global process or a purely local event remains problematic. In either case, the consequences for the prehistoric inhabitants of this area would have been equally significant. Changes in the local hydrology and vegetation of the river valley would have affected the kinds and quantities of resources available to the people during the Late Woodland period, an important era in the history of human occupation of the Pettaquamscutt.

Sometime around 600 B.P., the river valley took on the characteristics that have continued up to the present day. *Spartina* salt marsh became established as far north as the Campbell site at around 500–600 B.P. This slow progression of the salt marsh, 2 km in 1,200 years, is consistent with a reduced rate of sea level rise (2 m/1,000 years) in modern times (O'Donnell and Leatherman 1980:6). The head of the estuary, as measured by salt water mixing in the river, was certainly in advance of the salt marsh itself by as much as 200 years, according to differences in dates for the first organic deposits in the Narrows and the Upper Pond (Gaines 1975).

Beginning 600 B.P., freshwater wetlands expanded again, perhaps to the their former extent at 2500–1200 B.P. The area covered by the modern upland forest of red oak and white pine declined dramatically after European colonization 350 B.P. Groundcover species typical of old-field successions, ragweed in particular, increased when European settlers cleared the forest.

4. PRODUCTIVITY AND SITE LOCATIONS

It is axiomatic that the distribution of natural resources conditions the location of human settlements (Hagget *et al.* 1977). While accepting this basic premise, archaeologists are divided in their approaches to reconstructing the processes by which people made decisions on where to live in relation to the resources they required. This paper therefore focuses on the relationship between the availability of natural resources and the location of settlements within the Pettaquamscutt River valley during the last 2,000 years.

The basic measures of natural resource availability used here are *net primary* and *secondary biological productivity,* defined as the amount of organic matter involved in the growth of flora (primary producers) and of herbivorous fauna (consumers), respectively, per unit of area per unit of time (Whittaker and Marks 1975). These measurements represent the potential amount of plant and animal (excluding predators) tissue that could have been used by omnivorous human consumers.

Reconstructing or modeling primary and secondary productivity is obviously a very simplified approach to the complex issues of prehistoric subsistence systems. For example, some of the factors affecting land-use decisions not considered here include the differences between the potential and effective environment (Davidson 1972), the deliberate modification of the environment through domestication (Wenke 1980) or burning (Patterson and Sassaman, Chapter 5), and seasonal variations in resources (Binford 1980). These are certainly important issues, but in this paper we are seeking to establish only some of the basic characteristics of natural resources in the past and to determine how they may have influenced the location of prehistoric sites along the Pettaquamscutt River.

The estimated dates of occupation at ten sites along the Pettaquamscutt River are shown in Table 2. The sites are grouped by lower, middle, and upper sections of the river, from its outlet into Narragansett Bay to the upper pond, just below Carr Pond (Figure 2). Each of these arbitrary sections is 3.7 km in length. Occupation periods at these ten sites were dated by either radiocarbon assay or temporally diagnostic artifacts. The radiocarbon dates were taken from charcoal found in cultural features, except for one date that was taken from a quahog shell found in a midden. The diagnostic artifacts include Small-Stemmed, Atlantic, and Levanna projectile points, a Mansion Inn Blade, and ceramic fragments.

Most of the sites were occupied more than once during prehistoric times, although Table 2 shows only the best estimated date for each site. Neither

Table 2. Dates of Prehistoric Occupations from the Lower to Upper Pettaquamscutt River

Site	Estimated dates of occupation (B.P.)	Basis of estimates
Sprague I	(a) 4500–2500	Small stem projectile points
Campbell	(a) 4500–3000	Squibnocket triangle projectile points
	(b) 3500–2000	Orient projectile points
	(c) 1100–550	Radiocarbon dates
YWCA	(a) 4500–2500	Small stem projectile points
	(b) 1500–500	Levanna projectile points

radiocarbon dating nor use of diagnostic artifacts is error-free; for present purposes, however, the dates presented in Table 2 should be viewed as representing a non-deliberately selected sample of ten dates during which the river valley was occupied.

There are two important aspects to these dates. First, there are none prior to 4000 B.P. If there were any sites occupied before this date along the Pettaquamscutt, they must have been located along the lowest part of the river, which has since been inundated by the rising sea level. The second noteworthy aspect of these dates is the lack of discernible patterning in site location with respect to age. In other words, dated occupations are randomly scattered up and down the river. Sites were occupied from the Late Archaic to the Late Woodland periods all along the river. Even the Campbell site, situated in the middle section of the river basin, contained evidence of multiple occupations covering the Late Archaic to Late Woodland time span (Cox and Thorbahn 1982).

Other researchers have reported very different patterns of occupation along rivers in southeastern New England. Dincauze (1973) discerned a pattern of dated occupation along the Charles River that correlated with the dates of the advance of the head of tide caused by rising sea levels. Older occupations were generally found along lower sections of the river, and more recent occupations were found upstream, so that there was a positive correlation between the age of occupation and the date of salt water transgression. Barber (1979) observed much the same pattern along the estuary of the Merrimack River. In addition, sites with the longest time span of multiple occupations were found beside the downstream parts of the estuary that had been established earlier in the Holocene. Both Dincauze and Barber hypothesized that the estuaries forming after the advance of salt water up the Charles and Merrimack Rivers attracted prehistoric groups to these locations because of the abundance of food resources there, especially shellfish and anadromous fish.

In fact, the Charles and Merrimack River data support one of the basic tenets of New England archaeology, namely that from the Late Archaic period onward, coastal and estuarine environments played an increasingly important role in the settlement systems of the region (Braun 1974; Ritchie 1969; Snow 1980). The progress of the estuary up the Pettaquamscutt, however, seems to have had no effect on where or when prehistoric groups occupied this river valley.

Why should the pattern of settlements along the Pettaquamscutt River have been so different from other rivers in the region? One possibility might be that the local environment of the Pettaquamscutt was essentially different from that of other coastal rivers. The Pettaquamscutt runs parallel to the coastline for much of its course, and may have been affected by sea-level rise to a lesser degree than other drowned river valleys and embayments, such as those of the Charles and Merrimack Rivers. In other words, the Pettaquamscutt may have been more important as an interior river and less so as an estuary for longer periods of time. This would contrast with a river such as the Merrimack, where tidal incursion occurred earlier and progressed further upriver (Barber 1979).

The Pettaquamscutt is also a small river, only 12 km in length. At this scale, a settlement anywhere along the river would have been within a relatively short distance of the estuary. Even so, the freshwater sections of the Pettaquamscutt attracted settlements throughout the last 4,000 years, despite the fact that coastal and estuarine resources were no more than 1–2 km downriver, or across Boston Neck (Figure 2). This observation raises the question of the relative attractiveness of resources in different ecological zones along the river.

Table 3 shows the range of modern values for net primary productivity for a variety of ecosystems in southern New England. These data show that some freshwater wetlands and upland forests produce organic matter at a rate equal to or greater than estuary or tidal environments (see Nicholas, Chapter 10, for

Table 3. Net Primary Production of Ecosystems in Southern New England

System	Productivity[a]	Sources
Freshwater wetland	1500–1000	Bernard and Gorham 1978:48
Freshwater tidal marsh	1300–800	Wigham *et al.*, 1978:6–10
Upland forest	1000–800	Nixon and Oviatt 1973:470; S. W. Nixon, personal communication, 1985
Tidal salt marsh	900–300	S. W. Nixon, personal communication, 1985
Coastal bay[b]	700–400	S. W. Nixon, personal communication, 1985
Wooded bog	500–100	Reader 1978:56
Lake[b]	200	Brooks and Deevey 1966:152

[a]Above ground, dry organic matter g/m per year.
[b]Micro and macroflora in water column.

related discussion). Along the Pettaquamscutt, there were large areas of freshwater wetlands and upland forests in the immediate vicinity of the river bottom. Estuarine zones were limited by the Narrows of the river as it entered Narragansett Bay (see Figure 2). Not only were freshwater and upland environments equally as attractive per unit of land area, they were also more extensive than the estuary zone.

Were the freshwater and upland zones, in fact, more attractive than the estuary to the prehistoric inhabitants of the river? This would depend on the usable biomass of both plants and animals that existed in each of these environments. Estuarine vegetation communities in southern New England are dominated by essentially three kinds of plants: tussock sedge (*Cupercae* spp.) in the freshwater/tidal zone; saltgrass (*Spartina* spp.) in the tidal zone; and eelgrass (*Zostera marin*) in areas of highest salinity (Nixon and Oviatt 1973). Freshwater wetlands and upland forests have a much higher diversity of vegetation than does the estuarine zone, with dozens of species represented in any given community (Jorgenson 1978), so there would be a much greater range of usable plants in upland and wetland environments. It is particularly interesting to note that some of the upland species (e.g., sunflowers and other composites) that contributed to the early open-field pollen rain around 1200 B.P. are very productive and were domesticated by Native Americans in prehistoric times (Whittaker and Marks 1975).

More biomass and more kinds of plants were available away from the coast and estuary of the Pettaquamscutt, but what about animal protein sources? Shell middens are the most conspicuous feature of prehistoric sites along or near Narragansett Bay, and there is no question that this resource was repeatedly exploited near the Pettaquamscutt over the past 2,000–4,000 years. It is also clear that upland game, especially deer, was an important part of the prehistoric diet of the people in this region (Perlman 1976). Shell middens along Narragansett Bay typically contain remains of upland game animals. What might have been the relative rates of secondary production of usable animal species between the upland forest and estuary?

There is no doubt that estuaries and near-coastal waters would have been the most significant sources of marine resources for New England's prehistoric population (Braun 1974; Carlson 1986). However, ongoing research into rates of secondary production of commercially useful shellfish and fin fish along Long Island Sound during the nineteenth century indicates that all of these species represented less than 1% of the total primary production of plant matter (Scott Nixon, personal communication, 1985).

Figure 4 shows the relationship of the average dietary contribution of different forest tree species versus the number of species of game animals and birds that use these plants in their diet (Martin *et al.* 1961). In other words, Figure 4 could represent the relationship between the number of animal species that can

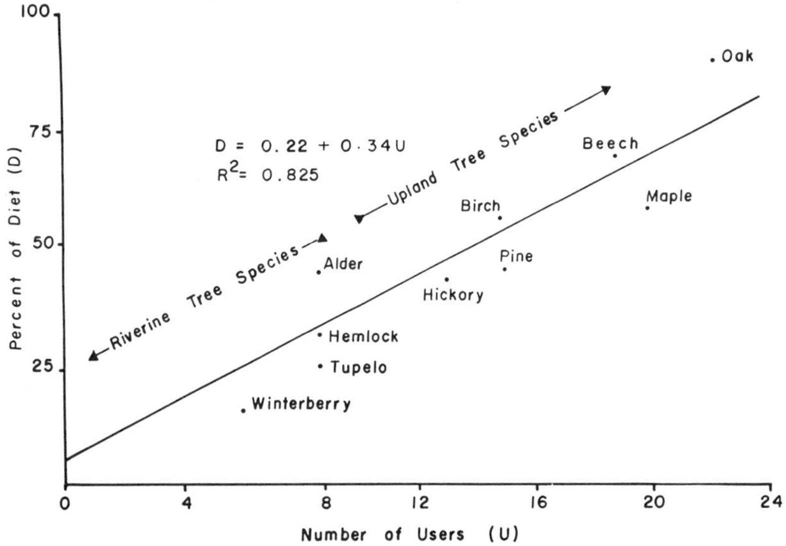

Figure 4. Diet contribution (D) versus number of mammal and bird game species users (U) for riverine and upland tree species.

be exploited by humans and the degree to which different plants contribute to the faunal biomass (secondary production) in the upland forests and wetlands.

It is clear from Figure 4 that the upland forest species (oak, beech, birch, pine, and hickory) support more kinds of animal, and also have a greater part in the secondary production of these animals, than do the flora of other zones. Wetland trees and shrubs are more likely to fall below the average values for number of animal species that use them. The wetlands, however, support many other usable animal sources (e.g., fish, turtles, small mammals, migratory waterfowl) that are not found in the forest.

Direct, quantifiable comparisons between the rates of estuarine/coastal and forest/wetland secondary biological productivity cannot be made using these data. Nonetheless, it appears that the upland forests and wetlands were a potential source of a great variety and high biomass of both plants and animals for the prehistoric inhabitants of the Pettaquamscutt River basin. It is also important to note that the spatial distribution of the plants and animals in all ecological zones of the river basin was changing with steadily rising sea levels and alternating periods of drier and moister climates throughout the Holocene, especially within the last 2,000 years.

The processes of environmental change within the Pettaquamscutt River basin were highly dynamic, so the processes by which prehistoric groups selected settlement locations had to have been very complex. Conventional knowledge stresses the abundance of resources in coastal and estuarine paleoenvironments (Braun 1974; Ritchie 1969; Snow 1980). It is only recently that researchers have begun to consider the role that diversity and dynamism play in the development of temperate forest ecosystems of prehistoric New England (Dincauze and Mulholland 1977; Mulholland 1979; Nicholas, Chapter 10).

5. CONCLUSIONS

The results of research presented here for the Pettaquamscutt River valley admit a different perspective on the relationship between environmental change and the location of prehistoric settlement. In this river valley, prehistoric groups appear to have selected locations from which they could take advantage of the resources in several environmental zones, as opposed to focusing soley on estuaries. Some of the upland zones were probably at least as productive as those of the estuary or the coast. This implies that the location of prehistoric settlements was conditioned by the mix of available resources in several areas, instead of by the productivity of a single environmental zone.

The environmental and archaeological data from the Pettaquamscutt basin suggest that other factors also had a significant effect on the location of prehistoric settlement. For example, fossil pollen of open-field species known to have been domesticated, such as sunflower, appeared at 1200 B.P. in the pollen spectra. In addition, a slate spade was recovered at the Browning site and grains of wild rice were found at the Campbell site. Perhaps the settlement system in the Pettaquamscutt basin was based on horticulture during the last 1,000 to 500 years.

Aboriginal horticulture in southern New England was probably never intensive and it was certainly supplemented by age-old patterns of hunting and gathering (Snow 1980). If horticulture was part of the Pettaquamscutt settlement system from 1200 B.P. onward, then the location of fertile, well-drained soils would have been added to the mix of upland forest and estuarine resources that prehistoric groups had to consider when choosing a site location. This would have further complicated the decision-making process and decreased the influence of any one environmental zone of the location of prehistoric settlements.

Acknowledgments

The data used in this study were collected during three cultural resource management studies of a construction easement for wastewater treatment facilities in the Town of Narragansett. The archaeological survey, site testing, and data recovery phases of this project were funded by the Environmental Protection

Agency, the Rhode Island Department of Environmental Management, and the Town of Narragansett. Additional research along the entire Pettaquamscutt River drainage was funded by the Rhode Island Historical Preservation Commission through the Survey and Planning Grant Program of the National Park Service, Department of the Interior. The support of these agencies is gratefully acknowledged. The authors assume sole responsibility for the conclusions and opinions presented here.

6. REFERENCES

Anderson, P. M., 1976, *The Historical and Palynological Record of Rhode Island in the Seventeenth through Nineteenth Centuries: A Methodological and Theoretical Consideration of Ecological Approaches in Archaeology,* M.A. Thesis, Brown University, Providence, Rhode Island, unpublished.

Barber, R. J., 1979, *Human Ecology and the Estuarine Ecosystem: Prehistoric Exploitation in the Merrimack Valley,* Ph.D. Dissertation, Harvard University, Cambridge, Massachusetts, unpublished.

Bernard, J. M., and Gorham, E., 1978, Life History Aspects of Primary Production in Sedge Wetlands, in: *Freshwater Wetlands* (R. E. Good, D. F. Whigham, and R. L. Simpson, eds.), Academic Press, New York, pp. 39–52.

Binford, L. R., 1980, Willow Smoke and Dogs' Tails: Hunter–Gatherer Settlement Systems and Archaeological Site Formation, *American Antiquity* 45:4–20.

Bradshaw, R. H. W., Nelson, S., and McGown, K. 1980, Paleoenvironmental Reconstructions in the Taunton Quadrangle, Ms. on file, Public Archaeology Laboratory, Providence, Rhode Island.

Braun, D. P., 1974, Explanatory Models for the Evolution of Coastal Adaptafion in Prehistoric Eastern New England, *American Antiquity* 39:582–596.

Brennan, L. A., 1976, Coastal Adaptation in Prehistoric New England, *American Antiquity* 41:112–113.

Brooks, J. L., and Deevey, E. S., Jr., 1966, New England, in: *Limnology in North America* (D. G. Frey, ed.), University of Wisconsin Press, Madison, Wisconsin, pp. 127–162.

Bryson, R. A., and Murray, T. J., 1977, *Climates of Hunger,* University of Wisconsin Press, Madison, Wisconsin.

Carlson, C. C., 1986, *Maritime Catchment Areas: An Analysis of Prehistoric Fishing Strategies in the Boothbay Region of Maine,* M.Sc. Thesis, Institute for Quaternary Studies, University of Maine, Orono, Maine, unpublished.

Cox, D. C., and Thorbahn, P. F., 1982, Prehistoric Archaeological Investigations at Narragansett, Rhode Island: Campbell and Sprague I Sites, Report submitted to Lee, Pare and Associates, Pawtucket, Rhode Island [Available from the Rhode Island Historical Preservation Commission, Providence, Rhode Island].

Cox, D. C., Thorbahn, P. F., and Leveillee, A., 1983, An Archaeological Assessment Survey of the Pettaquamscutt River Basin, Public Archaeology Laboratory, Report No. 12-1, Report submitted to the Rhode Island Historical Preservation Commission, Providence, Rhode Island.

Cronon, W., 1983, *Changes in the Land: Indians, Colonists and the Ecology of New England,* Wang and Hill, New York.

Davidson, D. A., 1972, Terrain Adjustment and Prehistoric Communities, in: *Man, Settlement and Urbanism* (P. J. Ucko, R. Tringham, and G. W. Dimbleby, ed.), Duckworth, London, pp. 17–22.

Dincauze, D. F., 1973, Prehistoric Occupation of the Charles River Estuary: A Paleogeographic Study, *Bulletin of the Archaeological Society of Connecticut* 38:25–39.

Dincauze, D. F., 1974, An Introduction to the Archaeology in the Greater Boston Area, *Archaeology of Eastern North America* 2:39–67.

Dincauze, D. F., 1980, Research Priorities in Northeast Prehistory, in: *Proceedings of the Conference on Northeastern Archaeology* (J. A. Moore, ed.), Research Reports, No. 19, Department of Anthropology, University of Massachusetts, Amherst, Massachusetts, pp. 29–48.

Dincauze, D. F., and Mulholland, M. T., 1977, Early and Middle Archaic Site Distributions and Habitats in Southern New England, *Annals of the New York Academy of Science* 288:439–456.

Emery, K. O., 1969, *A Coastal Pond,* Elsevier, New York.

Gaines, A., 1975, *Papers on the Geomorphology, Hydrography and Geochemistry of the Pettaquamscutt River Estuary,* Ph.D. Dissertation, Graduate School of Oceanography, University of Rhode Island, Kingstown, Rhode Island, unpublished.

Goldsmith, D., 1980, Report on Cores Taken from the Pettaquamscutt River, Ms. on file, Public Archaeology Laboratory, Providence, Rhode Island.

Gribben, J., and Lamb, H. H., 1978, Climatic Change in Historical Times, in: *Climate Change* (J. Gribben, ed.), Cambridge University Press, London, pp. 68–82.

Haggett, P., Cliff, A. D., and Frey, A., 1977, *Locational Analysis in Human Geography,* 2nd ed., Wiley, New York.

Jorgensen, N., 1978, *Naturalist's Guide to Southern New England,* Sierra Club Books, San Francisco, California.

Martin, A. C., Zim, H. S., and Nelson, A. L., 1961, *American Wildlife and Plants,* Dover, New York.

Mulholland, M. T., 1979, Forest Succession and Population Change in a Temperate Forest Environment, in: *Ecological Anthropology of the Middle Connecticut River Valley* (R. Paynter, ed.), Research Reports, No. 18, Department of Anthropology, University of Massachusetts, Amherst, Massachusetts, pp. 45–56.

Nelson, S., 1980, Palynological Studies in a River Estuary, Narragansett, Rhode Island, Ms. on file, Public Archaeology Laboratory, Providence, Rhode Island.

Nixon, S. W., and Oviatt, C. A., 1973, Ecology of a New England Salt Marsh, *Ecological Monographs* 43:463–498.

O'Donnell, P. A., and Leatherman, S. P., 1980, Generalized Maps and Geomorphic Reconstructions of Outer Cape Cod between 12,000 B.P. and 500 B.P. National Park Service Cooperative Research Unit Report 48, University of Massachusetts, Amherst, Massachusetts.

Orr, W. L., and Gaines, A., 1973, Observations on the Rate of Sulfate Reduction and Organic Matter Oxidation in the Bottom Water of an Estuarine Basin: The Upper Basin of the Pettaquamscutt River, in: *Advances in Organic Geochemistry: Proceedings of the Sixth International Meeting of Organic Geochemistry* (Rueil-Malmaison, France), Technip, Paris, pp. 791–812.

Perlman, S. M., 1976, *Optimum Diet Models and Prehistoric Hunter–Gatherers: A Test on Martha's Vineyard,* Ph.D. Dissertation, Department of Anthropology, University of Massachusetts, Amherst, Massachusetts, unpublished.

Reader, R. J., 1978, Primary Production in Northern Bog Marshes, in: *Freshwater Wetlands* (R. E. Good, D. F. Whigman, and R. L. Simpson, eds.), Academic Press, New York, pp. 53–62.

Ritchie, W. A., 1969, *The Archaeology of Martha's Vineyard,* Natural History Press, Garden City, New York.

Snow, D., 1980, The Archaeology of New England, Academic Press, New York.

Thorbahn, P. F., 1982, *The Prehistoric Settlement Systems of Southern New England: Final Report of the Interstate 495 Archaeological Data Recovery Program,* Vol. 1, Department of Anthropology, Brown University, Providence, Rhode Island, Report submitted to the Massachusetts Department of Public Works, Boston, Massachusetts.

Thorbahn, P. F., Loparto, L. W., Cox, D. C., and Simon, B., 1980, Prehistoric Settlement Processes in

Southern New England, Report submitted to the Massachusetts Historical Commission, Boston, Massachusetts.

Wenke, R. J., 1980, *Patterns in Prehistory,* Oxford University Press, New York.

Whigham, D. F., McCormick, J., Good, R. E., and Simpson, R. L., 1978, Biomass and Primary Production in Freshwater Tidal Wetlands of the Middle Atlantic Coast, in: *Freshwater Wetlands* (R. E. Good, D. F. Whigham, and R. L. Simpson, eds.), Academic Press, New York, pp. 3–20.

Whittaker, R. H., and Marks, P. L., 1975, Methods of Assessing Terrestrial Productivity, in: *Primary Productivity of the Biosphere* (H. Lieth and R. H. Whittaker, eds.), Springer-Verlag, New York, pp. 55–118.

Part III

Long-Term Perspectives

What happens to the human ecosystem over time? Many changes in human lifeways visible in the archaeological record are the result of repeated behavioral or technological responses to new social or environmental conditions, and represent long-term change in the way that climate represents long-term weather: on the day-to-day basis that individuals interact with their environment, they respond not to climate, but to weather. Anthropologists studying living populations cannot observe long-term environmental effects or cultural responses directly because those are either not observable by instrumentation or not perceived by the participants. One cycle of a settlement pattern may take a lifetime to complete, with changes in that pattern visible only after several generations, whereas cooling trends that significantly affect length of the growing season may become noticeable only after a decade and then only by objective measurement. Some social or demographic responses to relatively subtle ecological pressures may take an even longer period to appear. Archaeologists and paleoecologists have the advantage of time and space and are thus in a position to observe historical processes within the human ecosystem as populations responded to cultural and environmental pressures over millennia.

Rigorous explanations are required to confirm the influence of environmental factors upon human behavior during the prehistoric period. In a major revision of the larger history of interaction between human responses to environmental change during the early and middle Holocene, Joyce not only reviews the substantial environmental and archaeological literature for the region, but examines several important epistemological aspects of paleoecology. Like Kellogg, he reveals that most regional reconstructions are poorly constructed, outdated, or based upon inaccurate use of palynological and climatic data by nonspecialists. His revised reconstruction of the regional landscape follows recent trends in ecological theory in which change within the ecosystem is continuous, not discontinuous, and occurs primarily at the level of individual members of the biotic community. One implication of this study is that we should expect greater diversity in the Holocene archaeological record.

Gaudreau addresses the issue of dynamic equilibrium between vegetation, climate, and topography in her study of Holocene landscape development in the Northeast. This paper presents a comprehensive survey of the regional pollen data base, which she interprets and then uses to evaluate lags and feedback loops between the biotic and physical environments. The vegetational histories mapped by Gaudreau have obvious implications for prehistoric settlement, while her use of specific analytical methods to identify localized ecotonal contrasts may help archaeologists differentiate between cultural and environmental factors in long-term land-use studies. Two different processes of vegetational change are evidenced in the regional data base: colonization by trees in small populations, and climatic limitations on plant colonization. In each case, change within the ecosystem is primarily at the level of individuals and small aggregations, although the time scale required to shift the location or character of an ecotone may be substantial. Gaudreau's model of small invading populations implies greater environmental heterogeneity (i.e., patchiness) and more diverse biotic communities—two factors stressed by Thorbahn and Cox and by Nicholas in their studies of prehistoric spatial organization.

The opportunities provided by environmental heterogeneity at some locations may provide new ways of explaining diversity in early postglacial land-use behaviors and social relations. The interpretation by Nicholas of high archaeological site frequency and redundant use of space within a heterogeneous landscape emphasizes the importance of ecological contrasts in resource distribution patterns and the prehistoric use of space. By increasing the temporal parameters of this model to span the early and middle Holocene periods, Nicholas offers an historical perspective on the ecology of land-use behaviors at one place on the landscape.

Chapter 8

Early/Middle Holocene Environments in the Middle Atlantic Region

A Revised Reconstruction

Arthur A. Joyce

1. INTRODUCTION

The Late Archaic culture period has been characterized as a time of rapid and widespread changes in settlement–subsistence patterns, social organization, and technologies in the Middle Atlantic region (Catlin *et al.* 1983; Custer 1978, 1984a,b, 1986a,b; Gardner 1978, 1980; Mouer 1986; Mouer *et al.* 1981; Witthoft 1953; but see Cavallo and Joyce 1985). Recent explanatory models using a cultural–ecological theoretical perspective have focused on environmental change as the major cause of these shifts (Catlin *et al.* 1983; Custer 1978, 1984a,b, 1986a,b; Custer and Stewart 1983; Custer and Watson 1986). One of the major environmental factors viewed as being significant in generating Late Archaic cultural change is the mid-postglacial xerothermic. This climatic interval has been characterized as a period of extreme climatic oscillations, including maximum Holocene warm/dry conditions (Carbone 1976:77; Catlin *et al.* 1983:125; Custer 1978:2, 1984a:90, 1984b:33; 1986a,b; Custer and Stewart 1983:4; Custer and Wallace 1982:147; Custer and Watson 1986). It has been

ARTHUR A. JOYCE • Department of Anthropology, Rutgers University, New Brunswick, New Jersey 08903.

proposed that those climatic conditions caused changes in the abundance and distribution of resources exploited by human groups, resulting in an adaptive response by those populations that involved an increasing focus on estuarine and riverine sources and the emergence of long-range exchange networks (Catlin *et al.* 1983; Custer 1978, 1984a,b, 1986a,b; Custer and Stewart 1983).

The work of the late Victor A. Carbone has been used repeatedly to generate paleoenvironmental reconstructions for the early and middle Holocene period. Carbone's (1976) doctoral dissertation was the first study of late Pleistocene and Holocene environments in the Middle Atlantic region to be developed from a systematic synthesis of regional data. More recent reconstructions by other archaeologists for this period have differed little in structure or substance (see especially Custer 1984a,b; Custer and Watson 1986), and contrast with Carbone's only in the timing of Holocene climatic episodes (Carbone 1976; cf. Catlin *et al.* 1983; Curry and Custer 1982; Custer 1978, 1984a,b; Custer and Wallace 1982; Custer and Watson 1986). Palynological, geomorphological, and pedological data primarily from archaeological sites in the Middle Atlantic region have been used to support their reconstructions (Curry 1980; Curry and Custer 1982; Custer 1978, 1984a,b, 1986a,b; Custer and Watson 1986; Stewart 1983; although see Dent 1979, 1982). However, they have failed to consider adequately advances by paleoecologists in eastern North America generated since the completion of Carbone's dissertation. This paper reviews paleoenvironmental research by Middle Atlantic archaeologists in light of recent archaeological and ecological data and proposes a revised reconstruction more consistent with the data.

2. EARLY/MIDDLE HOLOCENE PALEOENVIRONMENTAL RECONSTRUCTIONS: PREVIOUS RESEARCH

Carbone (1976) adopted a modeling approach to the interpretation of late and postglacial environments and to the mechanisms of environmental change originally developed by Bryson and his colleagues (Bryson 1966, 1970; Bryson *et al.* 1970; Bryson and Wendland 1967; Wendland and Bryson 1974). They contend that the glacial to postglacial transition was characterized by abrupt climatic change that can be modeled in the form of a square wave (Bryson *et al.* 1970:56) (see Figure 1). Furthermore, various biogeomorphic responses to climatic change can be represented by variants in the basic square wave through the introduction of such factors as lag time into the system. Utilizing this approach, Carbone (1976) synthesized paleoenvironmental data from the Middle Atlantic region (e.g., Craig 1969; Gehris 1964; Maxwell and Davis 1972; Niering 1953; Walker and Hartman 1960) and portions of the Southeast (e.g., Frey 1953; Whitehead

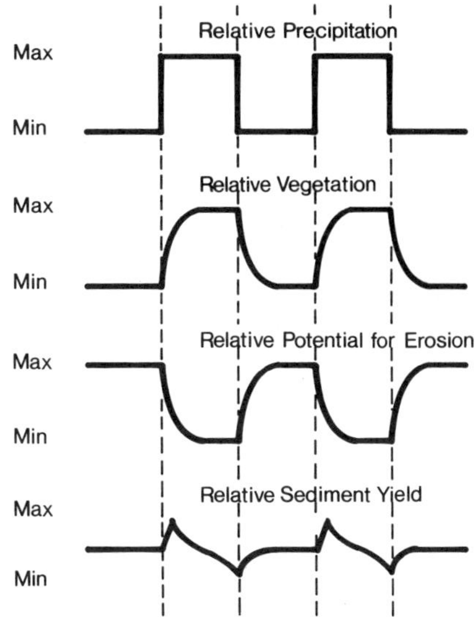

Figure 1. Idealized square wave model. After Carbone 1976.

1973), and organized the regional record of environmental change into a sequence of "quasi-steady"-state episodes. Changes in pollen percentages and maxima and minima in the percentages of different taxa were utilized to infer changing climatic conditions. He distinguished two climatic episodes during the early/middle Holocene in the Middle Atlantic: an earlier, mesic period and a later, xeric period.

The earlier episode, dating from approximately 8200 to 6500 B.P., is dominated by oak pollen, with low levels of pine and a hemlock maximum. These profiles were interpreted as indicating the establishment of a mesic forest environment and warm/wet climatic conditions (Carbone 1976:76, 200). The later episode is the mid-postglacial xerothermic and is dated between 6500 and 3200 B.P. Oak pollen remained dominant during this episode and hickory pollen attained a maximum (Carbone 1976:76). Pine pollen began to replace oak pollen, especially in the southern Middle Atlantic, as early as 5000 B.P. Carbone suggests that this may indicate the expansion of loblolly pine resulting from drier conditions (1976:55–56). Xeric conditions are postulated for this episode along with an increase in temperature over the previous period. This warm/dry climate extreme appears to culminate between 5000 and 4000 B.P., although pronounced climatic oscillations are suggested for the period as a whole (Carbone 1976:200). Carbone correlated the xerothermic with a gradual extension of oak–hickory forests and the eastward extension of the prairie peninsula from the Midwest. This represents the time of maximum warmth and moisture stress in the Middle

Atlantic region during the Holocene. Forested habitats supposedly decreased in extent, creating grassy openings, and "vegetation was probably at a minimum" (Carbone 1976:123).

Subsequent work by Middle Atlantic archaeologists has continued to rely on the square wave model and on the same maxima and minima in pollen percentages that Carbone used to infer climatic change (see Custer 1978, 1984a,b; Custer and Wallace 1982; Custer and Watson 1986). They rely primarily on the same palynological data discussed by Carbone (1976; see especially Custer 1978, 1984a:30–37, 1984b:Table 1), although Custer (1984a,b) has included some new palynological data, consisting mostly of pollen samples recently taken from archaeological sites in Delaware. Dent (1979) also used palynological data from north of Carbone's study area for his paleoenvironmental reconstruction for the upper Delaware valley. The most significant addition to the work of Carbone has been the collection of geomorphological and pedological data from archaeological sites in the Middle Atlantic region. These studies can be classified into four categories: (1) aeolian depositional sequences (Curry 1980; Curry and Custer 1982; Custer 1984b, 1986b; Custer and Watson 1986; McNamara 1981; Stewart 1983); (2) alluvial depositional sequences (Curry and Custer 1982; Custer 1978, 1981, 1984b; McNamara 1983; Stewart 1983); (3) roof fall deposits in rock shelters (Curry and Custer 1982; Custer 1984b); and (4) settlement patterns associated with relict ponds (Curry and Custer 1982).

Depositional sequences were interpreted using a model influenced by the work of Knox (1972, 1976), and based on a square wave model that specifies the relationship between climatic change, vegetation, soil properties, and geomorphological processes (Carbone 1982; Curry and Custer 1982; Custer 1978; cf. Knox 1972, 1976, 1983) (see Figure 1). This model has allowed researchers to relate depositional discontinuities to climatic shifts, such as the proposed wet/dry shift at the onset of the xerothermic and the dry/wet shift at its termination (Curry and Custer 1982; Custer 1984b; Custer and Watson 1986; Stewart 1983). Custer (1984b:33–34), however, notes that the links between climatic change and processes of soil deposition and erosion are complex and may vary with local vegetation and geology.

These researchers generally agree with Carbone's reconstruction concerning the nature and duration of climatic episodes in the Middle Atlantic region (Custer 1978, 1984a,b, 1986a; Custer and Wallace 1982; Custer and Watson 1986; Dent 1979). There is considerable disagreement, however, over the timing of these episodes (see Table 1). For example, Dent's (1979) reconstruction agrees with Carbone's, with the exception that the dates for transitions between similar climatic episodes are up to 2,000 years later. Furthermore, dates for the xerothermic have ranged from the interval 6950–3950 B.P., as proposed by Catlin *et al.* (1983), to the period 4200–2200 B.P., proposed by Curry and Custer (1982:1). The xerothermic is usually reconstructed as having lasted 2,000 years, although

Table 1. Dates Proposed for Xerothermic Interval

Reference	Date (B.P.)	Location
Carbone 1976	6500–3200	Middle Atlantic region
Custer 1978	4650–2150	Middle Atlantic region
Curry and Custer 1982	4200–2200	Middle Atlantic region
Custer and Wallace 1982	4950–2750	Piedmont Uplands
Catlin *et al.* 1983	6950–3950	Middle Atlantic region
Custer 1984a	4650–2150	Middle Atlantic region
Custer 1984b	5050–3550[a]	Middle Atlantic region

[a]Possibly as recent as 1350 B.P. in some areas

in papers published or delivered in the past 5 years, it can be found to occur in just about any 2,000–year interval between 7000 and 2000 B.P. (Table 1). Custer (1984b:33) notes that the effects of the transition between these climatic episodes is not consistent throughout the region and that a slight south to north time lag is evident. However, these factors are insufficient to explain the amount of variance in dates proposed for the xerothermic.

3. RECENT TRENDS IN PALEOENVIRONMENTAL RESEARCH: IMPLICATIONS FOR THE MIDDLE ATLANTIC REGION

In addition to paleoenvironmental studies conducted in the Middle Atlantic region by archaeologists, an abundance of research has been completed by paleoecologists in eastern North America during the last 10 years. Environments have been reconstructed for portions of the Middle Atlantic region (Brush 1986; H. Delcourt 1979; Watts 1979), New England (Davis *et al.* 1980; Likens and Davis 1975; Winkler 1982, 1985), the Southeast (Delcourt 1980; Delcourt and Delcourt 1977, 1979; Delcourt *et al.* 1980; Watts 1980), Appalachia (Watts 1979), and the northern Midwest (Bartlein and Webb 1982; Bernabo 1981; Webb *et al.* 1983). Other studies have emphasized broader regional perspectives, including most or all of eastern North America (Bernabo and Webb 1977; Davis 1976, 1983; Delcourt 1979; Delcourt and Delcourt 1985; Delcourt *et al.* 1983; Delcourt and Delcourt 1981, 1983; Gaudreau and Webb 1985; Webb 1980, 1981). Finally, a number of studies have focused on Holocene vegetational migration patterns for the eastern United States (Bernabo and Webb 1977; Davis 1976, 1981a, 1983; Delcourt and Delcourt 1985; Delcourt *et al.* 1983; Delcourt and Delcourt 1981, 1983; Gaudreau and Webb 1985). These studies have benefited from new theoretical developments coupled with improved analytical techniques and a greater number of sufficiently sampled and dated pollen cores not available to previous researchers (see Dincauze 1981; Ogden 1977; Webb 1980, 1981, 1982). This research will be used to evaluate and revise reconstructions now in use by archaeologists in the Middle Atlantic region (i.e., Carbone 1976; Custer 1984a,b, 1986a,b; Custer and Wallace 1982; Custer and Watson 1986; Dent 1979).

Palynologists have stressed the need for more rigorous methods to increase the accuracy of paleoenvironmental reconstructions (Ogden 1977; Overpeck *et al.* 1985; Prentice 1985, 1986; Webb 1980, 1981). These techniques include: multiple radiocarbon dates on pollen cores to increase temporal resolution; direct-counting methods, such as rates of absolute pollen influx and absolute pollen deposition (which requires multiple dates); decreasing the sampling interval of cores; and maximizing sample densities when measuring regional patterns of pollen deposition. Furthermore, paleoecologists recognize that the type of depositional basin from which pollen cores are taken influences the proportion of local, extralocal, and regional pollen sampled (Jacobson and Bradshaw 1981; Prentice 1985), making cores from small basins less relevant for regional paleoenvironmental reconstructions. Palynological data used to reconstruct Middle Atlantic region paleoenvironments suffer from a lack of pollen cores sampled with up-to-date techniques (Cavallo and Joyce 1985). Especially alarming is the fact that fewer than 50% of the cores used by archaeologists for reconstructions of Middle Atlantic paleoenvironments (see Carbone 1976; Custer 1984b:Table 1; Dent 1979) have been radiocarbon dated, and that pollen influx data have been collected from only two cores (Buckles Bog, western Maryland, and Tannersville Bog, northeastern Pennsylvania). In general, the density and quality of pollen cores for the Middle Atlantic region are less than those of other areas of northeastern North America.

Paleoenvironmental reconstructions in eastern North America are also at variance with those generated by archaeologists for the Middle Atlantic region (i.e., Carbone 1976; Custer 1984a,b; 1986a,b; Custer and Watson 1986; Dent 1979). Recent paleoenvironmental reconstructions throughout the eastern United States assert that the duration of warmer and drier conditions occurred between 9000 and 5000 B.P. (Davis *et al.* 1980; Delcourt 1979; Delcourt 1980; Webb 1980, 1981; Watts 1979). Furthermore, the maximum eastward extension of prairie conditions has been shown to have occurred between 7000 and 6000 B.P., reaching no further than southeastern Missouri (King and Allen 1977), and which was followed by a gradual westward retreat (Bartlein *et al.* 1984; Bernabo and Webb 1977; Webb 1981; Webb *et al.* 1983).

Reasons for the discrepancies between paleoenvironmental reconstructions generated by Middle Atlantic region archaeologists (e.g., Carbone 1976; Custer 1984a, 1986a,b; Custer and Watson 1986; Dent 1979) and those developed by paleoecologists in northeastern North America (e.g., Bernabo and Webb 1977; Davis 1983; Davis *et al.* 1980; Delcourt 1979; Delcourt and Delcourt 1985;

Delcourt 1980; Watts 1979) can be seen by comparing the epistemological foundations of these reconstructions. The most significant point of disagreement involves the theoretical linkage between the palynological record and hypothetical vegetational and climatic changes.

For example, Carbone (1976) and others (Custer 1984a,b; Dent 1979) have adopted an approach developed by Bryson and his colleagues in the late 1960s and early 1970s (see Carbone 1976:Chapter 2). In this case, the paleoenvironmental record is viewed as reflecting a sequence of stable (i.e., climax) vegetational communities in equilibriim with the dominant climate: "After all, the concept of a stable self-perpetuating floristic unit in equilibrium with the prevailing climate is a cornerstone of the palynological discipline" (Dent 1979:225). These periods of relative stasis are seen as being punctuated by periods of climatic–biotic disequilibrium resulting from rapid, steplike climatic change, followed by the reestablishment of new equilibrium conditions between climate and vegetation. This is the basic logic behind the square-wave model of climatic change.

The consequences of this position for reconstructing Holocene environmental change are clearly stated by Carbone:

> . . . the postglacial transition was characterized by abrupt climatic shifts, i.e., rapid, step-like transitions in the climatic regime, from one quasi-steady state to another. Thus, the paleoclimatologist's main task was to define both the character of the dominant pattern during each of the periods as well as the time interval in which the pattern was operative. (Carbone 1976:12)

Climate is considered the dominant factor controlling regional vegetational communities such that stable climatic regimes during the Holocene resulted in stable floristic communities (Carbone 1976:17). When significant changes are observed in the palynological record, usually represented as the transition between pollen zones, it is assumed to be the result of a rapid climatic change and a subsequent readjustment of climatic–biotic equilibrium conditions.

This position, excluding assumptions about the tempo of climatic change, is essentially identical to that of the earliest paleoenvironmental researchers in North America (i.e., Antevs 1948, 1955; Deevey 1939, 1949; Deevey and Flint 1957; Sears 1935, 1942a,b, 1948). These early researchers assumed that changes in forest communities, as reconstructed from the palynological record, were a direct result of climatic change (see Sears 1948:326; Harmon 1968:11). Pollen profiles were classified into zones that were viewed as temporal sequences of forest communities resulting from changing climatic conditions (Deevey and Flint 1957; Ogden 1965; Sears 1948). The optimal climatic conditions for each community, based on modern analogs, were assumed to have been the dominant climate for the period of time represented by each zone. Prior to the work of Bryson and his colleagues (Bryson 1970; Bryson *et al.* 1970; Bryson and Wendland 1967; Wendland and Bryson 1974), Holocene climatic change was assumed to have been gradual. Bryson (1970; Bryson and Wendland 1967) asserted that the nature of climatic change is episodic, so that abrupt rather than gradual transitions caused the variation in the pollen record.

In contrast to this approach, paleoecological specialists in eastern North America have recently taken a completely different position concerning paleoenvironmental reconstruction. These researchers assert that the establishment of forest communities must be explained by reference to the migrational histories of the individual species that make up the "community," rather than a superorganismic model of community development (West *et al.* 1981). This position is consistent with the recent trend in ecology away from community, or systems level, explanations, and toward an emphasis on the individual as the unit of adaptation (Colinvaux 1973; Orlove 1980; Richerson 1977).

This position dictates a different approach concerning inferences from the palynological record. The emphasis is on understanding migrational rates and patterns of particular taxa, rather than on defining pollen zones and broad vegetational communities (Bernabo and Webb 1977; Davis 1976, 1981a; Delcourt and Delcourt 1985). Correlations between the Blytt–Sernander sequence and North American sequences are considered unjustified (Dincauze 1981:79, 1985:14). Climate is considered to be only one of the significant influences on plant migrations. Other important factors include inter- and intraspecific competitive interactions, soils development, physical barriers to seed dispersion, the effects of pathogens, forest fires, and insolation values. Davis states:

> Climatic changes associated with the termination of the Wisconsin glacial phase were the ultimate cause of the expansion of tree species during the Holocene. However, much of the time the rate of spread was not controlled by climate, and the geographic distributions of many species were not in equilibrium with climate, depending instead on the availability of propagules and the ability of seedlings to survive in competition with plants already growing on the site. The idea that species ranges expanded northward in parallel with northward movements of isotherms as the climate gradually warmed is disproved because the migration routes for individual species differed greatly; some moved from east to west, others moved from west to east. Rates of movement also differed. (Davis 1981a:152)

Thus, forests mapped today as single communities (Braun 1950) may have had very different developmental histories. For example, oak–chestnut forests in the central Appalachians have contained chestnut as a dominant species for at least 5,000 years, while oak–chestnut forests in Connecticut have included chestnut for only 2,000 years (Davis 1981a:151).

These complex migrational patterns created vegetational communities without modern analogs. This makes it difficult to apply analog models that rely on projections of contemporary vegetational community/climate relationships to the past. Instead, a detailed analysis of the migrational patterns of individual species has been emphasized so that climatic effects can be separated from other factors using an understanding of the ecology of individual taxa in the context of their migrational histories. Thus, a more complex analog model is used that focuses on the ecological relationships of individual taxa rather than on analogs based on broad correlations between vegetational communities and climate.

Davis (1981a:152) concludes that the rate at which forests adjust to climate is far exceeded by the rate of climatic change (Brubaker 1986; Davis *et al.* 1986; Lewin 1985; Pennington 1986; Prentice 1986; Ritchie 1986; Webb 1986). While the rapid climatic change at the Pleistocene/Holocene transition was the impetus for subsequent vegetational migration in the Holocene, factors such as seed dispersion and the survival ability of invading seedlings were more important in influencing specific patterns and rates of plant migration than immediate climatic effects. Some species, such as hickory, did not reach their maximum possible range extension in the Northeast, because hickory reached New England after the time of maximum temperatures in the region (Davis 1981a:152–153). Climate did affect the rates and patterns of some taxa by limiting the areas favorable for invasion or by increasing or decreasing migration rates. Given these recent studies, plant migration must be seen as more than successive periods of climatic–biotic equilibrium punctuated by rapid climatic change and subsequent system readjustment.

Analyses of vegetational migration patterns have shown that many of the pollen trends asserted by Carbone (1976) and others (Custer 1984a,b, 1986a; Custer and Watson 1986; Dent 1979; Edwards and Merrill 1977; Sirkin 1977) to be climatic indicators are better explained by reference to other factors. Davis (1976, 1981a; Davis *et al.* 1980) has argued that the sudden increase in hickory pollen, as well as other pollen types in North America, is a time-transgressive phenomenon representing the spread of species northward following the end of the last glaciation, and cannot be considered as direct responses to climatic change:

> The changes in pollen frequency that were originally thought to represent the xerothermic interval in southern New England (Deevey 1939) are in younger sediment, precisely dated now at 2000–4500 years B.P. These changes seem most easily interpreted now as the reflection of successive immigrations. (Davis 1969:421)

Regardless of migrational lags, hickory is an especially poor indicator of climate because the many species of *Carya* present in eastern North America have widely divergent climatic and edaphic tolerance limits (Harlow and Harrar 1958). Davis (1981b) has also argued that the rapid decline of hemlock pollen throughout northeastern North America at approximately 4650 B.P. was the result of a pathogen rather than climatic change (see also Gaudreau and Webb 1985:254; Webb

1982). Furthermore, pollen cores from the Northeast and from the Middle and South Atlantic regions rarely indicate maxima in nonarboreal pollen during the middle Holocene (Bernabo and Webb 1977; Davis 1983; Watts 1979), and most of the exceptions are explainable as the result of edaphic factors or changing hydrological conditions (e.g., Russell 1985; Whitehead 1965:428).

The rise in pine during the middle Holocene in the Atlantic and Gulf Coastal Plains has been attributed to an increase, rather than decrease, in precipitation (Delcourt 1979; Delcourt and Delcourt 1985; Delcourt 1978, 1980; Watts 1979). Watts (1979:462–463) has presented a detailed analysis of the possible conditions necessary to cause this pine expansion. He concluded that the most plausible explanation is an increase in the frequency of lightning-caused fires and strong winds due to a stormier climate that increased the occurrence of light gaps in the forest, allowing the expansion of southern pines. Most other paleoecologists in the Southeast concur with Watts' (1979) conclusions (Delcourt 1979:277; Delcourt and Delcourt 1985:20; Delcourt 1980:384; Delcourt and Delcourt 1981:150; but see Whitehead 1979:351).

The differences previously discussed between the reconstructions of Carbone (1976) and Dent (1979) can be explained by considering vegetational migration rates and patterns. These differences result from the geographic placement of sites used for these reconstructions in relation to vegetational migration patterns. Carbone's (1976) study is centered in the Shenandoah Valley, Virginia, while Dent's (1979) reconstruction focuses on an area approximately 400 km to the northeast in the Upper Delaware Valley. The earlier occurrence in the Shenandoah Valley of the boreal pine, oak–hickory, and oak–chestnut pollen zones, and the hypothesized climatic episodes corresponding to these zones, can be attributed to the earlier expansion of white pine, hickory, and chestnut through Virginia. The differences in the timing of these zones for the two areas correspond to the rates and patterns of migration determined by Davis (1981a).

Unfortunately, many archaeologists in the Middle Atlantic region have not considered these trends in paleoecological research and therefore may not be aware of growing problems with their data and interpretations. For example, Custer (1984b:Table 1) continues to rely on undated pollen cores (e.g., Gehris 1964; Niering 1953; Owens *et al.* 1974; Rasmussen 1958) and on cores sampled prior to the development of modern sampling techniques (e.g., Niering 1953). The continued reliance on an epistemology of paleoenvironmental reconstruction that infers climatic episodes directly from pollen zones appears to have created confusion regarding the timing of a mid-postglacial xerothermic interval (see Table 1).

The proposed relationship between middle Holocene climatic change and depositional discontinuities in regional alluvial sequences should also be reevaluated. The model developed by Knox (1972, 1976) has influenced interpretations of alluvial sequences by archaeologists in the Middle Atlantic region (Curry and

Custer 1982; Custer 1978, 1984a,b, 1986a; Custer and Watson 1986). However, Knox's model of the relationship between climate and alluvial processes (Knox 1972, 1976, 1983, 1985) differs from those of Custer (Curry and Custer 1982; Custer 1978, 1984a,b, 1986b; Custer and Watson 1986) in regard to both theoretical underpinnings and resulting inferences. Recent applications of the Knox model to reconstruct paleoclimates from alluvial sequences in the eastern United States and throughout North America (Brakenridge 1980, 1981; Knox 1976, 1983, 1985) conflict with Middle Atlantic region reconstructions. The relationship between climatic patterns and alluvial depositional sequences is complex and it is usually impossible to relate depositional discontinuities to particular climatic trends without corroboration with other paleoenvironmental data (McDowell 1983:101; Ray Mueller, personal communication, 1985). These relationships should be reevaluated in the Middle Atlantic region because the paleoclimatic reconstructions used to interpret alluvial sequences there have been contradicted by recent research.

One major problem involves the use of geomorphological, pedological, and palynological data taken from archaeological contexts for the generation of paleoclimatic inferences. Analyses of alluvial and aeolian depositional sequences from archaeological sites in the Middle Atlantic region have assumed that climatic effects, most notable the mid-postglacial xerothermic, are responsible for depositional discontinuities (Curry 1980; Curry and Custer 1982; Custer 1978, 1984b, 1986b; Custer and Watson 1986; Stewart 1983). However, these studies neglect to consider the profound effects that human activities can have on geomorphological processes. Human disturbance of vegetational communities, even if confined to habitation areas, can have a significant impact on rates of alluvial and aeolian deposition. Butzer (1982:Chapter 8) discusses numerous archaeological examples of human impact on the landscape affecting alluvial processes. Most of Butzer's examples are from Europe and he notes that, "a systematic picture of prehistoric human impact on the landscape has not yet emerged for other continents" (Butzer 1982:140). The type of local vegetational disturbance necessary to affect geomorphological processes has been documented for aboriginal populations in the eastern United States (Cronon 1983; Delcourt *et al.* 1986; Russell 1983). Therefore, it is not unreasonable to consider human impact as a causative agent in alluvial depositional discontinuities from archaeological sites in the Middle Atlantic region.

Sites buried by aeolian sands (Curry 1980; Curry and Custer 1982; Custer 1986b; Custer and Watson 1986; Stewart 1983) represent an especially problematic source of paleoclimatic data. Evans (1978:86) notes that three conditions are necessary for aeolian deposition: (1) a supply of sand; (2) absence of surface vegetation; and (3) at least temporary dryness. It has been suggested that the warm/dry climatic conditions of the xerothermic caused upland forest openings, a decrease in surface vegetation in upland settings, and desiccating conditions necessary for aeolian erosion and deposition (Curry 1980:6–9; Custer and Watson 1986:8; Stewart 1983:46). However, large-scale climatic change is not required to generate these conditions. Drying conditions sufficient for aeolian deposition can be provided by normal periods of summer drought (Curry 1980:8; Ray Mueller, personal communication, 1985). Summer drought conditions or human activities can cause disruption of local vegetation at archaeological sites resulting in the removal of sufficient surface vegetation to mobilize sand. Furthermore, the use of diagnostic artifacts to date depositional sequences and calculate rates of deposition (Curry and Custer 1982; Custer and Watson 1986; Stewart 1983) is problematic because of questionable artifact integrity resulting from deflation of soils due to aeolian erosion and redeposition (Curry 1980:9). Diagnostic artifacts in some of these aeolian sites are confined to plow zone contexts, which adds to the difficulty of using these data for paleoclimatic interpretations (Curry 1980:3).

Human impact on local vegetation must also be considered when analyzing pollen samples from archaeological sites. Custer used pollen samples from three archaeological sites in Delaware to support a paleoenvironmental reconstruction based on the work of Carbone (1976), although he attributed changing pollen frequencies in these samples to climatic effects without considering human impact (Custer 1984a:34–37, 1984b:33–34). For example, Custer notes that "some pollen data [from the Delaware Park site] show plant communities indicative of dry conditions lasting as late as 600 A.D." (Custer 1984b:33). Unfortunately, these data consisted of ten pollen samples taken from dated archaeological refuse features (Thomas 1981), so that human selection is clearly indicated. Therefore these samples cannot be considered as valid unambiguous data for generating paleoclimatic inferences.

4. EARLY/MIDDLE HOLOCENE PALEOENVIRONMENTS: A REVISED RECONSTRUCTION

The following section presents a revised reconstruction for middle Holocene environments in the eastern United States that focuses on the Middle Atlantic region. This reconstruction relies on palynological, paleobotantical, geomorphological, and pedological data from this region, as well as from adjacent areas. It is necessary to expand the area of analysis both because the data base in the Middle Atlantic region is not as extensive as in areas to the north, west, and south, and because the processes that have caused environmental change are not confined to specific geographic regions. This approach also allows for hypothesis testing of large-scale paleoclimatic models developed elsewhere (e.g., Bryson and Wendland 1967; Bryson *et al.* 1970; Knox 1976).

Recent studies in the eastern United States suggest a sequence of middle Holocene environmental change significantly different from that reconstructed by archaeologists in the Middle Atlantic region. The duration of maximum Holocene warmth and dryness is interpreted as having occurred between approximately 9000–8000 and 5000–4000 B.P. A gradual adjustment to modern conditions occurred until approximately 2000 B.P., when the basic character of the present landscape was attained. This pattern parallels conditions reconstructed for the Midwest (Bartlein and Webb 1982; Bartlein *et al.* 1984; Webb *et al.* 1983).

Palynological data from the Southern and Middle Atlantic regions indicate maximum Holocene warmth and dryness during the period from approximately 9000–8000 to approximately 5000–4000 B.P. Palynological data from the central eastern United States (34°–42°N, 90°–74°W) indicate that mixed mesophytic forest taxa (Braun 1950; H. R. Delcourt 1979:255) became increasingly restricted in distribution and topographic position to the Cumberland Mountains of eastern Kentucky, suggesting warm/dry conditions (Delcourt 1979:277). Likewise, pollen in early Holocene deposits from several sites from Georgia to New Jersey have been damaged or destroyed by oxidation, while later deposits are well preserved Watts (1979:463), suggesting that summer drying of ponds was more extensive during the early Holocene than at present.

In the Southern and Middle Atlantic regions, a shift to climatic conditions similar to those of today appears to have occurred shortly after 5000 B.P. In general, vegetational patterns inferred from pollen data assume their modern character by approximately 5000 B.P. (Delcourt 1979:277; Delcourt 1980:384). Notable climatic indicators of a shift to cooler and wetter conditions are the pine expansion in the Atlantic and Gulf Coastal Plains and central Appalachia (Delcourt 1979:277) and the emergence of bog conditions suitable for northern conifers in Pennsylvania (Watts 1979:463–464). The pine expansion may have begun as early as 5500 B.P. (Watts 1979:462–463) and continued until shortly after 5000 B.P. (Delcourt 1979:277).

In the Chesapeake Bay area, Brush (1986) argues that from 4700 to approximately 3400 B.P., climatic conditions were wetter than in subsequent periods. Forests along the Chesapeake were dominated by black gum and sweet gum, with abundant river birch and *Osmunda* ferns. Pollen production was highest during this period, suggesting a greater abundance and size of trees. By approximately 1500 B.P., forests containing plants indicative of drier climate, such as holly, chestnut, and ericaceous shrubs, were the dominant taxa along the Chesapeake. Brush (1986:151) notes that climatic conditions during the late Holocene were generally similar to today, although oscillations toward drier conditions are indicated from 1500 to 1400 B.P. and again from 950 to 750 B.P. High charcoal and metal influxes during the latter period also suggest an increased fire frequency.

In New England, palynological and plant macrofossil data indicate similar trends. The most reliable data involve altitudinal movements of tree populations in mountainous areas of New England (Davis 1983:176; Davis *et al.* 1980; Jackson 1981; Likens and Davis 1975; Spear 1981; Whitehead *et al.* 1981). These data are especially important because elevational zonation of forest species appears to be climatically controlled (Davis *et al.* 1980), making them sensitive indicators of changing climatic conditions. Palynological and plant macrofossil data demonstrate that both white pine and hemlock occurred 300–400 m above their present elevation limit during the early to middle Holocene. The upward range extension of these species lasted from 9000 to at least 5000 B.P., when hemlock disappeared from its highest locations apparently as a result of biogenic events (Davis *et al.* 1980:248). This indicates that climatic conditions at 9000 B.P. were already warmer and drier than those of today. Estimates from modern lapse rates suggests a 2°C increase in mean annual temperature and a 400–125 mm decrease in mean annual precipitation (Davis 1983:176; Davis *et al.* 1980:218), although these estimates are rough approximations. Davis *et al.* (1980:249) argue that after 5000 B.P., a continuous change toward present conditions is implied, crossing a climatic threshold that allowed spruce to recolonize sites between 650 and 1000 m at 2000 B.P. This resurgence of spruce is consistent with observations from sites throughout the Northeast, indicating a southern migration of northern conifers beginning as early as 4000 B.P. (Bernabo and Webb 1977:80; Watts 1979:447). Needles of white pine and hemlock occur at high elevation sites until the Little Ice Age (A.D. 1450–1850), suggesting that conditions warmer than present persisted until this time. Davis *et al.* conclude: "The new evidence we have presented here indicates that a climatic optimum did occur during the Holocene in New England, that it was earlier than previously supposed and contemporaneous with the Prairie Period of the Middle West, and that the amplitude of change was similar to the Holocene temperature optimum in Europe" (Davis *et al.* 1980:250).

Data from early Holocene deposits in New England also suggest that drier conditions may have resulted in a higher fire frequency. Peaks in charcoal influx suggesting an increase in incidence of forest fires occur about 7500 B.P. at Mirror Lake, New Hampshire (Davis 1983:177) and between 9000 and 8200 B.P. at Duck Pond on Cape Cod (Winkler 1985:307). Charcoal influx rates are generally greater at these sites throughout the early Holocene relative to subsequent periods. The abundance of ironwood pollen in early Holocene deposits of New England is also compatible with a drier climate and a higher fire frequency (Davis 1983:177). Ironwood today is especially prevalent in woodlands experiencing frequent forest fires. Similarly, Winkler (1985) notes that palynological evidence from Duck Pond suggests a transition from a mixed mesophytic forest to a particularly xeric pine barrens association by approximately 8200 B.P., and a return toward somewhat more mesophytic conditions by about 5000 B.P. Pine barren communities are typically associated with high fire frequencies and xeric soils.

Davis (1969:419) has also noted a maximum for ragweed pollen dated to approximately 8000–7000 B.P. at Rogers Lake, Connecticut. She attributes this to forest openings caused by warmer and drier climatic conditions, which correlates with the maximum extension of the Prairie Peninsula in the Midwest. A slight decline in pollen sedimentation rates occurs at this time in both the Rogers Lake and Berry Pond (Massachusetts) profiles (Whitehead 1979:252), which is consistent with Davis' interpretation. However, increases in the percentage of nonarboreal pollen do not appear to be a general trend in northeastern profiles at this time (Bernabo and Webb 1977), suggesting that these changes may reflect local physiographic or hydrological conditions rather than climatic change.

In addition, sedimentological data from the southern basin of Lake George, New York, are consistent with the palynological data from New England. Hutchinson *et al.* (1981) inferred that a low-water stage occurred in Lake George between 10,000 and 7000 B.P. These data suggest that the lake may have been lower by as much as 20 m. While Hutchinson *et al.* (1981:59) fail to provide an explanation for this phenomenon and reject a climatic explanation, they cite only dated paleoclimatic sources (i.e., Connally and Sirkin 1969, 1971). The more recent data discussed here provide a climatic explanation consistent with their findings.

Changes in forest composition during the Holocene have been analyzed by the construction of isopleth, isopoll, and isochrone maps that synthesize a large number of pollen profiles (Bernabo and Webb 1977; Davis 1976, 1981a; Gaudreau and Webb 1985). These studies attempt to use only profiles sampled with modern techniques to avoid the problems inherent in undated or poorly sampled pollen cores. Bernabo and Webb (1977) constructed isopoll and isochrone maps which track changes in pollen percentages through time. Davis (1976, 1981a) has extended this approach to reconstruct the leading edge of the migrating populations of various arboreal taxa. These studies involve generalized regional migration patterns. Forest composition at particular locations will be complicated by a variety of factors, including soils properties, topography, hydrology, and human impact. For more specific paleoenvironmental reconstructions, local pollen profiles should be examined in the context of these broader regional patterns (e.g., Brush 1986; Delcourt 1979; Delcourt 1980; Russell 1985; Whitehead 1979).

Early Holocene forests generally had a low species diversity, which increased through time as successive migrations brought new taxa into Middle Atlantic forest regions (Bernabo and Webb 1977; Davis 1981a; Delcourt and Delcourt 1985; Gaudreau and Webb 1985; Watts 1979). Forests in the Middle Atlantic region have been dominated by oak species from approximately 9000 B.P. to the present. Oak rapidly migrated northward, replacing pine in the Middle Atlantic between approximately 10,000 and 9000 B.P. While forests were oak-dominated by 9000 B.P., they also contained significant populations of hemlock, maple, birch, and pine, with spruce still important at higher elevations, especially in the Appalachian Mountains. Beech was important in the southern and central Middle Atlantic at this time, but did not migrate into portions of the northern and western Middle Atlantic until 8000–7000 B.P. Similarly, hickory reached sites in the southern Middle Atlantic region by 10,000 B.P., but did not migrate into northern sites until approximately 6000 B.P. The majority of these early Holocene migrations into the Middle Atlantic was at the expense of northern pines and spruce, which were rapidly migrating northward in response to Holocene thermal conditions.

Changes in forest composition continued during the middle Holocene in the Middle Atlantic region (Bernabo and Webb 1977; Davis 1976, 1981a, 1983; Delcourt and Delcourt 1985; Gaudreau and Webb 1985; Watts 1979). Chestnut migrated northeastward across the central Appalachian states, reaching the northern Middle Atlantic by approximately 5000 B.P. and parts of New England by 2000 B.P. (Davis 1981a:150). Southern pine increased at the expense of oaks on the Atlantic and Gulf Coastal Plains beginning at approximately 5500 B.P. (Watts 1979). In New England and parts of the northern Middle Atlantic region, northern conifers, especially spruce, began to recolonize sites after 4000 B.P. (Bernabo and Webb 1977:80; Carbone 1976:76; Gaudreau and Webb 1985:262).

The most dramatic change in forest composition during the middle Holocene was the paneastern hemlock decline dated to approximately 4650 B.P. (Webb 1982). Prior to this, hemlock had been abundant throughout the Middle Atlantic region (Carbone 1976:76), and was the dominant taxon in profiles for the northern Middle Atlantic and New England (Dent 1979; Nicholas 1968; Whitehead 1979). The synchroneity of the hemlock decline throughout eastern North America and the lack of contemporaneous changes in other taxa have led Davis (1981b) to attribute it to the effects of a pathogen. The areas left vacant by hemlock were probably replaced by local taxa such as birch, oak, beech, ash, and red maple (see Davis 1983:177 for a summary of regional patterns of secondary succession following the hemlock decline). Hemlock eventually recovered its previous abundance in most areas by approximately 3000 B.P. (Davis 1981a:148).

5. EXPLANATORY MODELS OF HOLOCENE CLIMATE CHANGE

The timing of maximum Holocene warmth and dryness in the eastern United States is synchronous with the hypsithermal in the Midwest, so that mechanisms responsible for climatic change there may have also affected climate in the East (Davis *et al.* 1980:250). The midwestern hypsithermal has been inferred from palynological data indicating an eastward expansion of prairie environments beginning at approximately 10,000 B.P., reaching maximum eastward expansion at 7000–6000 B.P., and receding westward at a slower pace from 6000 to 2000 B.P. (Bartlein *et al.* 1984; Bernabo and Webb 1977; Webb *et al.* 1983). This suggests that warm/dry conditions increased in the Midwest from approximately 9000 to 6000 B.P. Palynological data indicate that maximum dryness occurred between 7000 and 6000 B.P., with precipitation estimated to have been 80% of its modern value (Bartlein *et al.* 1984:368). This interpretation has been supported by data indicating low midwestern lake levels at 7000 B.P. (Watts and Winter 1966). Recently, Dean *et al.* (1984) presented data from varved lake sediments suggesting that the dry interval in the northern Midwest was interrupted by an abrupt change to a moist period at approximately 5400 B.P., with a resumption of drying conditions between 4800 and 4000 B.P. Temperature appears to have reached its maximum value at approximately 6000 B.P. in the northern Midwest and at approximately 4000 B.P. in the southern Midwest (Bartlein *et al.* 1984:371; Winkler *et al.* 1986). Temperatures may have reached as high as 2°C above present conditions.

Explanatory models of climatic change in the Midwest have focused on changes in atmospheric circulation patterns (Bartlein and Webb 1982; Bartlein *et al.* 1984; Bryson and Wendland 1967; Webb and Bryson 1972). Bartlein *et al.* (1984:372) suggest that the trend toward warmer and drier conditions in the Midwest between 9000 and 6000 B.P. was related to the replacement of the retreating cooler Arctic air masses by the warmer and drier Pacific air masses. After 6000 B.P., the influence of Pacific air masses decreased and was replaced in the northern Midwest by Arctic air masses and in the southern Midwest by the warm Atlantic air masses.

These changing air mass patterns in the Midwest can be viewed as part of a trend in general circulation over North America (Bartlein *et al.* 1984; Bryson and Wendland 1967; Knox 1976). The early/middle Holocene increase in the duration of Pacific air masses and the accompanying warmer and drier conditions are consistent with a northward displacement and intensification of zonal (east–west) circulation and the axis of strongest westerlies over eastern North America. Thermal contrasts along the southern edge of the retreating Laurentide ice sheet would have strengthened the belt of westerlies and pulled them northward until the ice sheet collapsed at approximately 6000 B.P., after which time circulation patterns would have adjusted to more or less modern conditions characterized by greater meridional (north–south) circulation. This would have led to an increase in the influence of both Arctic and Atlantic air masses in the Midwest.

Hypothesized changes in atmospheric circulation patterns derived from midwestern palynological data are consistent with paleoenvironmental reconstructions from the eastern United States. The hypothesized hypsithermal and an intensified zonal circulation in the Midwest may have forced Pacific air masses eastward, creating the warm/dry climatic conditions hypothesized in the eastern United States for this period. The hypothesized increase in meridional circulation beginning at about 6000 B.P. is supported by palynological evidence, suggesting both an increasing influence of Atlantic air masses, such as the maritime tropical air mass, and Arctic air masses. H. R. Delcourt (1979:277; Delcourt and Delcourt 1985:20) argues that the pine rise, which begins at approximately 5500 B.P. on the Atlantic and Gulf Coastal Plains, was caused by an "increased dominance of the tropical maritime air mass from the Gulf of Mexico, and due to high fire frequency resulting from lightning strikes." The southward migration of northern boreal taxa (especially spruce), beginning at 4000 B.P., can be attributed to an increase in the influence of Arctic air in New England and the Middle Atlantic region.

This model of Holocene climatic change has been evaluated with geomorphological data using an approach developed by Knox (1976; see also Brackenridge 1980, 1981; Knox 1983, 1985; Knox *et al.* 1975). Knox (1976:182) asserts that the maintenance of a graded condition in fluvial systems depends to a large degree upon the recurrence intervals of large floods. Streams tend to maintain morphologic stability over time and are graded in the absence of large floods of high energy. Isolated large floods may temporarily cause a stream to become ungraded if thresholds of geomorphological stability are exceeded. In the absence of another large flood in the next several years, the stream will quickly regrade itself and maintain a stable channel morphology adjusted to the characteristics of moderate magnitude and relatively frequent floods. However, if the recurrence interval of large floods is sufficiently small, severe alterations in channel morphology can occur.

The recurrence interval of large floods is strongly influenced by patterns of upper atmospheric circulation. Knox (1976:183) argues that strong meridional circulation increases the variability of climatic and hydrological events, most notably large-scale floods, particularly in the middle latitudes. Meridional circulation favors deep penetration of polar air masses into low latitudes and equally deep penetration of tropical air masses into high latitudes. This frequently results in the development of intense cyclones and storms in the middle latitudes, followed by large floods. In contrast, the strong westerly component of zonal circulation diminishes the frequency of intense cyclones and storms in the middle latitudes. Streams will eventually adjust to particular climatic and hydrological conditions so that the occurrence of ungraded streams, as represented by major morphological adjustments and/or episodes of alluviation and degradation, should follow periods of change in upper atmospheric circulation patterns that exceed thresholds of geomorphological stability (see also Schumm 1976). Therefore, according to this model, these periods of geomorphological instability in stream channels should involve a shift toward an increased meridional component in atmospheric circulation (Brakenridge 1980:656, 1981:75; Knox 1976; Knox *et al.* 1975).

Knox (1976) analyzed 802 radiocarbon-dated stratigraphic horizons in floodplains, terraces, and alluvial fans for sites in the middle latitudes of the Northern Hemisphere. An effort was made not to include archaeological dates, especially those associated with deposits disturbed by humans (e.g., refuse pits, fire pits). His data suggest a period of stability between 8500 and 4600 B.P., ending with an episode of channel instability. This pattern is consistent with hypothesized changes in upper atmospheric circulation patterns during the Holocene (also see Brakenridge 1980, 1981).

Recent studies of alluvial sequences in the Midwest and in the eastern United States are consistent with the patterns found by Knox (1976). Sequences of early to middle Holocene alluvial stability, followed by periods of alluvial erosion and deposition dated between 6000 and 4500 B.P., have been noted throughout the Midwest (Brakenridge 1980, 1981; Butzer 1978; Knox 1983, 1985; Knox and Johnson 1974; McDowell 1983). In the eastern United States, Scully and Arnold (1981) have noted a period of channel stability dated between 8000 and 4700 B.P., followed by a period of vertical accretion and, by 3000 B.P., channel incision and lateral migration, in the upper Susquehanna River basin, New York. Carbone *et al.* (1982) found a thick paleo-B Horizon suggestive of a prolonged period of stability dating from the early to middle Holocene in the Savannah River Valley, which was followed by a period of fast deposition of up to 3 m of medium to fine sand during the middle Holocene. The variation in dates for this shift in alluvial depositional sequences may be due to variations in geomorphological thresholds (Schumm 1976) or to the geographic position of these sites in relation to changing air mass patterns.

The geomorphological data presented here are consistent with the explanatory model of climatic change generated from palynological data. However, it must be emphasized that alluvial depositional sequences may be difficult to interpret accurately and may not provide adequate information in the absence of other data to determine the causal factors responsible for the sequence. For example, three different factors may be responsible for gaps in an alluvial sequence when no dated deposits occur in a particular period: (1) the channel may have been relatively stable, with only low-level fluvial activity; (2) the gap may have been the result of continuous net erosion; or (3) intense net erosion may have occurred near the end of the gap (McDowell 1983:101). Furthermore, it has been suggested that the period of middle Holocene channel instability resulted from hemlock die-off, causing increased sediment loads due to deforestation and intensified upland erosion (Scully and Arnold 1981:341; see also Whitehead 1979:353). However, the consistency between the palynological and the geomorphological data, and the widespread occurence of this shift beyond the range of middle Holocene hemlock disruptions, argue for a climatic rather than a biotic cause.

6. CONCLUSIONS

The paleoenvironmental reconstruction generated by Carbone (1976) was an important step in gaining an understanding of past human/environment interactions in the Middle Atlantic region. The present paper has continued this effort by using recent paleoecological research to build on earlier work. The revised paleoenvironmental reconstruction presented here has benefited from recent developments in paleocological methodology and epistemology, as well as from a larger number of dated pollen cores and sedimentological profiles. Analysis of the migration patterns of individual taxa has been stressed, which allows the potential separation of climatic factors from other factors that affect vegetational distributions.

The reconstruction presented here differs from Carbone's primarily in the nature and timing of early/middle Holocene environmental change. The duration of warm/dry conditions in the Middle Atlantic region and throughout eastern North America is reconstructed as having occurred earlier than the xeriothermic hypothesized by Carbone (1976:76), which was dated to the period from 6500 to 3200 B.P. Instead, maximum Holocene warmth and dryness occurred in the Middle Atlantic region from approximately 9000 to 5000–4000 B.P., followed by an adjustment toward modern conditions. Vegetational communities were dominated by oaks throughout this period, with a general increase in diversity as taxa such as beech, hickory, and chestnut migrated into the region from the south. The Coastal Plain pine expansion and the paneastern hemlock decline at approximately 5000 B.P. were probably the most significant vegetation trends, and undoubtedly caused substantial changes in particular habitats within the region. There are no indications of increases in grasses or other nonarboreal species during this time period in the Middle Atlantic region, except as a result of local edaphic or hydrological conditions not directly related to climatic change. Palynological and geomorphological data in the East are consistent with hypotheses generated from an atmospheric model of Holocene climatic change in North America (i.e., Knox 1976). This reconstruction suggests that warm/dry conditions on the East Coast of the United States were contemporaneous with the hypsithermal in the Midwest. The conclusions reached here concerning early/middle Holocene paleoenvironmental change also suggest that archaeological models that rely on a midpostglacial xerothermic as a causative factor to explain Late Archaic culture change in the Middle Atlantic region (Catlin *et al.* 1983; Custer 1978, 1984a,b, 1986a,b; Custer and Stewart 1983; Custer and Watson 1986) will have to be revsed (Cavallo and Joyce 1985). Fortunately, Custer

(1986b) has begun to deemphasize the xerothermic in his model of Late Archaic culture change.

In the Middle Atlantic region, the collection of high-quality palynological data should be the most important short-term goal of paleoenvironmental research, especially cores with multiple radiocarbon dates. An approach should be used that emphasizes the integration of various forms of paleoenvironmental data and examines data from particular sites in terms of broader regional patterns. Research currently being conducted at the Abbott Farm National Landmark in Trenton, New Jersey, is achieving this type of integrated approach (Mueller 1985; Russell 1985). Furthermore, research aimed at identifying rigorous paleoclimatic indicators, such as the fluctuations of the Prairie Peninsula in the Midwest, must be conducted in the East (Watts 1979:464). Davis (Davis *et al.* 1980; Liken and Davis 1975) study of altitudinal movements of arboreal taxa on mountains in New England and Watts' (1979) examination of the Coastal Plain pine expansion are examples of promising research in this direction. This type of research requires that the relationship between palynological, geomorphological, and edaphic data must be more accurately defined.

The problems of paleoenvironmental reconstruction discussed here are not confined to the Middle Atlantic region. There are few areas where sufficient paleoecological data have been collected to produce a reasonable picture of Holocene paleoenvironmental change. Such locations are confined largely to the northern Midwestern United States (Bernabo and Webb 1977; Webb and Bryson 1972; Wright 1981) and northwestern Europe (Walker and West 1970), where Pleistocene glaciation created numerous depositional basins that preserve pollen. However, even in these regions, paleoenvironmental inferences remain problematic, making reconstructions tentative hypothetical models that require continued research (Bernabo 1981:158; Bernabo and Webb 1977:92; Webb 1980:769–771).

Recent developments in paleoecology are beginning to provide explanations for paleoclimatic change that integrate data from over wide areas, while allowing for increasingly more specific inferences. Atmospheric models (e.g., Bartlein *et al.* 1984; Bryson and Wendland 1967; Knox 1976; Otto-Bliesner *et al.* 1982) have proven useful for addressing long-term (i.e., millennial) trends in temperature and precipitation during the Holocene. Recent results from the application of the astronomical theory of climatic change (Bryson and Swain 1981; O. K. Davis 1984; Kutzbach 1981, 1983; Kutzbach and Otto-Bliesner 1982; Ritchie *et al.* 1983; Webb 1986) have also been highly promising, especially for the elucidation of long-term fluctuations in seasonal patterns of solar radiation affecting temperature.

By synthesizing theories such as these, which have been proven useful in understanding Holocene environments, a more comprehensive theory of Holocene climatic change may emerge that will allow researchers to discuss climatic fluctuations at a variety of temporal and spatial scales. Resolving debate over the degree to which migrational patterns can be explained by complex vegetational responses or by complex climatic fluctuations will allow researchers to make more specific and accurate inferences concerning the magnitude and complexity of climatic change at smaller temporal and spatial scales than are currently possible (Brubaker 1986; Bryson 1985; Davis and Botkin 1985; Davis *et al.* 1986; Lewin 1985; Pennington 1986; Prentice 1986; Ritchie 1986; Webb 1986). Methodological developments such as the use of phytoliths, diatoms, and chemical isotopes, as well as the use of computers to aid in paleoenvironmental reconstructions (Solomon and Webb 1985), will also add to our understanding of Holocene paleoclimatic change.

Archaeologists can contribute greatly to the development of more accurate paleoenvironmental reconstructions by continuing to collect and utilize paleoenvironmental data. However, it must be stressed that such data collected from archaeological sites may reflect human impact rather than climatic change. In regions where paleoenvironmental data are particularly sparse, data from archaeological contexts may be unwisely used for paleoclimatic reconstructions (e.g., Schoenwetter 1974; Schoenwetter and Flannery 1970). While these data may not be particularly useful for inferences concerning climatic change (Brown 1985:76), they can be invaluable for providing information concerning the type of human/environment interactions that interest ecologically oriented archaeologists. A classic example of this type of research is the Little Tennesee River Valley Project. In this study, Delcourt *et al.* (1986) integrated paleoethnobotanical, paleoecological, and archaeological data to provide a comprehensive picture of long-term human impact on the local environment:

> The Little Tennessee River Valley provides a case study of long-term human impact upon native ecosystems that is without precedent in eastern North America. Rather than a unique and localized situation, however, mid- and late-Holocene land clearance by Indians must have occurred along other major river valleys. (P. A. Delcourt *et al.* 1986:348)

This study should be used as a prototype for further, similarly focused research by archaeologists.

Paleoenvironmental reconstructions will continue to be revised as developments in paleoecological research continue and new data emerge. This does require that archaeologists remain flexible and open-minded in regard to the use of any paleoenvironmental reconstruction. Archaeologists must keep informed of recent developments within the discipline of paleoecology to utilize their data more effectively and accurately. The recent trend away from paleoenvironmental models that focus on broad climatic–vegetational correlations and "episodic change" and toward a concern with the migrational patterns of individual taxa and continuous change is a significant development in paleoecology that has not been adequately considered by most Middle Atlantic region archaeologists. Archaeologists must work together with paleoecologists to develop multidisciplinary research strategies that allow the strengths of each discipline to augment one another. Paleoenvironmental research must be conducted cautiously and deliberately, and paleoenvironmental reconstructions should be treated with appropriate skepticism. Only in this way can archaeologists generate rigorous explanations of prehistoric human–environment interactions at a variety of temporal and spatial scales.

ACKNOWLEDGMENTS

I would like to thank Robert Blumenschine, Dena Dincauze, William Patterson III, and Thompson Webb III for comments on earlier drafts of this paper. Gail Ashley, John Cavallo, Deborah Fimbel, Kurt Kalb, George Nicholas, and Emily Russell provided helpful comments and discussion throughout the preparation of this paper.

7. REFERENCES

Antevs, E., 1948, Climatic Changes and Pre-White Man, *University of Utah Bulletin* 38:168–191.

Antevs, E., 1955, Geologic-Climatic Dating in the West, *American Antiquity* 20:317–335.

Bartlein, P. J., and Webb, T., III, 1982, Holocene Climate Changes Estimated from the Northern Midwest, in: *Quaternary History of the Driftless Area,* Field Trip Guide Book, No. 5, University of Wisconsin Extension, Geological and Natural History Museum, Madison, Wisconsin.

Bartlein, P. J., Webb, T., III, and Fleri, E., 1984, Holocene Climate Change in the Northern Midwest: Pollen-Derived Estimates, *Quaternary Research* 22:361–374.

Bernabo, J. C., 1981, Quantitative Estimates of Temperature Changes over the Last 2700 Years in Michigan Based on Pollen Data, *Quaternary Research* 15:143–159.

Bernabo, J. C., and Webb, T., III, 1977, Changing Patterns in the Holocene Pollen Record of Northeastern North America: A Mapped Summary, *Quaternary Research* 8:64–96.

Brakenridge, G. R., 1980, Widespread Episodes of Stream Erosion during the Holocene and Their Climatic Cause, *Nature* (London) 283:655–656.

Brakenridge, G. R., 1981, Late Quaternary Floodplain Sedimentation along the Pomme de Terre River, Southern Missouri, *Quaternary Research* 15:62–76.

Braun, E. L., 1950, *Deciduous Trees of Eastern North America,* Blakiston, Philadelphia.

Brown, R. B., 1985, A Summary of Late-Quaternary Pollen Records from Mexico West of the Isthmus of Tehuantepec, in: *Pollen Records of Late-Quaternary North American Sediments* (V. M. Bryant and R. G. Holloway, eds.), American Association of Stratigraphic Palynologists Foundation, Dallas, Texas, pp. 71–94.

Brubaker, L. B., 1986, Responses of Tree Populations to Climatic Change, *Vegetatio* 67:119–130.

Brush, G. S., 1986, Geology and Paleoecology of Chesapeake Bay: A Long-Term Monitoring Tool for Management, *Journal of the Washington Academy of Science* 76:146–160.

Bryson, R. A., 1966, Air Masses, Streamlines and the Boreal Forest, *Geographical Bulletin* 8:228–269.

Bryson, R. A., 1970, The Character of Climatic Change, and the End of the Pleistocene, *AMQUA Abstract* (American Quaternary Association) 1:20–22.

Bryson, R. A., 1985, On Climatic Analogs in Paleoclimatic Reconstruction, *Quaternary Research* 23:275–286.

Bryson, R. A., and Swain, A. M., 1981, Holocene Variations of Monsoon Rainfall in Rajasthan, *Quaternary Research* 16:135–145.

Bryson, R. A., and Wendland, R. M., 1967, Tentative Climatic Patterns for Some Late Glacial and Postglacial Episodes in Central North America, in: *Life, Land, and Water* (W. Mayer-Oakes, ed.), University of Manitoba Press, pp. 271–298.

Bryson, R. A., Barreis, B. A., and Wendland, R. M., 1970, The Character of Late Glacial and Postglacial Changes, in: *Pleistocene and Recent Environments of the Central Great Plains* (W. Dort, Jr. and J. K. Jones, Jr., eds.), University of Kansas Press, Lawrence, Kansas, pp. 53–74.

Butzer, K. W., 1978, Changing Holocene Environments at the Koster Site: A Geo-Archaeological Perspective, *American Antiquity* 48:408–413.

Butzer, K. W., 1982, *Archaeology As Human Ecology,* Cambridge University Press, New York.

Carbone, V. A., 1976, *Environment and Prehistory in the Shenandoah Valley,* Ph.D. Dissertation, Catholic University of America [available from University Microfilms, Ann Arbor, Michigan].

Carbone, V. A., 1982, Environment and Society in Archaic and Woodland Times, in: *Practicing Environmental Archaeology: Methods and Interpretation* (R. W. Moeller, ed.), American Indian Archaeological Institute, Washington, Connecticut, pp. 39–52.

Carbone, V. A., Segoria, A., Foss, J., Sheehan, M., Whitehead, D., and Jackson, S., 1982, Paleoenvironmental Investigations along the Savannah River Valley, Richard B. Russell Dam and Lake Project, Paper presented at the 39th Annual Southeastern Archaeological Conference, Memphis, Tennessee.

Catlin, M., Custer, J. F., and Stewart, R. M., 1983, The Late Archaic in Virginia, *Quarterly Bulletin of the Archaeological Society of Virginia* 37:123–140.

Cavallo, J. A., and Joyce, A. A., 1985, Explanatory Models of Late Archaic Culture Change in the Middle Atlantic Region: An Evaluation, Paper presented at the 1985 Middle Atlantic Conference, Rehoboth Beach, Delaware.

Colinvaux, P. A., 1973, *Introduction to Ecology,* Wiley-Interscience, New York.

Connally, G. G., and Sirkin, L. A., 1969, Deglacial History of the Lake Champlain–Lake George Lowland, in: *Guidebook of the New York State Geological Association, 1969 Annual Meeting* (S. G. Barnett, ed.), New York State Geological Association, Plattsburg, New York, pp. 1–20.

Connally, G. G., and Sirkin, L. A., 1971, Luzerne Readvance near Glen Falls, New York, *Geological Society of America Bulletin* 82:989–1008.

Craig, A. J., 1969, Vegetational History of the Shenandoah Valley, Virginia, *Geological Society of America Special Paper* 123:283–296.

Cronon, W., 1983, *Changes in the Land: Indians, Colonists, and the Ecology of New England,* Hill and Wang, New York.

Curry, D. C., 1980, Burial of Late Archaic Coastal Plain Sites as a Result of Aeolian Deposition, Paper presented at the 1980 Middle Atlantic Archaeological Conference, Dover, Delaware.

Curry, D. C., and Custer, J. F., 1982, Holocene Climatic Change in the Middle Atlantic Area: Preliminary Observations from Archaeological Sites, Paper presented at the 1982 Middle Atlantic Archaeological Conference, Rehoboth Beach, Delaware.

Custer, J. F., 1978, Broadspears and Netsinkers, Paper presented at the 1978 Middle Atlantic Archaeological Conference, Rehoboth Beach, Delaware.

Custer, J. F., 1981, Report on Archaeological Research in Delaware FY-1981 by the Department of Anthropology, University of Delaware, Bureau of Archaeology and Historic Preservation, Division of Historical and Cultural Affairs, Dover, Delaware.

Custer, J. F., 1984a, *Delaware Prehistoric Archaeology,* University of Delaware Press, Newark, Delaware.

Custer, J. F., 1984b, The Paleoecology of the Late Archaic: Exchange and Adaptation, *Pennsylvania Archaeologist* 54(3–4):32–47.

Custer, J. F., 1986a, Prehistoric Use of the Chesapeake Estuary: A Diachronic Perspective, *Journal of the Washington Academy of Science* 76(3):146–160.

Custer, J. F., 1986b, Late Archaic Cultural Dynamics in the Central Middle Atlantic, Paper presented at the 1986 Meeting of the Eastern States Archaeological Federation, Wilmington, Delaware.

Custer, J. F., and Stewart, R. M., 1983, Maritime Adaptations in the Middle Atlantic Region of Eastern North America, Paper presented at the 1983 Meeting of the Society for American Archaeology, Pittsburgh, Pennsylvania.

Custer, J. F., and Wallace, E. B., 1982, Patterns of Resource Distribution and Archaeological Settlement Patterns in the Piedmont of the Middle Atlantic Region, *North American Archaeologist* 3(2):139–152.

Custer, J. F., and Watson, S. C., 1986, Making Cultural Paleoecology Work: An Example from Northern Delaware, Paper presented at the 1986 Middle Atlantic Archaeological Conference, Wilmington, Delaware.

Davis, M. B., 1969, Climate Changes in Southern Connecticut Recorded by Pollen Deposition at Rogers Lake, *Ecology* 50:409–422.

Davis, M. B., 1976, Pleistocene Biogeography of Temperate Deciduous Forests, *Geoscience and Man* 13:13–26.

Davis, M. B., 1981a, Quaternary History and the Stability of Forest Communities, in: *Forest Succession* (D. C. West, H. H. Shugart, and D. B. Botkin, eds.), Springer-Verlag, New York, pp. 132–153.

Davis, M. B., 1981b, Mid-Holocene Hemlock Decline: Evidence for a Pathogen or Insect Outbreak, in: *Geobotany II* (R. C. Romans, ed.), Plenum Press, New York, pp. 253–254.

Davis, M. B., 1983, Holocene Vegetational History of the Eastern United States, in: *Late-Quaternary Environments of the United States* (H. E. Wright, Jr., ed.), University of Minnesota Press, Minneapolis, Minnesota, pp. 166–181.

Davis, M. B., and Botkin, D. B., 1985, Sensitivity of Cool-Temperature Forest and their Fossil Pollen Record to Rapid Temperature Change, *Quaternary Research* 23:327–340.

Davis, M. B., Spear, R. W., and Shane, L. C. K., 1980, Holocene Climate of New England, *Quaternary Research* 14:240–250.

Davis, M. B., Woods, K. D., Webb, S. L., and Futyma, R. P., 1986, Dispersal Versus Climate: Expansion of *Fagus* and *Tsuga* into the Upper Great Lakes Region, Vegetatio 67(2):93–104.

Davis, O. K., 1984, Multiple Thermal Maxima during the Holocene, *Science* 225:617–619.

Dean, W. E., Bradbury, J. P., Anderson, R. Y., and Barnosky, C. W., 1984, The Variability of Holocene Climatic Change: Evidence from Varved Lake Sediments, *Science* 226:1191–1194.

Deevey, E. S., Jr., 1939, Studies on Connecticut Lake Sediments: A Postglacial Climatic Chronology for Southern New England, *American Journal of Science* 237:691–724.

Deevey, E. S., Jr., 1949, Biogeography of the Pleistocene: Part I. Europe and North America, *Geological Society of America Bulletin* 60:1315–1416.

Deevey, E. S., Jr. and Flint, R. F., 1957, Postglacial Hypsithermal Interval, *Science* 125:182–184.

Delcourt, H. R., 1979, Late Quaternary Vegetation History of the Eastern Highland Rim and Adjacent Cumberland Plateau of Tennessee, *Ecological Monographs* 49:255–280.

Delcourt, H. R., and Delcourt, P. A., 1985, Quaternary Palynology and Vegetational History of the Southeastern United States, in: *Pollen Records of Late-Quaternary North American Sediments* (V. M. Bryant and R. G. Holloway, eds.), American Association of Stratigraphic Palynologists Foundation, Dallas, Texas, pp. 1–38.

Delcourt, H. R., Delcourt, P. A., and Webb, T., III, 1983 Dynamic Plant Ecology: The Spectrum of Vegetational Change in Space and Time, *Quaternary Science Reviews* 1:153–175.

Delcourt, P. A., 1978, *Quaternary Vegetation History of the Gulf Coastal Plain,* Ph.D. Dissertation, University of Minnesota, Minneapolis, Minnesota [available from University Microfilms, Ann Arbor, Michigan].

Delcourt, P. A., 1980, Goshen Springs: Late Quaternary Vegetation Record for Southern Alabama, *Ecology* 61:371–386.
Delcourt, P. A., and Delcourt, H. R., 1977, Presettlement Magnolia-Beech Climax of the Gulf Coastal Plain: Quantitative Evidence from the Apalachicola River Bluffs, Northcentral Florida, *Ecology* 58:1085–1093.
Delcourt, P. A., and Delcourt, H. R., 1979, Late Pleistocene and Holocene Distributional History of the Deciduous Forest in the Southeastern United States, *Veroffentlichungen des Geobotanischen Innstitutes der ETH, Stiftung Rubel* (Zurich) 68:79–107.
Delcourt, P. A., and Delcourt, H. R., 1981, Vegetation Maps for Eastern North America: 40,000 Years B.P. to the Present, in: *Geobotany II* (R. C. Romans, ed.), Plenum Press, New York, pp. 123–165.
Delcourt, P. A., and Delcourt, H. R., 1983, Late-Quaternary Vegetational Dynamics and Community Stability Reconsidered, *Quaternary Research* 19:265–271.
Delcourt, P. A., Delcourt, H. R., Bristler, R. C., and Lackey, L. E., 1980, Quaternary Vegetation History of the Mississippi Embayment, *Quaternary Research* 13:111–132.
Delcourt, P. A., Delcourt, H. R., Cridlebaugh, P. A., and Chapman, J., 1986, Holocene Ethnobotanical and Paleoecological Record of Human Impact in the Little Tennessee River Valley, Tennessee, *Quaternary Research* 25:330–349.
Dent, R. J., 1979, *Ecological and Sociocultural Reconstruction in the Upper Delaware Valley,* Ph.D. Dissertation, American University, Washington, D.C. [available from University Microfilms, Ann Arbor, Michigan].
Dent, R. J., 1982, Amerind Society and the Environment: Evidence from the Upper Delware Valley, Unpublished.
Dincauze, D. F., 1981, Paleoenvironmental Reconstruction in the Northeast: The Art of Multidisciplinary Science, in: *Foundations of Northeast Archaeology* (D. Snow, ed.), Academic Press, New York, pp. 51–96.
Dincauze, D. F., 1985, Bravura Piece: A Review of *Delware Prehistoric Archaeology* by J. F. Custer, *Quarterly Review of Archaeology* 6(2):13–15.
Edwards, R. L., and Merrill, A. S., 1977, A Reconstruction of the Continental Shelf Areas of Eastern North America for the Times 9000 B.P. and 12,500 B.P., *Archaeology of Eastern North America* 5:1–42.
Evans, J. G., 1978, *An Introduction to Environmental Archaeology,* Cornell University Press, Ithaca, New York.
Frey, D. G., 1953, Regional Aspects of the Late Glacial and Post-Glacial Pollen Succession of Southeastern North Carolina, *Ecological Monographs* 23:289–313.
Gardner, W. M., 1978, Comparison of Ridge and Valley, Blue Ridge, Piedmont, and Coastal Plain Archaic Period Site Distributions: An Idealized Transect, Paper presented at the 1978 Middle Atlantic Archaeological Conference, Rehoboth Beach, Delaware.
Gardner, W. M., 1980, Settlement-Subsistence Strategies in the Middle and South Atlantic Portions of the Eastern United States during the Late Pleistocene and Early Holocene, Paper presented at the 1980 American Anthropological Association Meetings, Washington, D.C.
Gaudreau, D. C., and Webb, T., III, 1985, Late-Quaternary Pollen Sratigraphy and Isochrone Maps for the Northeastern United States, in: *Pollen Records of Late-Quaternary North American Sediments* (V. M. Bryant and R. G. Holloway, eds.), American Association of Stratigraphic Palynologists Foundation, Dallas, Texas, pp. 245–280.
Gehris, C. W., 1964, *Pollen Analysis of the Cranberry Bog Preserve, Tannersville, Monroe County, Pennsylvania,* Ph.D. Dissertation, Pennsylvania State University, State College, Pennsylvania [available from University Microfilms, Ann Arbor, Michigan].
Harlow, W. M., and Harrar, E. S., 1958, *Textbook of Dendrochronology,* McGraw-Hill, New York.
Harmon, K. P., 1968, *Late Pleistocene Forest Succession in Northern New Jersey,* Ph.D. Dissertation,

Rutgers University, New Brunswick, New Jersey [available from University Microfilms, Ann Arbor, Michigan].

Hutchinson, D. R., Ferrebee, W. M., Knebel, H. J., Wold, R. J., and Isachsen, Y. W., 1981, The Sedimentary Framework of the Southern Basin of Lake George, New York, *Quaternary Research* 15:44–61.

Jackson, S. T., 1981, Late-Glacial and Postglacial Local Vegetational Changes in the Adirondack Mountains, New York: The Macrofossil Record, *Ecological Society of America Bulletin* 62:126.

Jacobson, G. L., Jr. and Bradshaw, R. H. W., 1981, The Selection of Sites for Paleovegetational Studies, *Quaternary Research* 16:80–96.

King, J. E., and Allen, W. H., Jr., 1977, A Holocene Vegetation Record from the Mississippi River Valley, Southeastern Missouri, *Quaternary Research* 8:307–323.

Knox, J. C., 1972, Valley Alluviation in Southwestern Wisconsin, *Annals of the Association of American Geography* 62:401–410.

Knox, J. C., 1976, Concept of the Graded Stream, in: *Theories of Landform Development* (W. N. Melhorn and R. C. Flemal, eds.), Publications in Geomorphology, State University of New York, Binghampton, New York, pp. 169–198.

Knox, J. C., 1983, Responses of River Systems to Holocene Climates, in: *Late-Quaternary Environments of the United States* (H. E. Wright, Jr., ed.), University of Minnesota Press, Minneapolis, Minnesota, pp. 26–41.

Knox, J. C., 1985, Responses of Floods to Holocene Climatic Change in the Upper Mississippi Valley, *Quaternary Research* 23:287–300.

Knox, J. C., and Johnson, W. C., 1974, Late Quaternary Valley Alluviation in the Driftless Area of Southwestern Wisconsin, in: *Late Quaternary Environments of Wisconsin* (J. C. Knox and W. C. Johnson, eds.), American Quaternary Association Guidebook, 3rd Biennial Meeting, Madison, Wisconsin, pp. 134–162.

Knox, J. C., Bartlein, P. J., Hirschboeck, K. K., and Muckenhirn, R. J., 1975, The Response of Floods and Sediment Yields to Climatic Variation and Land Use in the Upper Mississippi Valley, Institute for Environmental Studies, Report 52, University of Wisconsin, Madison, Wisconsin.

Kutzbach, J. E., 1981, Monsoon Climate of the Early Holocene: Climate Experiment with the Earth's Orbital Parameters for 9000 Years Ago, *Science* 214:59–61.

Kutzbach, J. E., 1983, Modeling of Holocene Climates, in: *Late-Quaternary Environments of the United States* (H. E. Wright, Jr., ed.), University of Minnesota Press, Minneapolis, Minnesota, pp. 271–277.

Kutzbach, J. E., and Otto-Bliesner, B., 1982, The Sensitivity of the African–Asian Monsoonal Climate to Orbital Parameter Changes for 9000 Years B.P. in a Low Resolution General Circulation Model, *Journal of Atmospheric Science* 39:1177–1188.

Lewin, R., 1985, Plant Communities Resist Climatic Change, *Science* 228:165–166.

Likens, G. E., and Davis, M. B., 1975, Post-Glacial History of Mirror Lake and its Watershed in New Hampshire, USA: An Initial Report, *Verhanglungen Internationale Vereinigung für Theoretische und Angewandte Limnologie* 19:982–993.

Maxwell, J., and Davis, M. B., 1972, Pollen Evidence of Pleistocene and Holocene Vegetation in the Allegheny Plateau, Maryland, *Quaternary Research* 2:506–530.

McDowell, P. F., 1983, Evidence of Stream Response to Holocene Climatic Change in a Small Wisconsin Watershed, *Quaternary Research* 19:100–116.

McNamara, J., 1981, The Effects of Aeolian Processes on the Late Archaic Components at the Baldwin Site in Anne Arundel County, Maryland, Paper presented at the 1981 Middle Atlantic Archaeological Conference, Ocean City, Maryland.

McNamara, J., 1983, The Effects of Mid Post-Glacial Changes at Conowinga: A Stratified Late Archaic through Late Woodland Site in the Piedmont Floodplain of the Lower Susquehanna River Valley,

Paper presented at the 1983 Middle Atlantic Archaeological Conference, Rehoboth Beach, Delaware.

Mouer, L. D., 1986, The Lullabye of Broadspears: The Archaic/Woodland Transition in the James River Valley, Paper presented at the 1986 Meeting of the Eastern States Archaeological Federation, Wilmington, Delaware.

Mouer, L. D., Ryder, R. L., and Johnson, E. G., 1981, Down to the River in Boats: The Late Archaic/Transitional in the Middle James River Valley, Virginia, *Quarterly Bulletin of the Archaeological Society of Virginia* 36(1–2):29–48.

Mueller, R. G., 1985, Environmental Reconstruction: Natural Levee Position (23 Me 1-B), Ms. on file, Louis Berger and Associates, East Orange, New Jersey.

Nicholas, J., 1968, *Late Pleistocene Palynology of Southeastern New York and Northern New Jersey,* Ph.D. Dissertation, New York University, New York [available from University Microfilms, Ann Arbor, Michigan].

Niering, W. A., 1953, Past and Present Vegetation of High Point State Park, New Jersey, *Ecological Monographs* 23(2):127–148.

Ogden, J. G., 1965, Pleistocene Pollen Records for Eastern North America, *Botanical Review* 31:481–504.

Ogden, J. G., 1977, Limiting Factors in Paleoenvironmental Reconstruction, in: *Geobotany* (R. C. Romans, ed.), Plenum Press, New York, pp. 29–42.

Orlove, B. S., 1980, Ecological Anthropology, *Annual Review of Anthropology* 9:235–273.

Otto-Bliesner, B. L., Grantstator, G. W., and Houghton, D. D., 1982, A Global Low-Order Spectral General Circulation Model: Part I. Formation and Seasonal Climatology, *Journal of Atmospheric Science* 39:929–948.

Overpeck, J. T., Webb, T., III, and Prentice, I. C., 1985, Quantitative Interpretations of Fossil Pollen Spectra: Dissimilarity Coefficients and the Method of Modern Analogs, *Quaternary Research* 23:87–108.

Owens, J. P., Stefansson, K., and Sirkin, L. A., 1974, Chemical, Mineralogic, and Palynologic Character of the Upper Wisconsinan-Lower Holocene Fill in Parts of the Hudson, Delaware, and Chesapeake Estuaries, *Journal of Sedimentological Petrology* 44:390–408.

Pennington, W., 1986, Lags in Adjustment of Vegetation to Climate Caused by the Pace of Soil Development: Evidence from Britain, *Vegetatio* 67(2):105–118.

Prentice, I. C., 1985, Pollen Representation, Source Area, and Basin Size: Toward a Unified Theory of Pollen Analysis, *Quaternary Research* 23:76–86.

Prentice, I. C., 1986, Vegetation Responses to Past Climatic Variation, *Vegetatio* 67(2):131–141.

Rasmussen, W. C., 1958, *Geology and Hydrology of the "Bays" and Basins of Delaware,* Ph.D. Dissertation, Brywn Mawr College [available from University Microfilms, Ann Arbor, Michigan].

Richerson, P. J., 1977, Ecology and Human Ecology: A Comparison of Theories in the Biological and Social Sciences, *American Ethnologist* 4:1–26.

Ritchie, J. C., 1986, Climate Change and Vegetation Response, *Vegetatio* 67(2):65–74.

Ritchie, J. C., Cwynwar, L. C., and Spear, R. W., 1983, Evidence from North-West Canada for an Early Holocene Milankovitch Thermal Maximum, *Nature* 305:126–128.

Russell, E. W. B., 1983, Indian-Set Fires in the Forests of the Northeastern United States, *Ecology* 64(1):78–88.

Russell, E. W. B., 1985, Paleoecological Reconstruction of the Abbott Farm Site, Ms. on file, Louis Berger and Associates, East Orange, New Jersey.

Schoenwetter, J., 1974, Pollen Records of Guila-Naquitz Cave, *American Antiquity* 39(2):292–303.

Schoenwetter, J., and Flannery, K. V., 1970, Climate and Man in Formative Oaxaca, *Archaeology* 23:144–152.

Schumm, S. A., 1976, Episodic Erosion: A Modification of the Geomorphic Cycle, in: *Theories of*

Landform Development (W. N. Melhorn and R. C. Flemal, eds.), Publications in Geomorphology, State University of New York, Binghampton, New York, pp. 69–85.

Scully, R. W., and Arnold, R. W., 1981, Holocene Alluvial Stratigraphy in the Upper Susquehanna River Basin, New York, *Quaternary Research* 15:327–344.

Sears, P. B., 1935, Types of North American Pollen Profiles, *Ecology* 16:488–499.

Sears, P. B., 1942a, Xerothermic Theory, *Botanical Review* 8:708–736.

Sears, P. B., 1942b, Postglacial Migration of Five Forest Genera, *American Journal of Botany* 29:684–691.

Sears, P. B., 1948, Forest Sequence and Climatic Change in Northeastern North America Since Early Wisconsin Time, *Ecology* 29(2):326–333.

Sirkin, L. A., 1977, Late Pleistocene Vegetation and Environments in the Middle Atlantic Region, *Annals of the New York Academy of Science* 288:206–217.

Solomon, A. M., and Webb, T., III, 1985, Computer-Aided Reconstruction of Late-Quaternary Landscape Dynamics, *Annual Review of Ecology and Systematics* 16:63–84.

Spear, R. W., 1981, *History of Alpine and Subalpine Vegetation in the White Mountains, New Hampshire,* Ph.D. Dissertation, University of Minnesota, Minneapolis, Minnesota [available from University Microfilms, Ann Arbor, Michigan].

Stewart, R. M., 1983, Soils and the Prehistoric Archaeology of Abbott Farm, *North American Archaeologist* 4(1):27–49.

Thomas, R. A., 1981, Excavations at the Delaware Park Site (7NC-E-41), Delaware Department of Transportation, Dover, Delaware.

Walker, D., and West, R. G., (eds.), 1970, *Studies in the Vegetational History of the British Isles,* Cambridge University Press, Cambridge.

Walker, P. D., and Hartman, R. T., 1960, The Forest Sequence of the Hartsown Bog Area in Western Pennsylvania, *Ecology* 41:461–474.

Watts, W. A., 1979, Late Quaternary Vegetation of Central Appalachia and the New Jersey Coastal Plain, *Ecological Monographs* 49:427–469.

Watts, W. A., 1980, Late-Quaternary Vegetation History at White Pond on the Inner Coastal Plain of South Carolina, *Quaternary Research* 13:187–199.

Watts, W. A., and Winter, T. C., 1966, Plant Macrofossils from Krichner Marsh, Minnesota: A Paleoecological Study, *Geological Society of America Bulletin* 77:1339–1359.

Webb, T., III, 1980, The Reconstruction of Climatic Sequences from Pollen Data, *Journal of Interdisciplinary History* 10:749–772.

Webb, T., III, 1981, The Past 11,000 Years of Vegetational Change in Eastern North America, *Bioscience* 31:501–506.

Webb, T., III, 1982, Temporal Resolution in Holocene Pollen Data, in: *Proceedings of the Third North American Paleontological Convention,* Vol. 2, pp. 569–572.

Webb, T., III, 1986, Is the Vegetation in Equilibrium with Climate: How to Interpret Late-Quaternary Pollen Data, *Vegetatio* 67(2):75–92.

Webb, T., III, and Bryson, R. A., 1972, Late and Postglacial Climatic Change in the Northern Midwest, U.S.A.: Quantitative Estimates Derived from Fossil Pollen Spectra by Multivariate Statistical Analysis, *Quaternary Research* 2:70–115.

Webb, T., III, Cushing, E. J., and Wright, H. E., Jr., 1983, Holocene Changes in the Vegetation of the Midwest, in: *Late-Quaternary Environments of the United States* (H. E. Wright, Jr., ed.), University of Minnesota Press, Minneapolis, Minnesota, pp. 142–165.

Wendland, W., and Bryson, R. A., 1974, Dating Climatic Episodes of the Holocene, *Quaternary Research* 4:9–24.

West, D. C., Shugart, H. H., and Botkin, D. B., 1981, *Forest Succession,* Springer-Verlag, New York.

Whitehead, D. R., 1965, Palynology and Pleistocene Phytogeography of Unglaciated Eastern North

America, in: *The Quaternary of the United States* (H. E. Wright, Jr. and D. G. Frey, eds.), Princeton University Press, Princeton, New Jersey, pp. 417–432.

Whitehead, D. R., 1973, Late Wisconsin Vegetational History in Unglaciated Eastern North America, *Quaternary Research* 3:621–631.

Whitehead, D. R., 1979, Late-Glacial and Post-Glacial Vegetational History of the Berkshires, Western Massachusetts, *Quaternary Research* 12:333–357.

Whitehead, D. R., Sheehan, M. C., and Jackson, S. T., 1981, Late-Glacial and Postglacial Vegetational Changes in the Adirondack Mountains, New York: The Pollen Record, *Ecological Society of America Bulletin* 62:126.

Winkler, M. G., 1982, *Late-Glacial and Postglacial Vegetational History of Cape Cod and the Paleolimnology of Duck Pond, South Wellfleet, Massachusetts,* M.Sc. Thesis, University of Wisconsin, Madison, Wisconsin, unpublished.

Winkler, M. G., 1985, A 12,000 Year History of Vegetation and Climate for Cape Cod, Massachusetts, *Quaternary Research* 23:301–312.

Winkler, M. G., Swain, A. M., and Kutzbach, J. E., 1986, Middle Holocene Dry Period in the Northern Midwestern United States: Lake Levels and Pollen Stratigraphy, *Quaternary Research* 25:235–250.

Witthoft, J., 1953, Broadspear Points and the Transitional Period Cultures, *Pennsylvania Archaeologist* 23(1):4–31.

Wright, H. E., Jr., 1981, Vegetation East of the Rocky Mountains 18,000 Years Ago, *Quaternary Research* 15:113–125.

Chapter 9

The Distribution of Late Quaternary Forest Regions in the Northeast

Pollen Data, Physiography, and the Prehistoric Record

DENISE C. GAUDREAU

1. INTRODUCTION

Fossil pollen are the primary data on paleovegetational history for the last 100,000 years. Interfacing pollen data with data from other disciplines augments their usefulness for paleoenvironmental interpretation; maps are convenient for interfacing data bases when these data are scaled similarly in space and time (Webb and McAndrews 1976). For example, Dincauze and Mulholland (1977) mapped both pollen and archaeological data for New England to show how past vegetation influenced prehistoric settlement patterns. This paper updates pollen maps for this type of interdisciplinary study. Vegetational distributions for the last 14,000 years are summarized in contoured maps of pollen percentages. The pollen types mapped represent the major vegetational regions in northeastern United States and southeastern Canada. The maps show how pollen distributions reflect vegetational contrasts associated both with latitudinal/longitudinal gra-

DENISE C. GAUDREAU • Geology Department, Southampton College, Southampton, New York 11968.

dients and with topographic gradients. The topographic patterns in the pollen contours can be used to address paleoecological issues, including aspects of postglacial plant colonization. The topographic patterns also have significance for archaeological studies on human prehistory.

Features of vegetational history are better displayed in pollen maps, in contrast to pollen diagrams. A pollen diagram emphasizes temporal variation, while a map displays spatial features in pollen data. A series of pollen maps shows both temporal and spatial variation, and in a format that can be translated directly into features of past vegetation. Szafer's (1935) maps for Europe and Sears' (1942) maps for the midwestern United States represent early mapping efforts. Mapped summaries of vegetational history for the Northeast include isopoll maps (Bernabo and Webb 1977; Gaudreau 1986; Webb *et al.* 1983b), isochrone maps (Davis 1976, 1981b, 1983; Gaudreau and Webb 1985; Webb *et al.* 1983a,b), or isophyte maps (Davis and Jacobson 1985; Delcourt and Delcourt 1985; Delcourt and Delcourt 1981), which show, respectively, contours of pollen frequencies, time, or inferred vegetation.

Vegetational distributions at different temporal and spatial scales reflect the influence of different proximate factors. For example, continental or macroscale vegetation distributions reflect the broad latitudinal and longitudinal gradients in climate; in contrast, meso- and microscale distributional patterns reflect physiographic contrasts associated with substrate and with subclimatic effects imposed by topography or water bodies. The scale of pollen maps (site grid size and geographic area projected) can be varied to display and emphasize either the macroscale or mesoscale aspects of vegetational distribution. The scale of the pollen map thus influences the paleoenvironmental interpretations possible from pollen data (Delcourt *et al.* 1983).

Sufficient radiocarbon-dated pollen stratigraphies for the Northeast now permit mapped summaries of pollen data at the mesoscale. The mesoscale maps shown in this paper illustrate whether pollen can serve as a remote sensor (*sensu* Webb *et al.* 1978) of the vegetational contrasts associated with topographic contrasts in the Northeast or whether pollen transport blurs the mesoscale vegetational contrasts. This issue is not trivial, given the small size of the physiographic features of the Northeast relative to pollen-dispersal distances.

Resolution of mesoscale pattern in pollen distributions aids new comparisons of palynological data with the data of other disciplines. For example, archaeologists use physiography to differentiate human habitats in the prehistoric record; studies have compared the types and diversity of resources available for human subsistence in riverine versus highland environments (Curran and Dincauze 1977). Such inventories have served as a basis for archaeological models that attempt to correlate prehistoric settlement patterns in the Northeast with physiographic features and their associated resources (Dincauze and Mulholland 1977; Nicholas 1980, 1983). These studies highlight the need for a vegetational history of the Northeast at the mesoscale.

Maps of pollen distributions at the mesoscale display new features that can be used to address paleoecological issues (Gaudreau 1986); one current issue is whether exogenous factors (e.g., climate) or endogenous factors (e.g., inherent seed-dispersal rate) control plant distributions at 10^3-year intervals within 10^4-year periods. This issue and other paleoecological inferences relevant to archaeology are considered in the Discussion.

2. THE STUDY AREA

A major physiographic feature of northeastern North America is the Appalachian Mountain system, which extends from the Allegheny Mountains in the

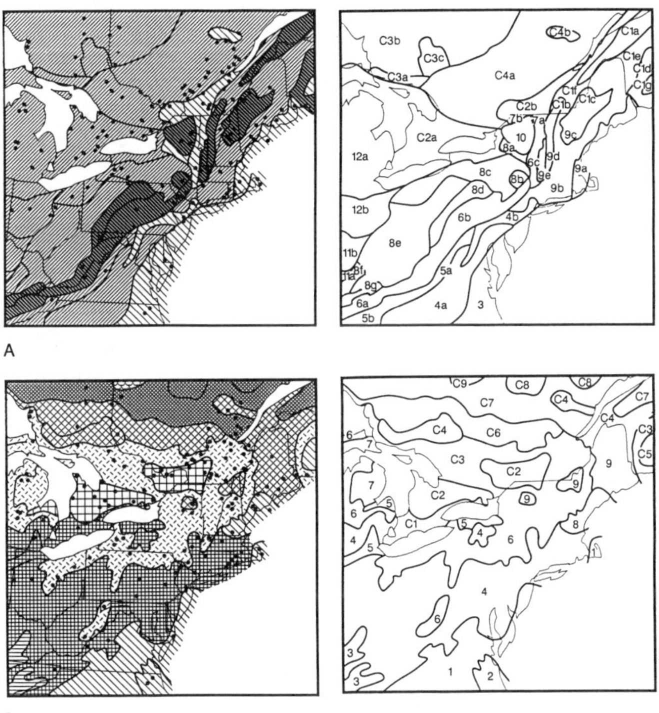

Figure 1. Map of pollen sites, physiographic regions, and vegetation regions for northeastern North America. (A) Map of the major physiographic regions of northeastern North America after Fenneman (1946) for the United States and after Fremlin (1974) for Canada. Numbers correspond to codes in Table 1. (B) Map of the major vegetation regions of northeastern North America after Eyre (1980) for the United States and after Fremlin (1974) for Canada. The principal forest regions are: deciduous forest, mixed conifer/northern-hardwood forest (including pine forests), boreal forest, and transitional regions. Numbers correspond to codes in Table 2.

southeastern United States to the Laurentide highlands in Canada. With the Adirondacks, the Appalachians comprise the major upland area, which grades into lowlands along river valleys, the Great Lakes, and the coastal plain (Figure 1 and Table 1).

The forest types of northeastern North America can be classified broadly as deciduous forests, mixed forests of conifers and hardwoods, and boreal forests (Braun 1950; Eyre 1980; Fremlin 1974; Küchler 1964; Lull 1968; Nichols 1935) (Figure 1 and Table 2). The maps in Figure 1 show that at the macroscale, the distributions of these vegetational regions reflect the latitudinal and longitudinal gradients associated with climatic gradients; and at the mesoscale, the distributions of vegetation types reflect the topographic and edaphic differences associated with major physiographic regions in northeastern North America. For example, the deciduous forests (e.g., nos. 4, 5, and C1 in Figure 1 and Table 2) dominate in the southern half of the Northeast; their distributions extend northward in the lowlands around the Great Lakes and in lowlands of major rivers such as the Hudson and St. Lawrence. Mixed forests of central or northern hardwoods (nos. 6, 7, 8, and C3) are distributed across the middle latitudes of northeastern North America and extend southernmost in the Appalachian highlands.

Mixed forests dominated by various pine species are found along the coast (nos. 1, 2, 8, and C5), in the southeastern United States (no. 3), and in Canada just northwest of the mapped region. In these mixed forests, pine co-occurs with oak in the southeast and along the coast and with northern hardwoods in the Canadian pine forests. Boreal forests (nos. C7 and C8) occur to the north of the mixed forests and predominate to the north of the region mapped in Figure 1. Transitional forests intergrade with the deciduous, mixed, and boreal forest regions; these transitional forests include taxa from deciduous and mixed forests (e.g., no. C2), or taxa from mixed and boreal forests (e.g., nos. 9, C4, C5, and C6).

The physiography of the Northeast has changed over the last 14,000 years due to the wasting of Wisconsinan ice sheets (Denton and Hughes 1981; Hughes *et al.* 1985; Quigley 1983). The ice reached a southern limit in the Northeast about 21,000–17,000 B.P. and sea level was lowered, exposing areas of the continental shelf. By 14,000 B.P., the ice had retreated to northern areas in New England and the Midwest; by about 6000 B.P., the ice sheet had disappeared. Proglacial lakes flooded lowlands of the Northeast during the glacial recession. In addition, between 12,000 and 8000 B.P., marine transgressions inundated recently deglaciated coasts (e.g., in Maine) and also inland areas (e.g., in the St. Lawrence and Champlain lowlands).

The ice front is the only paleogeographic feature mapped in Figures 3 and 4; its position is estimated from figures in Quigley (1983), Davis and Jacobson (1985), and Hughes *et al.* (1985).

Table 1. Physiographic Regions for Northeastern North America

United States (Fenneman 1946)					Canada (Fremlin 1974)				
Major division	Code[a]	Province	Code[a]	Section		Code[a]	Region	Code[a]	Division
Laurentian Upland	1	Superior Upland			The Borderlands	C1	Appalachian	C1a	Notre Dame Mountains
Atlantic Plain	2	(Continental Shelf)						C1b	Sutton Mountains
	3	Coastal Plain		Embayed				C1c	Megantic Hills
Appalachian Highlands	4	Piedmont	4a	Piedmont Upland				C1d	New Brunswick Highland
			4b	Piedmont Lowlands				C1e	Chaleur Uplands
	5	Blue Ridge	5a	Northern				C1f	Eastern Quebec Uplands
			5b	Southern				C1g	Maritime Plain
	6	Valley and Ridge	6a	Tennessee		C2	St. Lawrence	C2a	Western St. Lawrence Lowlands
			6b	Middle			Lowlands	C2b	Central St. Lawrence Lowlands
			6c	Hudson Valley	The Shield	C3	James	C3a	Penokean Hills
	7	St. Lawrence Valley	7a	Champlain				C3b	Abitibi Upland
			7b	Northern				C3c	Cobalt Plain
	8	Appalachian	8a	Mohawk		C4	Laurentian	C4a	Laurentian Highlands
		Plateaus	8b	Catskill				C4b	Lake St. Jean Lowland
			8c	Southern New York					
			8d	Allegheny Mountain					
			8e	Kanawha					
			8f	Cumberland Plateau					
			8g	Cumberland Mountain					
	9	New England	9a	Seaboard Lowland					
			9b	New England Upland					
			9c	White Mountain					
			9d	Green Mountain					
			9e	Taconic					
	10	Adirondack							
Interior Plains	11	Interior Low	11a	Highland Rim					
		Plateaus	11b	Lexington Plain					
	12	Central Lowland	12a	Eastern Lake					
			12b	Till Plains					

[a]Codes refer to labels in Figure 1.

Table 2. Vegetation Regions for Northeastern North America

United States (Eyre 1980)				Canada (Fremlin 1974)			
Vegetation region[a]	Code[b]	Forest type	Important taxa	Vegetation region[a]	Code[b]	Forest type	Important taxa
Southern forest (Southeastern conifer forest)	1	Loblolly–shortleaf pine	*Pinus taeda*, *P. echinata*	Southern broadleaf forest (Deciduous forest)	C1	Sugar maple–beech–hickory–oaks	*Acer saccharum*, *Fagus grandifolia*, *Carya* spp. (incl. *C. ovata*, *C. cordiformis*), *Quercus* spp. (incl. *Q. alba*, *Q. velutina*, *Q. rubra*)
	2	Oak–gum–cypress	*Quercus* spp. (incl. *Q. stellata*, *Q. falcata*, *Q. coccinea*, *Q. marilandica*), *Nyssa* spp. (incl. *N. sylvatica*), Cupressaceae (incl. *Taxodium distichum*)	Southeastern mixed forest (Transitional: deciduous/mixed)	C2	Sugar maple–beech–oak–hemlock–white pine	*Acer saccharum*, *Fagus grandifolia*, *Quercus* spp. (incl. *Q. alba*, *Q. rubra*), *Tsuga canadensis*, *Pinus strobus*
	3	Oak–pine	*Quercus* spp. (incl. *Q. palustris*, *Q. bicolor*, *Q. phellos*, *Q. prinus*), *Pinus* spp. (incl. *P. echinata*)	(Mixed conifer/northern-hardwood forest)	C3	Sugar maple–yellow birch–hemlock–spruce–white pine	*Acer saccharum*, *Betula alleghaniensis*, *Tsuga canadensis*, *Picea mariana*, *P. glauca*, *Pinus strobus*
Central forest (Deciduous forest)	4	Oak–hickory	*Quercus* spp. (incl. *Q. alba*, *Q. velutina*, *Q rubra*)				

			Carya spp. (incl. *C. tomentosa*, *C. glabra*, *C. ovata*)	(Transitional: mixed/boreal)	C4	Sugar maple–yellow birch–spruce–fir	*Acer saccharum*, *Betula alleghaniensis*, *Picea mariana*, *P. glauca*, *Abies balsamea*
	5	Elm–ash–cottonwood	*Ulmus americana*, *Fraxinus* spp. (incl. *F. nigra*), *Populus* spp. (incl. *P. deltoides*, *P. balsamifera*)		C5	Red spruce–balsam fir–hemlock–pine–maple–birch	*Picea rubens*, *Abies balsamea*, *Tsuga canadensis*, *Pinus resinosa*, *P. strobus*, *Acer rubrum*, *A. saccharum*, *Betula papyrifera*, *B. alleghaniensis*
Northern forest (Mixed conifer/northern-hardwood forest)	6	Maple–beech–birch	*Acer* spp. (incl. *A. saccharum*, *A. rubrum*), *Fagus grandifolia*, *Betula* spp. (incl. *B. alleghaniensis*, *B. populifolia*)	Boreal forest (Transitional: mixed/boreal)	C6	Balsam fir–spruce–birch–white pine	*Abies balsamea*, *Picea mariana*, *P. glauca*, *Betula papyrifera*, *B. Alleghaniensis*, *Pinus strobus*
	7	Aspen–birch	*Populus* spp. (incl. *P. tremuloides*), *Betula* spp. (incl. *B. populifolia*)	(Boreal)	C7	Spruce–balsam fir	*Picea mariana*, *P. glauca*, *Abies balsamea*
	8	White–red–jack pine	*Pinus strobus*, *P. resinosa*, *P. banksiana*		C8	Black spruce	*Picea mariana*
Boreal forest (Transitional: mixed/boreal)	9	Spruce–fir	*Picea mariana*, *P. glauca*, *Abies balsamea*	Bogs—organic terrain	C9	Sphagnum moss–sedges–black spruce–tamarack	*Sphagnum* spp., Cyperaceae, *Picea mariana*, *Larix laricina*

[a]Names of vegetation regions are from Eyre (1980) for the United States and from Fremlin (1974) for Canada. Descriptions in parentheses indicate a uniform terminology for the vegetation regions of the United States and Canada as follows: southeastern conifer forest, deciduous forest, mixed conifer/northern-hardwood forest, boreal forest, and transitional forests.
[b]Codes refer to labels in Figure 1.

3. DATA AND METHODS

3.1. Map Data Base

The data base for the maps in Figures 2–4 consists of 166 sites with pollen stratigraphies from lakes and bogs between 35°–50°N and 65°–85°W (Figure 2 and Table 3). The sites range in elevation from 2 m in the coastal plain to 1,542 m in the White Mountains; these sites provide radiocarbon-dated pollen profiles from major physiographic features of the study area, including the Allegheny, Catskill, Taconic, White, and Laurentide Mountains of the Appalachian Mountain system, the Adirondack Mountains, and lowlands associated with the coastline, the Great Lakes, and the Hudson and St. Lawrence Rivers.

3.2. Map Preparation

The pollen data and chronostratigraphic data are stored as computer files at Brown University. Data analysis was automated at all stages from the calculation of pollen percentages to the production of maps. Webb *et al.* (1983b) and

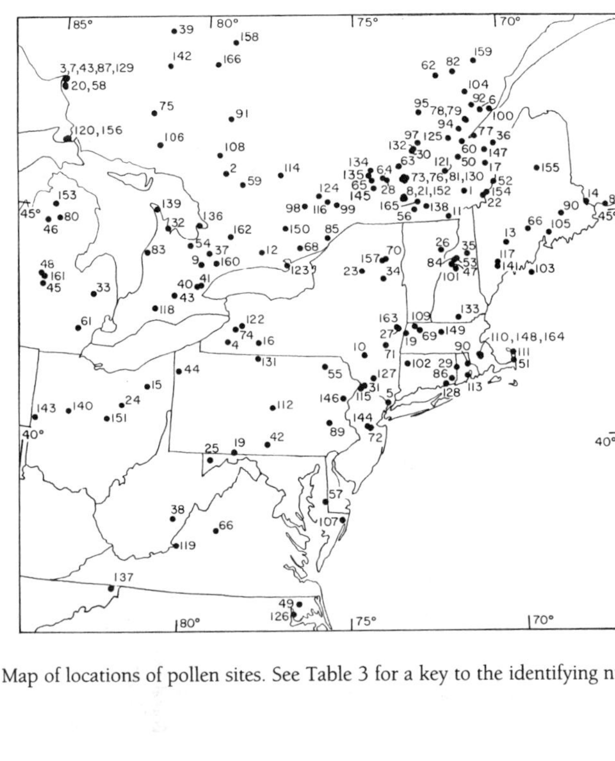

Figure 2. Map of locations of pollen sites. See Table 3 for a key to the identifying number for each site.

Gaudreau and Webb (1985) described the mapping procedure, which includes the following steps:

1. The radiocarbon dates and depths were plotted on scatter diagrams to aid the choice of age model for each site.
2. Dates were assigned to each pollen sample using the interpolation methods outlined in Overpeck and Fleri (1982). For most of the sites, age models were determined by linear interpolation between dates in the chronostratigraphic file for each site. Topmost and bottommost dates mapped at each site were limited by extrapolation of up to 500 years from the youngest and oldest dates in the chronostratigraphic file for each site.
3. Pollen percentages were calculated from a sum of tree, shrub, and herb pollen with pollen from aquatic plants, unknown and unidentifiable pollen, and spores excluded from the sum. For each site, tables of pollen percentages (of up to 15 pollen types per table) were produced.
4. The percentages to be plotted on maps at even 1,000-year intervals were determined by interpolation of pollen percentages between the two pollen samples at each site that, according to the age model, bracketed the target map date.
5. Interpolated percentages were plotted by computer on line-printer maps.

Figures 3 and 4 show isopoll maps for 2,000-year intervals for 0–14,000 radiocarbon years before present. Isofrequency contours, or isopolls, were hand-drawn on the maps. In order to control for the introduction of mesoscale pattern in the isopolls, certain contouring criteria were applied: (1) Because the major topographic features in the Northeast trend north/south, the isopolls were drawn with a latitudinal bias. The contours were deflected meridionally only as specifically forced by the data. (2) The contours are literal representations of the data. The contours deflect to record all spatial anomalies in the data; and all isolated pollen values that differ from regional pollen values are delimited by a circle (for single sites) or an ellipse (for two or a few neighboring sites).

3.3. Pollen Types Mapped

Three arboreal pollen types (*Quercus,* oak; *Picea,* spruce; and *Pinus,* pine) and one arboreal pollen group (BAFT: *Betula,* birch; *Acer,* maple; *Fagus,* beech; and *Tsuga,* hemlock) were mapped. Oak dominates in the deciduous forest, spruce in the boreal forest, and BAFT or pine in the mixed conifer/hardwood forest. These pollen types/groups were used by Bernabo and Webb (1977) to summarize the vegetation history of eastern North America.

Table 3. Location, Dating Information, and References for Pollen Sites in Northeastern North America

Site name[a]	Latitude	Longitude	State[b]	Elevation(m)	Type[c]	14C[d]	Source[e]
1. Albion	45°40′N	71°19′W	PQ	320	L	4	Richard 1977
2. Alderdale Bog	46° 3′N	79°12′W	ON	240	B	1	Terasmae 1968
3. Alfies Lake	47°53′N	84°52′W	ON	288	L	3	Saarnisto 1974
4. Allenberg Bog	42°15′N	78°52′W	NY	494	B	0	Miller 1973
5. Alpine Peat Bog	40°57′N	73°54′W	NJ	30	B	5	Nickmann, R.J., unpublished
6. Ange, lac a l' (Baie St. Paul)	47°28′N	70°41′W	PQ	640	L	7	Labelle and Richard 1981
7. Antoine Lake	47°44′N	84°54′W	ON	271	L	1	Saarnisto 1975
8. Atocas, Lac des	45°32′N	73°18′W	PQ	120	L	4	Gauthier 1981
9. Ballycroy Bog	43°57′N	79°52′W	ON	297	B	2	Karrow *et al.* 1975
10. Balsam Lake	42° 1′N	74°36′W	NY	762	L	8	Ibe 1982, 1985
11. Barnston Lake	45° 7′N	71°53′W	PQ	415	L	1	Mott 1977
12. Barry Lake	44°18′N	77°55′W	ON	183	L	1	McAndrews, J.H., unpublished
13. Basin Pond	44°28′N	70° 3′W	ME	124	L	0	Gajewski 1983
14. Basswood Road Lake	45°15′N	67°20′W	NB	106	L	3	Mott 1975
15. Battaglia Bog	41° 8′N	81°19′W	OH	318	B	3	Shane 1975
16. Belmont Bog	42°15′N	77°55′W	NY	497	B	6	Spear and Miller 1976
17. Benjamin, St. (Lac a Busque)	46°17′N	70°36′W	PQ	330	L	1	Richard 1977
18. Berry Pond	42°30′N	73°19′W	MA	600	L	16	Whitehead 1979
19. Big Pond	39°46′N	78°33′W	PA	634	L	1	Watts 1979
20. Blackington Lake	47°54′N	84°52′W	ON	261	L	2	Saarnisto 1975
21. Bouleaux, Lac	45°33′N	73°19′W	PQ	126	L	2	Mott, R.J., unpublished
22. Boundary Pond	45°34′N	70°41′W	ME	603	L	5	Mott 1977
23. Brandreth Lake	43°55′N	74°41′W	NY	583	B	5	Overpeck 1985
24. Browns Lake Bog	40°41′N	82° 3′W	OH	290	B	11	Ogden, J.G., III, unpublished
25. Buckles Bog	39°34′N	79°16′W	MD	814	B	6	Maxwell and Davis 1972
26. Bugbee Bog	44°22′N	72° 9′W	VT	398	B	4	McDowell *et al.* 1971
27. Burden Lake	42°36′N	73°34′W	NY	192	L	4	Gaudreau, D.C., unpublished
28. Calixte, St.	45°57′N	73°52′W	PQ	261	L	4	Richard, P. unpublished
29. Carbuncle Pond	41°42′N	71°47′W	RI	342	L	0	Bernabo 1977

30. Castor, Lac	46°36′N	72°59′W	PQ	220	L	6	Richard, P., unpublished
31. Cedar Bog	41°20′N	74°40′W	NJ	137	B	0	Niering 1953
32. Charles Lake	44°44′N	81° 1′W	ON	219	L	1	Bailey, R.E., unpublished
33. Chippewa Bog	43° 7′N	83°15′W	MI	262	B	4	Bailey and Ahearn 1981
34. Clear Pond	43°45′N	74° 1′W	NY	510	L	0	Swain, A.M., unpublished
35. Clouds, Lake of the	44°16′N	71°19′W	NH	1542	L	5	Spear 1981
36. Colin, Lac	46°43′N	70°18′W	PQ	658	L	8	Mott 1977
37. Cookstown Bog	44°13′N	79°37′W	ON	234	B	1	Karrow *et al.* 1975
38. Cranberry Glades	38°12′N	80°17′W	WV	1029	B	3	Watts 1979
39. Crates Lake	49°11′N	81°16′W	ON	259	L	6	Liu 1982
40. Crawford Bog	43°28′N	79°57′W	ON	279	B	0	McAndrews, J.H., unpublished
41. Crawford Lake	43°30′N	79°50′W	ON	102	L	0	McAndrews 1976
42. Criders Pond	39°58′N	77°33′W	PA	290	L	4	Watts 1979
43. Crozier Lake	47°54′N	84°52′W	ON	223	L	1	Saarnisto 1975
44. Crystal Lake	41°33′N	80°22′W	PA	313	L	1	Walker and Hartman 1960
45. Crystal Lake	43°15′N	84°55′W	MI	260	L	2	Kapp, R.O., and Lay, D., unpublished
46. Cub Lake	44°42′N	84°57′W	MI	370	L	5	Rasmussen, J., and Bailey, R.E., unpublished
47. Deer Lake Bog	44° 2′N	71°50′W	NH	1325	B	6	Spear 1981
48. Demont Lake	43°29′N	85° 0′W	MI	248	L	5	Kapp, R.O., unpublished
49. Dismal Swamp	36°35′N	76°27′W	VA	6	B	2	Whitehead 1972
50. Dosquet	46°27′N	71°30′W	PQ	140	B	1	Richard 1977
51. Duck Pond	41°50′N	70° 0′W	MA	2	L	6	Winkler 1985
52. Dufresne, Lac	45°51′N	70°21′W	PQ	650	L	6	Mott 1977
53. Eagle Lake Bog	44°10′N	71°40′W	NH	1275	B	3	Spear 1981
54. Edward Lake	44°22′N	80°15′W	ON	518	L	3	McAndrews 1981
55. Ely Lake	41°46′N	75°50′W	PA	385	L	0	Swain, A.M., unpublished
56. Farnham bog	45°17′N	72°59′W	PQ	55	B	4	Morassa, N., Jette, H., and Richard, P., unpublished
57. Federalsburg	38°42′N	75°46′W	MD	15	B	1	Sirkin *et al.* 1977
58. Fenton Lake	47°52′N	84°52′W	ON	253	L	1	Saarnisto 1975
59. Found Lake	45°48′N	78°38′W	ON	488	L	7	McAndrews 1981
60. Foy, Ste., Base de Plein-Air de	46°47′N	71°20′W	PQ	16	B	2	Larouche A., and Richard, P. unpublished

(Continued)

Table 3. (*cont.*)

Site name[a]	Latitude	Longitude	State[b]	Elevation(m)	Type[c]	14C[d]	Source[e]
61. Frains Lake	42°20′N	83°38′W	MI	271	L	9	Kerfoot 1974
62. Francois de Sales, St.	48°17′N	72° 8′W	PQ	358	L	4	Richard, P., unpublished
63. Gabriel, St.	46°16′N	73°28′W	PQ	250	L	5	Richard 1977
64. Geai, Lac (St.-Hyppolyte)	45°59′N	73°59′W	PQ	365	L	1	Richard, P., unpublished
65. Germain, St., Lac a	45°56′N	74°22′W	PQ	473	L	5	Savoie and Richard 1979
66. Gould Pond	44°44′N	69°19′W	ME	89	L	13	Jacobson, G.L., unpublished
67. Hack Pond	37°59′N	79° 0′W	VA	451	L	2	Craig 1969
68. Harrowsmith Bog	44°25′N	76°42′W	ON	145	B	1	Terasmae 1968
69. Hawley Bog Pond	42°34′N	72°53′W	MA	549	L	5	Patterson, W.A., III unpublished
70. Heart Lake	44°11′N	73°58′W	NY	664	L	8	Whitehead, D.R., unpublished
71. Hearts Content Bog	42°14′N	73°58′W	NY	107	B	0	Ibe 1982
72. Helmetta Bog	40°23′N	74°26′W	NJ	15	B	1	Watts 1979
73. Henri, St. (Lanoraie)	45°59′N	73°18′W	PQ	18	B	6	Comtois 1982
74. Houghton Bog	42°32′N	78°40′W	NY	428	L	1	Miller 1973
75. Jack Lake	47°19′N	81°46′W	ON	430	L	3	Liu 1982
76. Jean, St. (Lanoraie)	46° 0′N	73°13′W	PQ	20	B	4	Comtois 1982
77. Jean, St., Ile d'Orleans	46°56′N	70°56′W	PQ	68	B	3	Richard 1977
78. Joncas Bog	47°16′N	71°10′W	PQ	747	B	2	Richard 1977
79. Joncas Lake	47°15′N	71°10′W	PQ	780	L	1	Richard, P., unpublished
80. Jones Lake	44°46′N	84°36′W	MI	365	L	3	Bernabo 1981
81. Joseph, St. (Lanoraie)	45°59′N	73°18′W	PQ	18	B	3	Comtois 1982
82. Kenogami	48°22′N	71°34′W	PQ	166	B	2	Richard 1977
83. Kincardine Bog	44° 9′N	81°39′W	ON	198	B	4	Karrow *et al.* 1975
84. Kinsman Pond	44° 8′N	71°44′W	NH	1140	L	3	Spear 1981
85. Lambs Pond	44°39′N	75°48′W	ON	107	L	4	Anderson, T.W., unpublished
86. Lantern Hill Pond	41°27′N	71°57′W	CT	36	L	5	Trent, K.M., unpublished
87. Last Lake	47°53′N	84°52′W	ON	265	L	2	Saarnisto 1975
88. Little Lake	45° 9′N	66°43′W	NB	64	L	4	Mott 1975
89. Long Swamp	40°29′N	75°40′W	PA	192	B	6	Watts 1979

90. Loon Pond	45° 2′N	68°12′W	ME	110	L	10	Jacobson, G.L., unpublished
91. Louis, Lac	47°17′N	79° 7′W	PQ	300	L	3	Vincent 1973
92. Malbaie	47°36′N	70°58′W	PQ	800	B	2	Richard 1977
93. Maplehurst Lake	43°13′N	80°39′W	ON	300	L	6	Mott and Farley-Gill 1978
94. Marcotte, Lac	47° 4′N	71°25′W	PQ	503	L	5	Labelle and Richard 1981
95. Martini, Lac	47°28′N	72°45′W	PQ	242	L	4	Richard, P., unpublished
96. Mashapaug Pond	41°47′N	71°26′W	RI	12	L	1	Bernabo 1977
97. Mauricie, Lac du Noyer	46°47′N	72°50′W	PQ	270	L	5	Richard 1977
98. McLaughlan Lake	45°21′N	76°33′W	ON	196	L	1	Anderson, T.W., unpublished
99. Mer Bleue Peat Bog	45°24′N	75°30′W	ON	69	B	1	Mott and Camfield 1969
100. Mimi, Lac	47°30′N	70°22′W	PQ	411	L	6	Richard 1977
101. Mirror Lake	43°57′N	71°42′W	NH	200	L	9	Likens and Davis 1975
102. Mohawk Pond	41°49′N	73°17′W	CT	360	L	6	Gaudreau, D.C., unpublished
103. Monhegan Island Meadow	43°46′N	69°18′W	ME	3	B	17	Bostwick 1978
104. Montagnais	47°54′N	71°10′W	PQ	800	B	4	Richard 1977
105. Moulton Pond	44°37′N	68°38′W	ME	143	L	16	R.B. Davis *et al.* 1975
106. Nina Lake	46°36′N	81°30′W	ON	380	L	3	Liu 1982
107. Ninepin 24	38°18′N	75°17′W	MD	15	B	2	Sirkin *et al.* 1977
108. North Bay Bog	46°27′N	79°28′W	ON	369	B	1	Terasmae 1968
109. North Pond	42°39′N	73° 3′W	MA	586	L	13	Whitehead and Crisman 1978
110. Nunkets Pond	41°58′N	71° 3′W	MA	18	L	4	Bradshaw, R.H.W., unpublished
111. Pamet Cranberry Bog	42° 0′N	70° 2′W	MA	3	B	3	Patterson, W.A., III, unpublished
112. Panther Run Pond	40°48′N	77°25′W	PA	634	L	2	Watts 1979
113. Pasacaco Pond	41°31′N	71°27′W	RI	3	L	3	Bernabo 1977
114. Perch Lake	46° 2′N	77°22′W	ON	160	L	1	McAndrews, J.H., unpublished
115. Pine Bog	41°17′N	74°45′W	NJ	137	B	0	Niering 1953
116. Pink Lake	45°28′N	75°49′W	PQ	162	L	3	Mott and Farley-Gill 1981
117. Poland Spring Pond	44° 2′N	70°21′W	ME	94	L	11	Jacobson, G.L., unpublished
118. Pond Mills Pond	42°55′N	81°15′W	ON	274	L	1	McAndrews 1981
119. Potts Mountain Pond	37°36′N	80° 8′W	VA	840	L	2	Watts 1979
120. Prince Lake	46°34′N	84°33′W	ON	290	L	2	Saarnisto 1974

(Continued)

Table 3. (*cont.*)

Site name[a]	Latitude	Longitude	State[b]	Elevation(m)	Type[c]	14C[d]	Source[e]
121. Princeville, Lac Brule a	46° 8′N	71°56′W	PQ	135	L	0	Richard 1977
122. Protection Bog	42°37′N	78°28′W	NY	430	B	3	Miller 1973
123. Quinte, Bay of (Lake Ontario)	44° 2′N	77° 5′W	ON	75	L	4	Anderson, T.W., unpublished
124. Ramsay Lake	45°36′N	76° 6′W	PQ	200	L	5	Mott and Farley-Gill 1981
125. Raymond, St.	46°53′N	71°48′W	PQ	160	B	2	Richard 1977
126. Rockyhock Bay	36°10′N	76°41′W	NC	6	B	13	Whitehead 1981
127. Rogerine, Lake	41°30′N	74°20′W	NJ	137	L	2	Nicholas 1968
128. Rogers Lake	41°22′N	72° 7′W	CT	91	L	53	M.B. Davis 1969
129. Roller Lake	47°54′N	84°52′W	ON	213	L	2	Saarnisto 1975
130. Romer, Lac (Lanoraie)	45°58′N	73°20′W	PQ	20	L	4	Comtois 1982
131. Rose Lake	41°55′N	77°55′W	PA	631	L	2	Cotter and Crowl 1981
132. Sam, Lac a	46°39′N	72°58′W	PQ	240	L	6	Richard, P., unpublished
133. Sandogardy Pond	42°50′N	71°40′W	NH	100	L	4	M.B. Davis 1978
134. Sav-I (Ste. Agathe)	46° 3′N	74°28′W	PQ	454	L	3	Savoie and Richard 1979
135. Sav-II (Lac aux Quenouilles)	46°10′N	74°23′W	PQ	403	L	3	Savoie and Richard 1979
136. Second Lake	44°50′N	79°59′W	ON	196	L	1	McAndrews, J.H., unpublished
137. Shady Valley Peat	36°31′N	81°56′W	TN	383	B	1	Barclay 1957
138. Shefford, Mont	45°21′N	72°35′W	PQ	282	B	10	Richard 1977
139. Shouldice Lake	45° 9′N	81°25′W	ON	177	L	2	McAndrews, J.H., unpublished
140. Silver Lake	40°26′N	83°40′W	OH	341	L	15	Ogden 1966
141. Sinkhole Pond	43°58′N	70°21′W	ME	95	L	12	Jacobson, G.L., unpublished
142. Six, Lake	48°24′N	81°19′W	ON	305	L	3	Liu 1982
143. Stotzel-Leis Site	40°13′N	84°41′W	OH	320	B	11	Shane 1976

144. Szabo Pond	40°24′N	74°29′W	NJ	29	L	3	Watts 1979
145. Tania Lac	45°46′N	74°18′W	PQ	305	L	6	Richard, P., unpublished
146. Tannersville Bog	41° 2′N	75°16′W	PA	277	B	5	Watts 1979
147. Terrien, Petit Lac	46°35′N	70°37′W	PQ	404	L	1	Mott 1977
148. Titicut Swamp	41°57′N	71° 2′W	MA	18	B	3	Nelson 1984
149. Tom Swamp	42°31′N	72°13′W	MA	229	B	8	Gaudreau, D.C., unpublished
150. Tonawa Lake	44°51′N	77°10′W	ON	305	L	2	McAndrews, J.H., unpublished
151. Torrens Bog	40°21′N	82°28′W	OH	302	B	13	Ogden and Hay 1967
152. Tortue, Lac a la	45°32′N	73°19′W	PQ	137	L	4	Gauthier 1981
153. Twenty-seven, Lake	45° 4′N	84°47′W	MI	378	L	3	Bernabo 1981
154. Unknown Pond	45°37′N	70°38′W	ME	489	L	4	Mott 1977
155. Upper South Branch Pond	46° 5′N	68°54′W	ME	300	L	7	Anderson, 1979
156. Upper Twin Lake	46°33′N	84°35′W	ON	302	L	3	Saarnisto 1974
157. Upper Wallface Pond	44° 9′N	74° 3′W	NY	945	L	6	Whitehead, D.R., unpublished
158. Val St. Gilles Bog	49° 1′N	79° 5′W	PQ	290	B	2	Terasmae and Anderson 1970
159. Valin, Mont	48°36′N	70°50′W	PQ	891	L	4	Richard, P., unpublished
160. Van Nostrand Lake	44° 0′N	79°23′W	ON	297	L	2	McAndrews 1970
161. Vestaburg Bog	43°25′N	84°53′W	MI	255	L	4	Gilliam *et al.* 1967
162. Victoria Road Bog	44°37′N	78°57′W	ON	274	B	1	Terasmae 1968
163. West Sand Lake Peat Bog	42°38′N	73°36′W	NY	170	B	1	Gaudreau, D.C., unpublished
164. Winneconnet Pond	41°58′N	71° 8′W	MA	20	L	12	Suter 1985
165. Yamaska, Mont	45°28′N	72°52′W	PQ	265	L	7	Richard, P., unpublished
166. Yelle, Lac	48°30′N	79°38′W	PQ	355	L	3	Richard 1980

[a]Site name: numbers refer to labels in Figure 2.
[b]U.S. state or Canadian province.
[c]Type: site type, either lake (L), or bog or swamp (B).
[d]14C: number of radiocarbon dates.
[e]References in the Appendix.

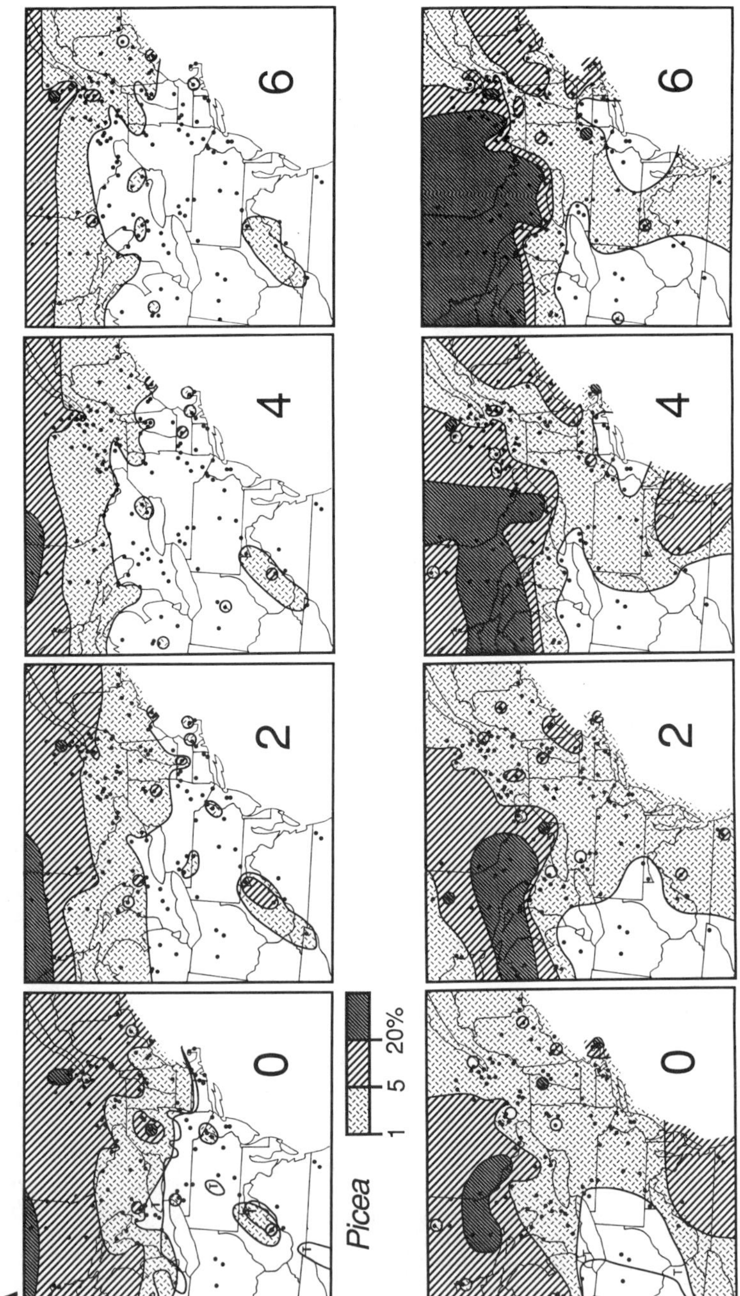
A
Picea
1 5 20%
0
2
4
6
Pinus
5 20 40%
0
2
4
6

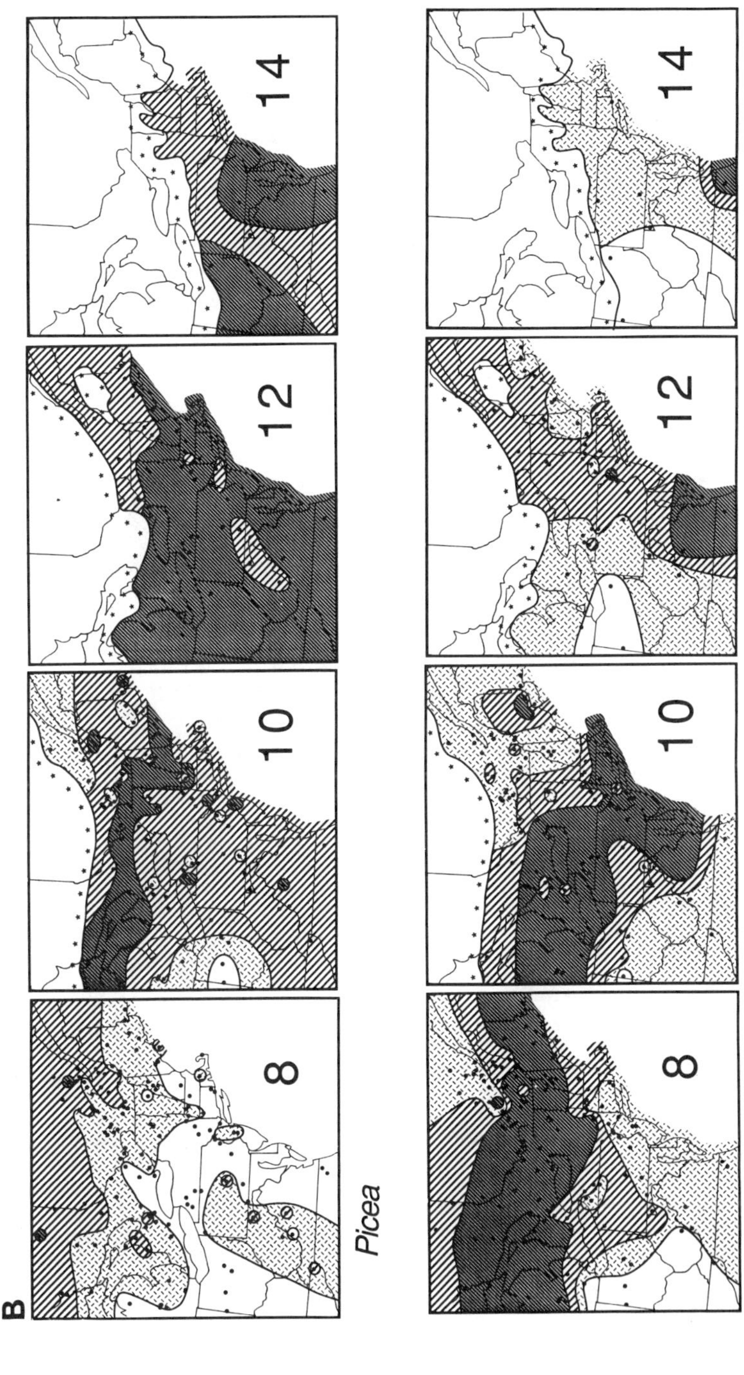

Figure 3. Isopoll maps for spruce and pine (0–14,000 B.P.). (A) Maps with isopolls for spruce and for pine at 2,000-year intervals from 0 to 6000 B.P. The modern range for spruce (three species; from Little 1971) is north of the line labeled with "T" and within the ellipses labeled with "T" on the 0 B.P. map for spruce. The modern range for pine (ten species; from Little 1971) is east of the line labeled with "T" on the 0 B.P. map for pine. (B) Maps with isopolls for spruce and pine at 2,000-year intervals from 8000 to 14,000 B.P. The southern margin of the ice sheet is delimited by a line labeled with asterisks.

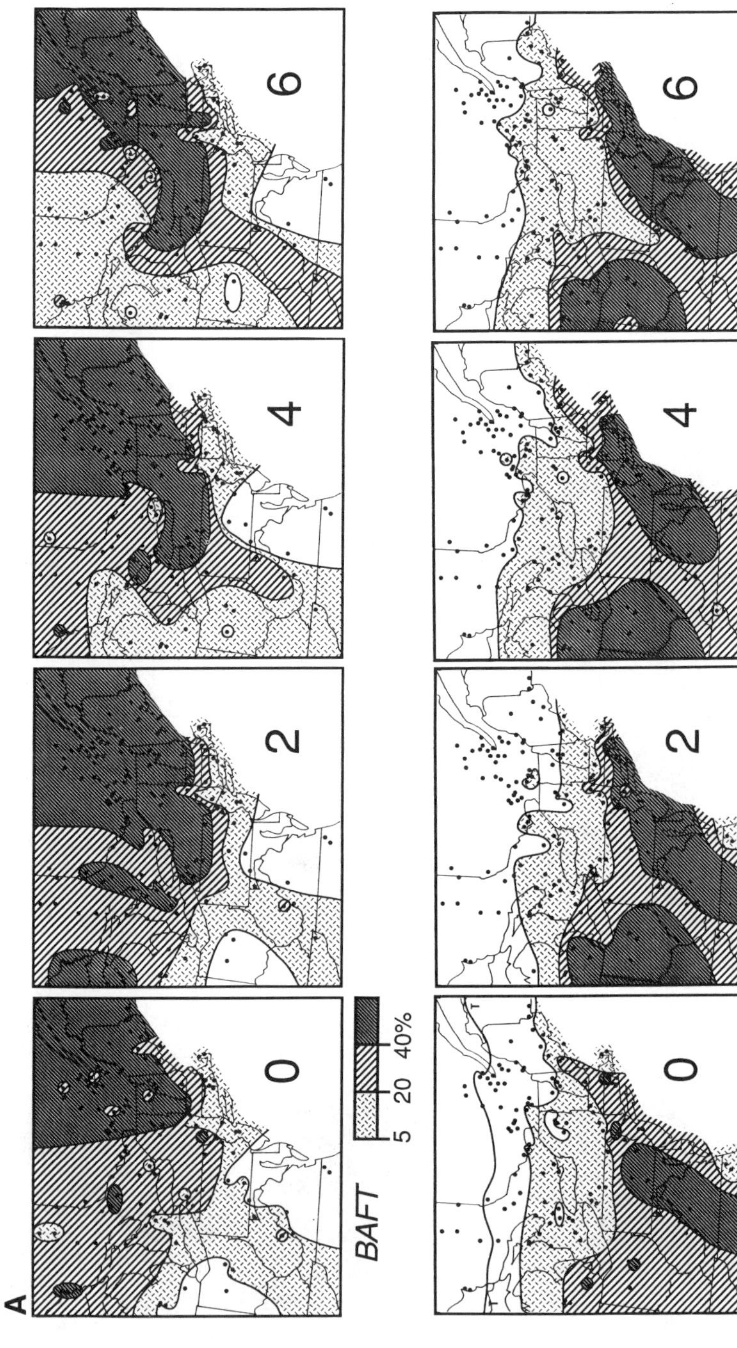
A
BAFT
5 20 40%
Quercus
5 20 40%
0
2
4
6

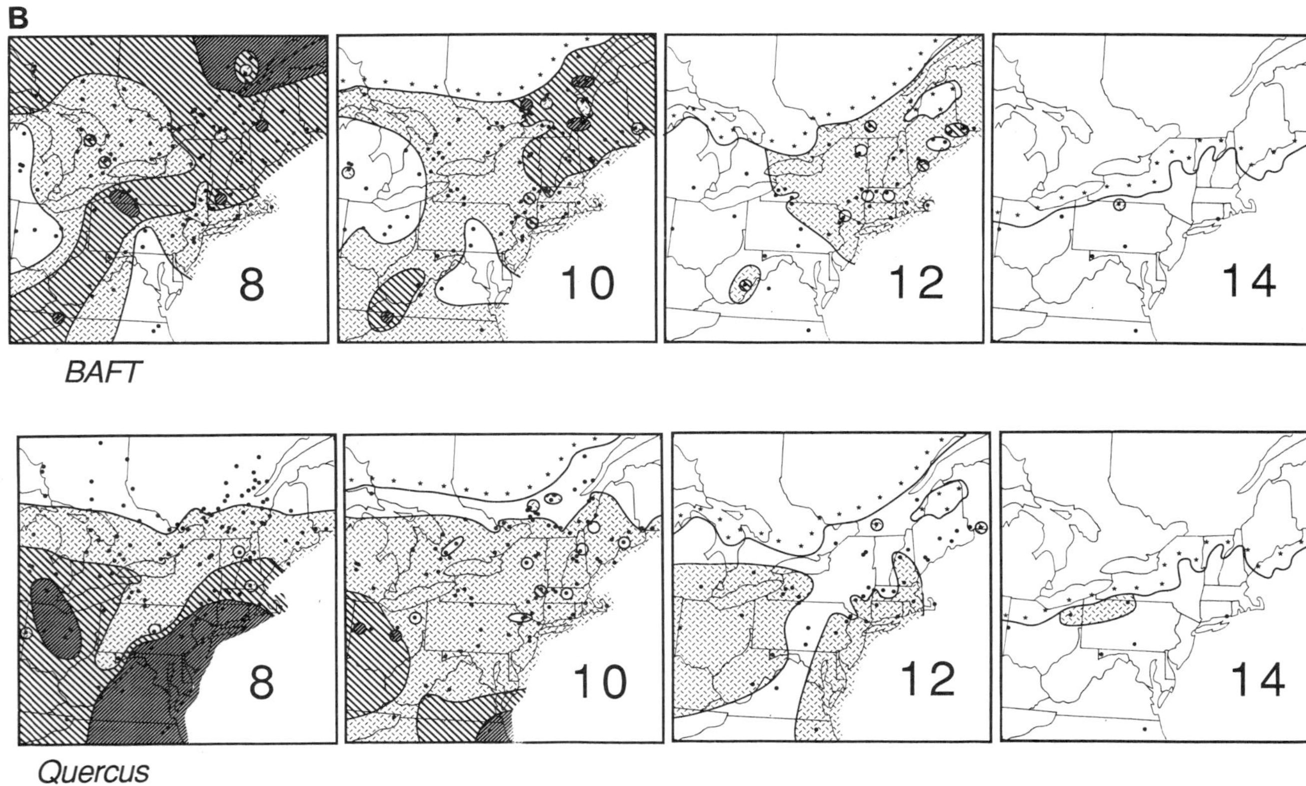

Figure 4. Isopoll maps for BAFT and oak (0–14,000 B.P.). (A) Maps with isopolls for BAFT and for oak at 2,000-year intervals from 0 to 6000 B.P. The modern range for oak (20 species; from Little 1971) is south of the line labeled with "T" on the 0 B.P. map for oak. (B) Maps with isopolls for BAFT and for oak at 2,000-year intervals from 8000 to 14,000 B.P.

3.4. Paleovegetational Inferences from the Isopolls

Three isopolls were mapped for each of the four pollen types/groups. As described below, the lowest valued isopoll approximates the range boundary of a taxon; the intermediate and highest valued isopolls delimit either population centers for a taxon or vegetation regions in which the taxon dominates. Together the three isopolls show changing gradients in plant populations.

Deciduous and mixed conifer/hardwood forests predominate in the Northeast today (Figure 1). Figures 1 and 4 show that, at present, the 20% oak isopoll corresponds with the northern limit of the deciduous forest for which oak pollen is numerically important; and the 20% isopoll of BAFT corresponds with the southern limit of the mixed forest in which birch, beech, and hemlock are important pollen types. The 40% isopolls for oak and BAFT delimit the population centers of these taxa. The distribution of the 5% oak isopoll approximates conservatively the range limit of species of oak (Figure 4). Trees of the BAFT group range throughout the Northeast, and the 5% isopoll delimits the distributions of small populations within the range of these taxa. Because birch, beech, and hemlock pollen are abundant pollen types in the BAFT group and each has a unique distributional history, isochrone maps of the 10% isopolls for each of these taxa (from Gaudreau and Webb 1985) are shown in Figure 5. [Although maple is a dominant taxon in the BAFT-rich mixed forest, its pollen is underrepresented (Delcourt *et al.* 1984)].

In the early Holocene and the late-glacial, spruce and pine pollen were dominant in the Northeast (Figure 3). During the Holocene, the abundances of spruce and pine pollen declined regionally; and at present, spruce and pine dominate in vegetation regions beyond the study area. The 20% spruce isopoll approximates conservatively the southern boundary of the spruce-rich boreal forest of Canada (Figures 1 and 3). The 5% isopoll aligns with the southern limit of boreal forest elements in transitional forests of the Northeast. The 1% isopoll closely approximates the range boundary for spruce, which extends southernmost in the highlands of the Northeast (Figure 3). The 20% and 40% pine isopolls delimit pine-rich mixed forests that dominate along the coast and also in

Figure 5. Isochrone maps of the northeastern United States for birch, hemlock, and beech (14,000–0 B.P.). From Gaudreau and Webb 1985; reproduced by permission. Note: Contours are in 10^3-year intervals; isolated numbers indicate the presence of pollen at the mapped abundance level for single sites. Percentages of each pollen type increase in the direction of the hachures on each isochrone. (A) Maps with isochrones at 2,000-year intervals showing the position of the 10% isopolls of birch from: (a) 14,000 to 10,000 B.P., (b) 8000 to 4000 B.P., (c) 4000 to 2000 B.P., and (d) 2000 to 0 B.P. (B) Isochrone maps of the 10% isopolls for hemlock from: (a) 14,000 to 4000 B.P. and (b) 2000 to 0 B.P. (C) Isochrone maps of the 10% isopolls for beech from: (a) 14,000 to 4000 B.P. and (b) 2000 to 0 B.P.

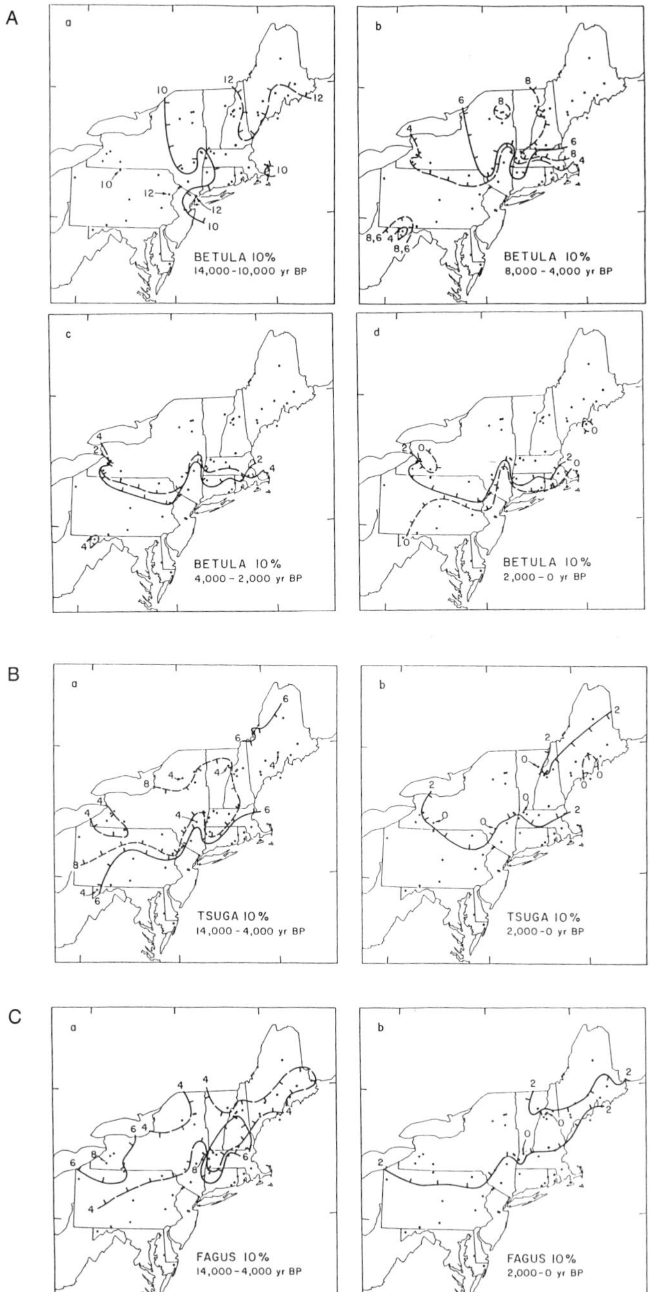
A
a
BETULA 10%
14,000-10,000 yr BP
b
BETULA 10%
8,000 - 4,000 yr BP
c
BETULA 10%
4,000 - 2,000 yr BP
d
BETULA 10%
2,000 - 0 yr BP
B
a
TSUGA 10%
14,000 - 4,000 yr BP
b
TSUGA 10%
2,000-0 yr BP
C
a
FAGUS 10%
14,000-4,000 yr BP
b
FAGUS 10%
2,000-0 yr BP

Canada to the northwest of the region mapped (Figures 1 and 3). The position of the 5% pine isopoll approximates the range boundary of pine taxa (Figure 3).

3.5. Paleoecological Inferences from Isopoll Patterns

Two features in the isopoll patterns have special application in inferring plant range boundaries (and thus postglacial plant colonization) and also positions of ecotones. The two features discussed below include (1) isopoll gradients and (2) isopoll patterns that reflect topographic contrasts.

3.5.1. Gradients

Over the last 14,000 years, gradients in the three isopolls for each pollen type changed in steepness and in geographic location. These changes in abundance gradients suggest that the isopoll patterns reflect vegetational changes, as opposed to noise due to long-distance pollen transport, because such diffusion would likely result in constant gradients in pollen abundances from plant population centers. Thus, the lowest valued isopoll that shows patterns of changing abundance gradients with adjacent isopolls is used as a proxy for the range boundary of its taxon (Gaudreau 1986; see also the criterion in Topographic Patterns below). In addition, weak gradients between the low-valued (i.e., range-proxy) and higher valued isopolls suggest that plants colonized as small populations and at great distances beyond population centers.

An ecotone is a transition region between two vegetation regions each dominated by different taxa. The location of and nature (e.g., abruptness) of this transition zone can vary in time. Steepened gradients in the isopolls of two taxa/groups suggest abrupt vegetational transitions and are used to trace the histories of spruce/pine, oak/BAFT, oak/pine, and pine/BAFT ecotones over that last 14,000 years (see Results and Discussion).

3.5.2. Topographic Patterns

The patterns in the isopolls show anomalies that reflect the broad-scale physiographic contrast imposed by the Appalachian Mountain system and also the local topographic contrast of lowlands relative to adjacent highlands. These anomalies are most evident in Figures 3 and 4 as: (1) broad-scale patterns in isopolls reflecting the topographic relief imposed by the Allegheny Mountains; (2) localized deflections in isopolls across the central New England/New York area associated with the topographic relief imposed by the Adirondack, Catskill, Green, White, and Taconic Mountains in contrast to the Hudson River lowlands or the coastal lowlands; and/or (3) localized deflections in isopolls that reflect the elevational contrast between lowlands of the Great Lakes and the St. Lawrence River relative to adjacent highlands.

Topographic pattern in the distribution of an isopoll provides evidence that the isopoll signals a vegetational distribution, as opposed to indicating noise due to long-distance pollen transport. Thus, the lowest valued isopoll that reflects topographic patterns is used as a range proxy for its taxon (Gaudreau 1986; see also the criterion in Gradients above). Further, the topographic patterns in isopoll distributions are used as evidence that the distribution of a taxon is influenced by climate (see Discussion, p. 242).

3.6. Summary of Terminology for Vegetational Inferences from Isopoll Distributions

The *range boundary* for a taxon is approximated by the position of the lowest valued isopoll for the taxon mapped in Figures 3 and 4; *range extension* or *population establishment* (i.e., "immigration") in a region is inferred from the initial appearance of this isopoll. *Population increase* is inferred from the appearance within a region of the intermediate or highest abundances of pollen for the taxon; and *population centers* are delimited by the highest valued isopolls. *Population expansion* is observed as enlargement in distribution of either the low- or intermediate-/high-valued isopolls that indicate, respectively, population establishment or increase. Loose gradients between the range-boundary proxy and adjacent isopolls show establishment of taxa in *small* populations, whereas tighter gradients suggest establishment via *larger* populations, or "migrating fronts" or "waves." Steep gradients in isopolls for different taxa are used to infer *ecotones* between vegetational regions dominated by the respective taxa.

4. RESULTS

4.1. Major Features in the Distributions of the Four Pollen Groups

4.1.1. Spruce

In the late-glacial, population centers of spruce (*Picea;* Figure 3) moved from southern to northern locations across the Northeast, and by 8000 B.P., a southern range limit developed in the Northeast. By 6000 B.P., spruce populations had their most restricted distribution in the Northeast. Then, by 4000 B.P. and continuing until 0 B.P., spruce populations expanded southward.

Pattern in the spruce isopolls reflects the topographic contrasts in the Northeast from 14,000 B.P., when population centers were in the south and abundances declined with elevation in the southern Appalachians. Populations of spruce expanded regionally in the Northeast by 12,000 B.P. and extended north near the retreating ice margin. Before 10,000 B.P., herb pollen (including Cyperaceae, Gramineae, and *Artemisia*) was an important element in pollen assemblages of the Northeast. The co-occurrence of herb and spruce pollen suggests open woodland or parkland vegetation, physiognomically and compositionally different from the closed boreal forests found today in Canada and in the northeastern United States (Davis and Jacobson 1985; Gaudreau and Webb 1985).

By 10,000 B.P., spruce populations had declined across the Northeast and the distributional center of spruce populations shifted northward and westward into Canada. By 8000 B.P., spruce populations had a southern range boundary in the Northeast; and in contrast to the location of population centers in the lowlands around the Appalachians in the late-glacial, the southern range limit extended southernmost in the highlands.

The large decrease in the distribution of 20% spruce pollen observed from 12,000 to 10,000 B.P. was repeated in the decrease in the distribution of 5% spruce pollen from 10,000 to 8000 B.P., and again in a subsequent decrease in the distribution of 1% spruce pollen from 8000 to 6000 B.P. After 4000 B.P., this pattern of decline in spruce populations was reversed; populations of spruce increased and the range limit for spruce shifted southward. The range limit extended southernmost in the highlands of New England and in the southern Appalachians. The increase in spruce populations continued to 0 B.P. The increase in spruce populations and their southward expansion are seen in both the southward movement of isopolls and in changes in isopoll gradients.

4.1.2. Pine

Population centers of pine (*Pinus;* Figure 3) expanded rapidly from the southeast between 14,000 and 12,000 B.P. Like spruce populations, pine extended north to the retreating ice sheet. Large populations of pine were regionally extensive by 10,000 B.P.; and these population centers then retreated northwestward during the Holocene. As a proxy for the range limit of pine, the 5% isopoll shows, over the last 14,000 years, a persistent western limit for pine populations. The gradient between the range proxy and adjacent isopolls changes through time, but the gradient is steeper in the west than in the east.

Pine populations expanded along the East Coast from 14,000 to 12,000 B.P. By 10,000 B.P., pine populations also increased in the midwestern United States and Canada, but these populations did not increase in the intervening southern Appalachians. From 10,000 to 6000 B.P., pine population centers moved northward; and by 6000 B.P., populations in the east declined. The isopolls show that between 6000 and 4000 B.P. three disjunct population centers appeared. Complex changes in isopoll gradients accompanied the development of these three regions. The three pine-rich forest regions (each dominated by different species) persist today: (1) in southern and southeastern United States; (2) in the coastal plain of the northern United States and Canada; and (3) in interior Canada, northwest of the mapped region.

From 6000 B.P. to the present, the highest pine abundances are found in Canada; during that time, pine population centers decreased in areal extent and shifted west into Ontario. Also during that time, this western distribution of pine population centers was complemented by increases in populations of trees of the BAFT group in the east, as described below.

4.1.3. BAFT

Besides birch, both beech and hemlock are important pollen indicators of the mixed forest, and these taxa show spatially and temporally independent histories; the use of the distribution of 20% BAFT pollen to approximate the major distributional features of the mixed forest does not imply that the BAFT taxa behaved as an intact association over the last 14,000 years.

As areas supporting large pine populations shrank northwestward during the Holocene, populations of BAFT taxa (Figure 4) increased in the Northeast and subsequently maintained their northeastern-centered distribution that extended south along the southern Appalachians and in the highlands of New England. Over the last 14,000 years, the distribution of BAFT pollen showed a strong east–west gradient; a distributional feature shared with pine, though in the opposite sense.

By 10,000 B.P., populations of trees in the BAFT group had established a range that remained relatively stable through the Holocene, and by 10,000 B.P. their populations increased in two regionally disjunct areas: in northeastern New England and Canada, and in the southern Appalachians. Increases in birch populations [probably gray and white birch, as suggested by pollen-grain measurements or macrofossils (e.g., Watts, 1983)] accounted for the increase in BAFT values in the northeastern United States and in Canada (Figure 5). Increases in birch (possibly yellow birch) and hemlock populations accounted for the increase in BAFT values in the southern Appalachians (Figure 5) (Gaudreau and Webb 1985). The initial increase in birch populations from the north contrasts with the initial increase from the south observed for spruce and also pine populations. This pattern of increase in spruce and pine populations corresponds intuitively with an expected direction of colonization of deglaciated areas from south to north; the pattern of increase in birch populations suggests decoupling of colonization from population increase.

By 8000 B.P., BAFT populations had increased within the range that was established by 10,000 B.P. At this time, hemlock populations were centered across New England and the Midwest (Figure 5), and these populations decreased in size to the north and south; populations of birch (probably white birch) generally increased northward (Figure 5). By 6000 B.P., population centers of birch (probably yellow birch) expanded in the Northeast (Figure 5), and hemlock and beech populations expanded in New England and New York (Figure 5) and the Midwest. At 4800 B.P., hemlock pollen decreased in abundance, possibly the result of a pathogen that decimated hemlock populations (M. B. Davis 1981a). This "decline" is not reflected in the BAFT map for 4000 B.P., showing that populations of other taxa in the BAFT group (e.g., birch and beech; Figure 5) expanded as hemlock populations declined.

As a result of the population expansions described above, the distribution of 40% BAFT pollen in the north and east complemented that of 40% pine pollen in the north and west; the distribution of 40% BAFT pollen also complemented the distribution of 40% oak pollen to the south. The isopolls for BAFT, pine, and oak show the development of ecotones that persisted from the early Holocene to 0 B.P., though the vegetational gradients in these transition zones changed.

4.1.4. Oak

Populations of oak (*Quercus;* Figure 4) expanded from the south and remained centered south of central New England and New York during the Holocene within a range established by the start of the Holocene.

The 5% isopoll shows that between 12,000 and 10,000 B.P., populations of oak achieved their northern limit for the Holocene. By 10,000 B.P., oak populations increased in the south, to the east and west of the Appalachians, and along the eastern slope. Throughout the Holocene, oak populations decreased to the north and also decreased with elevation in the Appalachians. At 10,000 B.P., this elevational contrast was evident in deflections in the 20% isopoll, and by 8000 B.P., in the 40% isopoll as well. The distributions of 40% oak pollen during the Holocene show expansions of oak populations.

The deciduous forest is dominated by oak trees and oak pollen is well-dispersed and well-represented (Webb *et al.* 1981); hickory and chestnut are also important in the deciduous forest, but their pollen is not as abundant as that of oak. Because oak is the dominant pollen indicator of deciduous forest, the distribution of high abundances of oak pollen was used to summarize the major features in the history of the deciduous forest; however, oak, hickory, and chestnut populations did have independent histories (Davis 1976; Gaudreau 1986).

4.2. The Development of Present Forest Regions and Their Ecotones

From 14,000 to 12,000 B.P., the arboreal vegetation of the Northeast was spruce-rich. By 10,000 B.P., pine populations were areally dominant and an ecotone existed between spruce-dominated vegetation to the north centered in Canada and pine vegetation to the south in the United States. This ecotone weakened by 8000 B.P.

During the Holocene, the distributions of oak-dominated forests and BAFT-dominated forests expanded, as pine populations became centered northward and westward and spruce population centers moved even farther northward. Since the early Holocene, oak-, pine-, and BAFT-dominated forest regions have developed complementary distributions in northeastern North America. Steepening abundance gradients between oak, BAFT, and pine pollen show the development of the ecotones between these forest regions in the early Holocene, and also the subsequent stability in the locations of the ecotones during the middle and late Holocene.

During the Holocene, pine- and BAFT-rich mixed forests developed strong east/west gradients; oak-rich forests developed strong north/south gradients with the mixed forests dominated by BAFT or pine taxa. Besides strong latitudinal and longitudinal gradients, these forest regions showed topographic gradients in their distributions. For example, at present, the positions of the 20% oak and BAFT isopolls coincide across New England and in the Appalachians and mark the location of the ecotone between oak-rich deciduous forests to the south and the BAFT-rich mixed forests to the northeast; patterns in the isopolls of BAFT and oak reflect the topographic contrasts of the Appalachians and also the contrast between adjacent highlands and lowlands of southern/central New England. Steepening gradients between BAFT and oak isopolls show that the ecotone developed in New England by the early Holocene; this transition zone became more pronounced between 8000 and 6000 B.P. The level of resolution shown in the isopoll patterns suggests that more pollen sites should allow more accurate tracking of changes in the ecotone during the Holocene.

While the position of the transition zone between oak-dominated deciduous forests and mixed forests of conifers and hardwoods remained relatively stable, population centers of associated taxa, including hemlock, beech, hickory, and chestnut, changed distributions during the Holocene (Gaudreau and Webb 1985). The temporal range of existing forest regions is an unresolved issue. Although the development of the deciduous/mixed forest ecotone is a prominent feature in the maps of Figures 3 and 4, it is difficult to identify the temporal limit for valid usage of classificatory terms for major vegetation regions such as deciduous forest or mixed conifer/northern-hardwood forest, or for forest types such as oak–hickory.

The histories of ecotones described above suggest that the maps for 10,000 to 0 B.P. show one vegetational regime, whereas late-glacial maps show another for northeastern North America. The late-glacial maps do not show patterns interpretable as ecotones for the pollen types mapped. This may be a function of the pollen abundances mapped, or of the low density of pollen sites. Also, pollen/tree relationships may have differed for the late-glacial relative to the Holocene because of changes in pollen dispersal and representivity. (Dispersal and representivity can change in response to changes in vegetational physiognomy and in sedimentological regimes of pollen catchment basins.) Alternatively, the development of forest regions similar to present forest regions in northeastern North America appears to be a Holocene phenomenon.

4.3. Summary

The distribution of isopolls for spruce, pine, BAFT, and oak show the following patterns: (1) broad-scale latitudinal and longitudinal changes that varied in rate of change over the last 14,000 years, (2) anomalies in the broad-scale distributions that reflect regional and local topographic gradients, (3) changes in steepness and location of gradients between adjacent isopolls for a pollen type, indicating changing distributions of population centers, and (4) steep gradients between isopolls of different pollen types, indicating ecotones. The maps show an ecotone between spruce-/pine-rich vegetation at 10,000 B.P.; and from 10,000 to 0 B.P., ecotones developed between oak-/BAFT-/pine-rich vegetation regions.

5. DISCUSSION

The paleoecological consequences of the patterns described above and their implications for archaeology are considered below.

5.1. Resolution in Pollen Distributions of Vegetational Contrasts Associated with Physiography

Pollen data convey information on vegetational distribution related to regional gradients of latitude and longitude in the Northeast (Bernabo and Webb 1977; Davis 1976; Davis and Webb 1975). However, the question of how well pollen distributions in the New England area reflect vegetational gradients associated with the local gradients of topography has been largely unaddressed. Davis *et al.* (1980) showed that pollen transport blurs the resolution of vegetation zones on mountain slopes in New Hampshire and that this effect is intensified near and beyond tree-line. The elevational gradients and corresponding vegetational gradients shown in the maps of Figures 1, 3, and 4 are less steep than those on mountain slopes. The isopoll patterns in the maps of Figures 3 and 4 show that, in addition to resolving regional distributions of vegetation, pollen distributions can resolve the mesoscale vegetational distributions associated with the physiographic features of the Northeast. This resolution allows certain statements about pollen dispersal and vegetational source regions.

In general, basins collect pollen primarily from a radius approximating 20 m to 20 km (Prentice 1983). This radius varies with basin attributes (Jacobson and Bradshaw 1981) and with pollen type (Webb *et al.* 1981) and can extend for orders of magnitude beyond the 20-km endmember, especially in treeless landscapes. Many of the physiographic features of the southern and central New England area have a minimum dimension within 20 km; for example, the north/south-trending Hudson lowlands are approximately 15 km in breadth, as are the Taconic Highlands/Berkshire Mountains. The isopoll patterns in Figures 3 and 4 illustrate synoptically that pollen dispersal is skewed toward the source (Prentice 1985; Solomon and Silkworth 1986) to the degree that pollen assemblages at a site represent mesoscale patterns in the vegetation in addition to regional patterns.

Pollen distributions can resolve a given vegetational distribution if the plants cover a critical area within the vegetational source region (for a given pollen-catchment basin) so that the plants produce pollen above a certain threshold. The threshold is that amount of pollen needed to overcome background noise, where "noise" is pollen derived from long-distance transport and from vegetational features other than those of interest. This threshold is not likely exceeded where elevational gradients are steep and therefore vegetational zones are narrow, as on mountain slopes. For localities with steep elevational and vegetational gradients, a pollen signature of mesoscale vegetational distribution is therefore unlikely; the pollen derived from regional and long-distance sources overcomes any mesoscale signal. The average elevational gradients and their corresponding vegetational gradients in the maps of Figures 1–5 are less steep than those on mountain slopes. Figures 3–5 show that, for these average topographic contrasts, the pollen can record vegetational contrasts associated with topographic contrasts superimposed on the pollen record of long-distance transport and regional vegetation.

Further, the maps in Figures 3–5 show that the signal of mesoscale vegetational distributions in the pollen distributions is robust, because the topographic contrasts in the pollen distributions are preserved in a network of sites that differ in collecting basin attributes, including site type (e.g., lake, bog), size, and morphometry. These attributes cause differences in the size of vegetational source regions among catchment basins.

Additional evidence that the signal in pollen distributions of mesoscale vegetational patterns is robust is the observation that mesoscale patterns are preserved in the pollen distributions at two spatial scales: (1) in the regional maps of Figures 3–5 referred to above, and (2) in pollen distributions at nearby sites. For example, a paired-site study by Gaudreau (1986) [using the methods of Brubaker (1975) and Jacobson (1979)] showed that pollen distributions at sites within 20 km can resolve the vegetational contrasts associated with 400 m of elevational contrast between a lowland and adjacent highland in the central New England area.

5.1.1. Application to Archaeology of Physiographic Pattern in Pollen Distributions

Bernabo and Webb (1977) mapped the percentages of the major pollen types from 62 sites across northeastern North America for 11,000 B.P. to the present. The maps in Figures 3–5 show broad-scale features that update Bernabo and Webb (1977) and add mesoscale pattern to their maps, which provided a framework for the interpretation of broad-scale vegetation history during the Holocene. Dincauze and Mulholland (1977) produced maps that showed both the locations of archaeological sites in New England and the oak isopolls used by Bernabo and Webb (1977) to indicate the geographic extent of deciduous forest. Dincauze and Mulholland (1977) used these oak pollen percentages to indicate habitat attractiveness, and they related the distributions of Archaic sites to the Holocene movements of the oak-dominated deciduous forests. Their "riverine model" serves as an hypothesis on prehistoric settlement patterns in New England. This model correlated habitat selection and subsistence patterns with the local environmental differences between highlands and river lowlands. Dincauze and Mulholland's (1977) maps included Early Archaic (10,000–9000 B.P.) sites and Middle Archaic (8000–7000 B.P.) sites in New York and New England within the region bounded to the north and west by the headwaters of the Hudson River and extending south and east across New England. Their maps showed sites clustered in the Hudson and Connecticut River lowlands and in coastal areas. However, the pollen data of Bernabo and Webb (1977) are not constrained at this local geographic scale. The location of the 20% oak isopoll from the continental-scale maps of Bernabo and Webb (1977) was interpolated from pollen percentages at two sites in southern New England: Rogers Lake in the coastal plain of Connecticut (Davis 1969) and Berry Pond in the Berkshire Mountains, Massachusetts (Whitehead 1979). The Bernabo and Webb (1977) maps therefore did not show vegetational differences associated with physiographic contrasts and thus do not allow adequate comparison of the pollen and archaeological data bases.

The maps in Figures 3–5 show isopolls constrained by a network of sites across New York and southern/central New England at the present location of the ecotone between deciduous and mixed conifer/northern-hardwood forests. These sites sample the diverse physiography of the region (and its associated vegetation); the sites include coastal and river lowlands and also highlands and mountains. Pollen data from this network of sites show patterns that can be used to infer the history of the ecotone and they show how the position of the ecotone reflects the physiographic contrasts of the New England area.

No-analog pollen assemblages and changes in pollen representation and dispersal and in the sedimentary environments of catchment basins make vegetational interpretations of late-glacial pollen assemblages less certain relative to Holocene assemblages; pollen dispersal and representivity may have differed in the late-glacial because of changes in vegetational physiognomy (which can affect pollen production and dispersal) and also changes in characteristics of the sedimentary basin (which can affect pollen catchment and deposition). The small size of the pollen data base for the late-glacial and poorer radiocarbon dating control contribute to the uncertainty in vegetational inferences. The coarser site grid for the late-glacial maps (14,000 and 12,000 B.P. in Figures 3 and 4) shows only general distributional features for spruce and pine, but their isopolls do show patterns that reflect physiography. Recent discussions argue to reverse the conceptions that the late-glacial landscape in the Northeast was vegetationally uniform, low in diversity, unproductive, and therefore regionally uniform in resources and habitats for animals and humans (Bonnichsen *et al.* 1985; R. B. Davis and Jacobson 1985; Nicholas 1983). These authors argue that inferences of environmental uniformity for the late-glacial are a function of site density.

For example, Davis and Jacobson's (1985) paleovegetational history for the late-glacial of northern New England reflects the philosophy that environments of the late-glacial were heterogeneous: they compensated for the low number of existing pollen sites by extrapolating the vegetation inferred from one pollen site to all other areas that lacked pollen data but were similar in elevation to that site. Their vegetation maps for 13,000–10,000 B.P. show mosaics of treeless vegetation ("tundra") coexisting with assemblages of trees. The trees occurred in mosaics that varied in taxonomic associations and in physiognomy (e.g., woodlands or forests).

Nicholas' (1980, 1983) "glacial lake mosaic" model likewise postulates heterogeneity in habitats and resources associated with postglacial lake environments common to late-glacial New England; he states that much of the early postglacial cultural hiatus may be an artifact of the discovery and recognition of sites. The vegetational inferences of Davis and Jacobson (1985) and the habitat assessment of Nicholas (1980, 1983) show heterogeneity for the late-glacial at scales as fine as those described from Holocene palynological and archaeological data. These studies suggest that the number of recovered sites limits present levels of resolution of paleoenvironments for the late-glacial relative to the Holocene.

5.1.2. General Considerations of Spatial and Temporal Scale in Palynology and Archaeology

The data discussed in this paper show that pollen distributions can resolve, over 10^3-year intervals, the vegetational differences associated with highland versus lowland environments and ecotonal versus nonecotonal environments.

This level of resolution is expected, given (1) vegetational source areas on the order of 10^0 to 10^1-km radii for the pollen recovered from lake or marsh sediments (Prentice 1983) and pollen dispersal skewed toward the vegetational source (Prentice 1985) as discussed above; and (2) chronostratigraphic resolution on the order of 10^2 years [200–500 years is the limit of resolution for radiocarbon dating (Webb 1982)]. Archaeological sites record human activity at these temporal and spatial scales, so archaeological and palynological data can be compatible for such scales of inquiry.

Archaeological sites also contain records of activity over radii of 10^{-2} or 10^{-3} km and for intervals of 10^{-3} years (Anderson 1982). These latter site-specific activities are best recorded in the pollen record from isolated samples in special sedimentary settings, such as in association with artifacts, in trash pits, burials, or feces; but such pollen samples yield little paleoecological information (Bryant and Holloway 1983; King 1985). Palynological and archaeological data bases have a scale incompatibility at this latter level: stratigraphically continuous pollen records from lakes and bogs yield a paleoecological signal, but cannot be used for inferring site-specific cultural activities (King 1985).

5.2. Paleoecological Implications of Isopoll Patterns

Response functions quantify how macroscale distributions of individual tree taxa are correlated with environmental variables (Bartlein *et al.* 1986). As a consequence of the correlation between plant distribution and climatic variables, plant distributional changes should reflect changes in the distributions of climatic variables. However, at certain temporal and spatial scales, biological processes, such as seed-dispersal rate, become a limiting factor on plant distributional changes, so that the plant distributions may be out of equilibrium with climate (e.g., Davis 1976). There is general agreement that for a 10^5-year period, the average time of a Pleistocene glacial–interglacial cycle, plant distributions on continents ($> 10^3$ km radius) and in large regions (10^2 km radius) reflect climatic control (Davis 1976). Debate has centered on the relative importance of climatic versus biological controls on plant distributions within regions over 10^3-year intervals in the postglacial (10^4-year period) (Davis 1981b, 1983; Davis *et al.* 1986; Delcourt *et al.* 1983; Delcourt and Delcourt 1983; Gaudreau, 1986; Prentice 1983, 1986; Webb 1986).

The isopoll patterns in Figures 3 and 4 can be used to address the issue of endogenous versus exogenous control of plant population establishment. The isopoll patterns are consistent with environmental control of vegetational distributions on the 10^3-year scale. Though it is difficult to demonstrate climatic control when the climate of the past is unknown, some patterns in the maps appear to demonstrate climate control. These include patterns in the pollen data that reflect, as discussed below, (1) macroscale distributional changes of plants and (2) mesoscale distributions, namely topographic gradients, in plant distributions.

5.2.1. Macroscale Patterns in the Isopolls

The isopoll maps illustrate a sequence of dominance first for spruce pollen (14,000–12,000 B.P.), then pine pollen (10,000 B.P.), and finally for pine, oak, and BAFT pollen (8000–0 B.P.). The high pollen values for these types represent increased populations of the taxa, and the maps show that the spatial and temporal patterns of population expansion and increase for the individual tree taxa varied over the last 14,000 years. The broad-scale distributional changes of the taxa relative to each other show that all four groups had similar rapid expansions of population centers and also showed relative stability at other times over the last 14,000 years. These patterns are consistent with climatic control of population increase.

While the two higher valued isopolls mapped for each pollen type in Figures 3 and 4 indicate increased tree populations, the low-valued isopolls in each map approximate the range boundaries for the taxa. Patterns in Figures 3 and 4 as well as patterns in detailed maps for the low-valued isopolls of hemlock, beech, hickory, chestnut, spruce, pine, birch, and oak (Gaudreau 1986) show that (1) range extension for these taxa can occur via small populations at great distances from population centers, (2) the establishment of these taxa can precede, by 10^3-year intervals, their population increases, and (3) rates of range extension for these taxa vary over the last 14,000 years, but relative to each other, the taxa share similar maximum rates of range extension. These observations support the idea that at the mesoscale (1) inherent differences in seed-dispersal rate were not a limitation on relative rates of establishment for different plant taxa for 10^3-year intervals, and (2) that population establishment and subsequent increases in population size were influenced by climate. These observations contrast with an alternative interpretation (1) that low or stratigraphically sporadic abundances of pollen indicate noise rather than a signal of plant distribution, (2) that the initial postglacial increase in abundance of a pollen type beyond the low or sporadic levels corresponds with colonization of its taxon, and further (3) that the increased pollen abundances indicate colonization by migrating fronts of trees which show delays in "immigration" of one taxon relative to another due to differences in inherent seed-dispersal rate (e.g., Davis 1976, 1983).

5.2.2. Mesoscale Patterns in the Isopolls

Because elevation emplaces local gradients in temperature and precipitation, which affect plant distributions, patterns in fossil pollen distributions that reflect topographic gradients are evidence that past plant distributions reflect climatic contrasts (Davis *et al.* 1980).

5.2.2.1. Topographic Pattern in High Pollen Values. The high-valued isopolls in Figures 3 and 4 showed changing distributional patterns for the four pollen groups that correlated with elevational contrasts. These patterns are consistent with the view that population increase is correlated with climatic contrasts. Figure 4 shows that during the Holocene, the position of ecotones (e.g., the deciduous/mixed forest ecotone) reflected the topographic contrasts of the Northeast; this pattern likewise suggests that ecotones are influenced by climatic factors.

5.2.2.2. Topographic Pattern in Low (Range-Proxy) Pollen Values. The isopolls that were used to approximate range boundaries for the individual taxa showed distributions that reflect topographic gradients. Gaudreau (1984, 1986) proposed that range-proxy abundances of pollen distributions that reflect topographic patterns (among other criteria) provide evidence for climatic control of plant-population establishment. Patterns in isopoll maps of Figures 2 and 3 and also in isopoll maps for oak, birch, pine, spruce, hemlock, beech, hickory, and chestnut (Gaudreau 1986) are consistent with climatic control of range extension for these taxa at 10^3-year intervals.

5.3. Significance to Archaeology

Two issues from the above discussion on plant colonization are directly applicable for archaeological models of human prehistory. These issues include: (1) colonization of trees in small populations versus as "migrating fronts," and (2) climatic versus seed-dispersal limitation of plant colonization over thousands of years.

As described above, recent interpretations of pollen distributions suggest that certain taxa extend their ranges by establishment of small populations (Bennett 1985; Gaudreau 1984, 1986). Later these populations increase as a result of changes in environmental conditions (e.g., climate and ecological relationships mediated by climate). "Migrating fronts" of trees limited by seed-dispersal rate imply regional waves of invasion by populations lagging their climatic limits and thus these trees are not in marginal environments at their invading front; therefore, they likely occur in large populations. The alternative model of small invading populations can imply greater patchiness and environmental mosaicism, and taxonomically and physiognomically more diverse communities; such attributes can increase the heterogeneity of resources and habitats available to animal and human populations.

The model of range extension of trees in small populations is analogous to the observation of Bonnichsen *et al.* (1985) that between-site differences in toolmaking technologies in northern New England "suggest that the region may have been colonized initially by quite distinct and separate groups of people, and not by a uniform advancing front of Paleo-Indians." Further study of the nature of colonizing populations is necessary for useful comparisons of archaeological and palynological records.

The broad-scale patterns and topographic patterns in Figures 3–5 show evidence that vegetational distributions at the mesoscale are under climatic control (i.e., are in equilibrium with climate) over 10^3-year intervals within the postglacial. Models of the dynamics of human settlement and subsistence must take into account lags and feedbacks with both the biotic environment (including plant and animal distributions) and the physical environment (including climate and physiography). Demonstration of dynamic equilibrium between vegetation and climate on the 10^3-year scale has implications for the dynamic relationship of humans with their biotic and physical environment and speaks to the question of whether climatic change causes cultural change (Dincauze 1981).

5.4. Discussion Summary

This paper has reviewed general features of the vegetational history for northeastern North America over the last 14,000 years by showing maps of selected pollen types. Patterns in the maps show (1) macroscale distributional histories for plant taxa, (2) mesoscale features that reflect the vegetational contrasts associated with physiography, and (3) changing abundance gradients that show histories of plant populations and their ecotones. Discussion focused on the paleoecological implications of these paleovegetational features and their application to archaeology:

1. The isopoll maps show that pollen distributions can resolve the vegetational differences associated with the physiographic contrasts of highlands and lowlands in the Northeast. The mesoscale patterns in the maps facilitate new attempts to correlate vegetational history with archaeological models that seek to differentiate habitats and resources between lowland and highland environments and between ecotonal and nonecotonal areas.

2. The isopoll patterns that show changing gradients in pollen abundances and those patterns that reflect topographic gradients aid in identifying proxies for plant range boundaries. The patterns in the range proxies and their gradients with adjacent isopolls are consistent with the view that certain plant taxa recolonized the deglaciated Northeast in small populations and not as migrating fronts of large populations. How and when plant populations recolonized deglaciated areas have implications both for environmental heterogeneity and resource diversity; both of the latter environmental attributes affect distributions of animal and human populations.

3. The topographic and macroscale isopoll patterns suggest climatic control of the inferred plant distributions. Such a relationship between plant distributions and climate has implications for paleoecological models that consider the dynamic relationships between humans, animals, plants, and climate.

Acknowledgments

This research was supported by grants from the NSF Climate Dynamics Program to T. Webb III at Brown University and by a grant from the Southampton College/Long Island University Academic Research Committee. I thank the many people whose contributions of published and unpublished pollen data to the data base at Brown allowed this research to assume its broad perspective. I thank J. Avizinis for technical assistance. R. Webb helped in compiling the paleogeographic base maps. J. W. Gephart provided advice on the use of Macintosh graphics for drafting the figures. I thank T. Webb III, S. Suter, and G. Nicholas for valuable discussions. D. F. Dincauze, W. A. Patterson III, and T. Webb III read versions of the manuscript and offered helpful criticism.

6. APPENDIX: REFERENCES FOR MAP DATA BASE

Anderson, R. S., 1979, *A Holocene Record of Vegetation and Fire at Upper South Branch Pond in Northern Maine,* M.Sc. Thesis, University of Maine, Orono, Maine, unpublished.

Bailey, R. E., and Ahearn, P. J., 1981, A Late- and Postglacial Pollen Record from Chippewa Bog, Lapeer Co., MI: Further Examination of White Pine and Beech Immigration into the Central Great Lakes Region, in: *Geobotany II* (R. C. Romans, ed.), Plenum Press, New York, pp. 53–74.

Barclay, F. H., 1957, *The Natural Vegetation of Johnson County, Tennessee Past and Present,* Ph.D. Dissertation, University of Tennessee, Knoxville, Tennessee, unpublished.

Bernabo, J. C., 1977, *Sensing Climatically and Culturally Induced Environmental Changes with Palynological Data,* Ph.D. Dissertation, Brown University, Providence, Rhode Island, unpublished.

Bernabo, J. C., 1981, Quantitative Estimates of Temperature Changes over the Last 2700 Years in Michigan Based on Pollen Data, *Quaternary Research* 15:143–159.

Bostwick, L. G., 1978, *An Environmental Framework for Cultural Change in Maine: Pollen Influx and Percentage Diagrams from Monhegan Island,* M.Sc. Thesis, University of Maine, Orono, Maine, unpublished.

Comtois, P., 1982, Histoire Holocène du Climat et de la Végétation à Lanoraie (Québec), *Canadian Journal of Earch Sciences* 19:1938–1952.

Cotter, J. F. P., and Crowl, G. H., 1981, The Paleolimnology of Rose Lake, Potter Co., Pennsylvania: A Comparison of Palynologic and Paleo-Pigment Studies, in: *Geobotany II* (R. C. Romans, ed.), Plenum Press, New York, pp. 91–116.

Craig, A. J., 1969, Vegetational History of the Shenandoah Valley, Virginia, *Geological Society of America Special Paper* 123:283–296.

Davis, M. B., 1969, Climatic Changes in Southern Connecticut Recorded by Pollen Deposition at Rogers Lake, *Ecology* 50:409–422.

Davis, M. B., 1978, Climatic Interpretation of Pollen in Quaternary Sediments, in: *Biology and*

Quaternary Environments (D. Walker and J. C. Guppy, eds.), Australian Academy of Science, Canberra, Australia, pp. 35–51.

Davis, R. B., Bradstreet, T. E., Stuckenrath, R., Jr., and Borns, H. W., Jr., 1975, Vegetation and Associated Environments during the Past 14,000 Years Near Moulton Pond, Maine, *Quaternary Research* 5:436–465.

Gajewski, K., 1983, *On the Interpretation of Climatic Change from the Fossil Record: Climatic Change in Central and Eastern United States over the Past 2000 Years Estimated from Pollen Data,* Ph.D. Dissertation, University of Wisconsin, Madison, Wisconsin, unpublished.

Gauthier, R., 1981, *Histoire de la Colonisation Vegetale Postglaciaire des Monteregiennes: Deux Sites du Mont Saint-Bruno,* Memoire M.Sc., Université de Montréal, Québec, Canada, unpublished.

Gilliam, J. A., Kapp, R. O., and Bogue, R. D., 1967, A Post-Wisconsin Pollen Sequence from Vestaburg Bog, Moncalm County, Michigan, *Michigan Academy of Science, Arts, and Letters* 52:3–17.

Ibe, R. A., 1982, *Quaternary Palynology of Five Lacustrine Deposits in the Catskill Mountain Region of New York,* Ph.D. Dissertation, New York University, New York, unpublished.

Ibe, R. A., 1985, Postglacial Montane Vegetational History Around Balsam Lake, Catskill Mountains, New York, *Bulletin of the Torrey Botanical Club* 112:176–186.

Karrow, P. F., Anderson, T. W., Clarke, A. H., Delorme, L. D., and Steenivasa, M. R., 1975, Stratigraphy, Paleontology, and Age of Lake Algonquin Sediments in Southwestern Ontario, Canada, *Quaternary Research* 5:49–87.

Kerfoot, W. C., 1974, Net Accumulation Rates and the History of Cladoceran Communities, *Ecology* 55:51–61.

Labelle, C., and Richard, P. J. H., 1981, Végétation Tardiglaciaire et Postglaciare au Sud-Est du Parc des Laurentides, Québec, *Géographie Physique et Quaternaire* 35:345–359.

Likens, G. E., and Davis, M. B., 1975, Post-Glacial History of Mirror Lake and Its Watershed in New Hampshire, U.S.A.: An Initial Report, *Verhandlungen der Internationalen Vereinigung für Theoretische und Angewandte Limnologie* 19:982–993.

Liu, K. B., 1982, *Postglacial Vegetational History of Northern Ontario: A Palynological Study,* Ph.D. Dissertation, University of Toronto, Toronto, Ontario, unpublished.

Maxwell, J. A., and Davis, M. B., 1972, Pollen Evidence of Pleistocene and Holocene Vegetation on the Allegheny Plateau, Maryland, *Quaternary Research* 2:506–530.

McAndrews, J. H., 1970, Fossil Pollen and Our Changing Landscape and Climate, *Rotunda* 3:30–37.

McAndrews, J. H., 1976, Fossil History of Man's Impact on the Canadian Flora: An Example from Southwestern Ontario, *Canadian Botanical Association Bulletin Supplement* 9:1–6.

McAndrews, J. H., 1981, Late Quaternary Climate of Ontario: Temperature Trends from the Fossil Pollen Record, in: *Quaternary Paleoclimate* (W. C. Mahaney, ed.), Geo Abstracts, Norwich, England, pp. 319–333.

McDowell, L. L., Dole, R. M., Jr., Howard, M., Jr., and Farrington, R. A., 1971, Palynology and Radiocarbon Chronology of Bugbee Wildflower Sanctuary and Natural Area, Caledonia County, Vermont, *Pollen et Spores* 13:73–91.

Miller, N. G., 1973, Late-Glacial and Postglacial Vegetation Change in Southwestern New York State, *Bulletin of the New York State Museum and Science Service* 420:1–102.

Mott, R. J., 1975, Palynological Studies of Lake Sediment Profiles from Southwestern New Brunswick, *Canadian Journal of Earth Sciences* 12:273–288.

Mott, R. J., 1977, Late-Pleistocene and Holocene Palynology in Southeastern Quebec, *Géographie Physique et Quaternaire* 31:139–149.

Mott, R. J., and Camfield, M., 1969, Palynological Studies in the Ottawa area, *Geological Survey of Canada Paper* 69-38:1–16.

Mott, R. J., and Farley-Gill, L. D., 1978, A Late-Quaternary Pollen Profile from Woodstock, Ontario, *Canadian Journal of Earth Sciences* 15:1101–1111.

Mott, R. J., and Farley-Gill, L. D., 1981, Two Late Quaternary Pollen Profiles from Gatineau Park, Quebec, *Geological Survey of Canada Paper* 80-31:1–10.

Nelson, S., 1984, Upland and Wetland Vegetational Change in Southeastern Massachusetts: a 12,000 Year Record, *Northeastern Geology* 6:181–191.

Nicholas, J., 1968, *Late Pleistocene Palynology of Southeastern New York and Northern New Jersey,* Ph.D. Dissertation, New York University, New York, unpublished.

Niering, W. A., 1953, The Past and Present Vegetation of High Point State Park, New Jersey, *Ecological Monographs* 23:127–148.

Ogden, J. G., III, 1966, Forest History of Ohio: Radiocarbon Dates and Pollen Stratigraphy of Silver Lake, Ohio, *Ohio Journal of Science* 66:387–400.

Ogden, J. G., III, and Hay, R. J., 1967, Ohio Wesleyan University Natural Radiocarbon Measurements III, *Radiocarbon* 9:316–332.

Overpeck, J. T., 1985, A Pollen Study of a Late-Quaternary Peat Bog: South-Central Adirondack Mountains, New York, *Geological Society of America Bulletin* 96:145–154.

Richard, P., 1977, *Histoire Post-Wisconsinienne de la Végétation du Québec Méridional par l'Analyse Pollinique,* Service de la Recherche, Direction Generale des Forêts, Ministère des Terres et Forêts, Quebec, Canada.

Richard, P., 1980, Histoire Postglaciaire de la Végétation au Sud du Lac Abitibi, Ontario et Québec, *Géographic Physique et Quaternaire* 34:77–94.

Saarnisto, M., 1974, The Deglaciation History of the Lake Superior Region and its Climatic Implications, *Quaternary Research* 4:316–339.

Saarnisto, M., 1975, Stratigraphic Studies on the Shoreline Displacement of Lake Superior, *Canadian Journal of Earth Sciences* 12:300–319.

Savoie, L., and Richard, P., 1979, Paléophytogéographie de l'Épisode de Saint-Narcisse dans la Région de Sainte-Agathe, Québec, *Géographie Physique et Quaternaire* 33:175–188.

Shane, L. C. K., 1975, Palynology and Radiocarbon Chronology of Battaglia Bog, Portage County, Ohio, *Ohio Journal of Science* 75:96–102.

Shane, L. C. K., 1976, *Late-Glacial and Postglacial Palynology and Chronology of Drake County, West-Central Ohio,* Ph.D. Dissertation, Kent State University, Kent, Ohio, unpublished.

Sirkin, L. A., Denny, C. S., and Rubin, M., 1977, Late Pleistocene Environment of the Central Delmarva Peninsula, Delaware–Maryland, *Geological Society of America Bulletin* 88:139–142.

Spear, R. W., 1981, *The History of High-Elevation Vegetation in the White Mountains of New Hampshire,* Ph.D. Dissertation, University of Minnesota, Minneapolis, Minnesota, unpublished.

Spear, R. W., and Miller, N. G., 1976, A Radiocarbon Dated Pollen Diagram from the Allegheny Plateau of New York State, *Journal of the Arnold Arboretum of Harvard University* 57:369–403.

Suter, S. M., 1985, Late-Glacial and Holocene Vegetational History in Southeastern Massachusetts: A 14,000 Year Pollen Record, *Current Research in the Pleistocene* 2:87–89.

Terasmae, J., 1968, A Discussion of Deglaciation and the Boreal Forest in the Northern Great Lakes Region, *Proceedings of the Entomological Society of Ontario* 99:31–43.

Terasmae, J., and Anderson, T. W., 1970, Hypsithermal Range Extension of White Pine (*Pinus strobus* L.) in Quebec, Canada, *Canadian Journal of Earth Sciences* 7:406–413.

Vincent, J. S., 1973, A Palynological Study for the Little Clay Belt, Northwestern Quebec, *Naturaliste Canadien* 100:59–69.

Walker, P. C., and Hartman, R. T., 1960, The Forest Sequence of the Hartstown Bog Area in Western Pennsylvania, *Ecology* 41:461–474.

Watts, W. A., 1979, Late Quaternary Vegetation of Central Appalachia and the New Jersey Coastal Plain, *Ecological Monographs* 49:427–469.

Whitehead, D. R., 1972, Development and Environmental History of the Dismal Swamp, *Ecological Monographs* 42:301–315.

Whitehead, D. R., 1979, Late-Glacial and Postglacial Vegetational History of the Berkshires, Western Massachusetts, *Quaternary Research* 12:333–357.

Whitehead, D. R., 1981, Late-Pleistocene Vegetational Changes in Northeastern North Carolina, *Ecological Monographs* 51:451–471.

Whitehead, D. R., and Crisman, T. L., 1978, Paleolimnological Studies of Small New England (U.S.A.) Ponds. Part I. Late-Glacial and Postglacial Trophic Oscillations, *Polskie Archiwum Hydrobiologii* 25:471–481.

Winkler, M. G., 1985, A 12,000-Year History of Vegetation and Climate for Cape Cod, Massachusetts, *Quaternary Research* 23:301–312.

7. REFERENCES

Anderson, P. M., 1982, *Reconstructing the Past: The Synthesis of Archaeological and Palynological Data, Northern Alaska and Northwestern Canada,* Ph.D. Dissertation, Brown University, Providence, Rhode Island, unpublished.

Bartlein, P. J., Prentice, I. C., and Webb, T., III, 1986, Climatic Response Surfaces from Pollen Data for Some Eastern North America Taxa, *Journal of Biogeography* 13:35–57.

Bennett, K. D., 1985, The Spread of *Fagus grandifolia* across Eastern North America during the Last 18000 Years, *Journal of Biogeography* 12:147–164.

Bernabo, J. C., and Webb, T., III, 1977, Changing Patterns in the Holocene Pollen Record from Northeastern North America: A Mapped Summary, *Quaternary Research* 8:64–96.

Bonnichsen, R., Jacobson, G. L., Jr., Davis, R. B., and Borns, H. W., Jr., 1985, The Environmental Setting for Human Colonization of Northern New England and Adjacent Canada in Late Pleistocene Time, in: *Late Pleistocene History of Northeastern New England and Adjacent Quebec* (H. W. Borns, Jr., P. LaSalle, and W. B. Thompson, eds.), *Geological Society of America Special Paper,* No. 197, pp. 151–159.

Braun, E. L., 1950, *Deciduous Forests of Eastern North America,* Blakiston, Philadelphia, Pennsylvania.

Brubaker, L. B., 1975, Postglacial Forest Patterns Associated with Till and Outwash in Northcentral Upper Michigan, *Quaternary Research* 5:499–527.

Bryant, V. M., Jr., and Holloway, R. G., 1983, The Role of Palynology in Archaeology, in: *Advances in Archaeological Method and Theory,* Vol. 6 (M. Schiffer, ed.), Academic Press, New York, pp. 191–224.

Curran, M. L., and Dincauze, D. F., 1977, Paleoindians and Paleo-Lakes: New Data from the Connecticut Drainage, *Annals of the New York Academy of Sciences* 288:333–348.

Davis, M. B., 1969, Climatic Changes in Southern Connecticut Recorded by Pollen Deposition at Rogers Lake, *Ecology* 50:409–422.

Davis, M. B., 1976, Pleistocene Biogeography of Temperate Deciduous Forests, *Geoscience and Man* 13:13–26.

Davis, M. B., 1981a, Outbreaks of Forest Pathogens in Quaternary History, in: *Proceedings of the Fourth International Palynological Conference, Lucknow, India (1976–1977),* pp. 216–227.

Davis, M. B., 1981b, Quaternary History and the Stability of Forest Communities, in: *Forest Succession* (D. C. West, H. H. Shugart, and D. B. Botkin, eds.), Springer-Verlag, New York, pp. 132–177.

Davis, M. B., 1983, Holocene Vegetational History of the Eastern United States, in: *Late Quaternary Environments of the United States.* Vol. 2. *The Holocene* (H. E. Wright, Jr., ed.), University of Minnesota Press, Minneapolis, Minnesota, pp. 166–181.

Davis, M. B., Spear, R. W., and Shane, L. C. K., 1980, Holocene Climate of New England, *Quaternary Research* 14:240–250.

Davis, M. B., Woods, K. D., Webb, S. L., and Futyma, R. P., 1986, Dispersal versus Climate: Expansion of *Fagus* and *Tsuga* in the Upper Great Lakes Region, *Vegetatio* 67:93–103.

Davis, R. B., and Jacobson, G. L., Jr., 1985, Lateglacial and Early Holocene Landscapes in Northern New England and Adjacent Areas of Canada, *Quaternary Research* 23:341–368.

Davis, R. B., and Webb, T., III, 1975, The Contemporary Distribution of Pollen from Eastern North America: A Comparison with the Vegetation, *Quaternary Research* 5:395–434.

Delcourt, H. R., and Delcourt, P. A., 1985, Quaternary Palynology and Vegetational History of the Southeastern United States, in: *Pollen Records of Late-Quaternary North American Sediments* (V. M. Bryant, Jr., and R. G. Holloway, eds.), American Association of Stratigraphic Palynologists Foundation, Dallas, Texas, pp. 1–37.

Delcourt, H. R., Delcourt, P. A., and Webb, T., III, 1983, Dynamic Plant Ecology: The Spectrum of Vegetational Change in Space and Time, *Quaternary Science Reviews* 1:153–175.

Delcourt, P. A., and Delcourt, H. R., 1981, Vegetation Maps for Eastern North America, in: *Geobotany II* (R. C. Romans, ed.), Plenum Press, New York, pp. 123–165.

Delcourt, P. A., and Delcourt, H. R., 1983, Late-Quaternary Vegetational Dynamics and Community Stability Reconsidered, *Quaternary Research* 19:265–271.

Delcourt, P. A., Delcourt, H. R., and Webb, T., III, 1984, *Atlas of Mapped Distributions of Dominance and Modern Pollen Percentages for Important Tree Taxa of Eastern North America,* American Association of Stratigraphic Palynologists Contributions Series, No. 14.

Denton, G. H., and Hughes, T. J. (eds.), 1981, *The Last Great Ice Sheets,* Wiley-Interscience, New York.

Dincauze, D. F., 1981, Paleoenvironmental Reconstruction in the Northeast: The Art of Multidisciplinary Science, in: *Foundations of Northeast Archaeology* (D. Snow, ed.), Academic Press, New York, pp. 51–96.

Dincauze, D. F., and Mulholland, M. T., 1977, Early and Middle Archaic Site Distributions and Habitats in Southern New England, *Annals of the New York Academy of Sciences* 288:439–456.

Eyre, F. H. (ed.), 1980, *Forest Cover Types of the United States and Canada,* Society of American Foresters, Washington, D.C.

Fenneman, N. M., 1946, *Physical Divisions of the United States,* U. S. Department of the Interior, Geological Survey.

Fremlin, G. (ed.), 1974, *The National Atlas of Canada,* Canada Department of Energy, Mines and Resources, Ottawa, Canada.

Gaudreau, D. C., 1984, Features of Holocene Plant Population Dynamics Inferred from Mapped Pollen Data from Northeastern North America, *Bulletin of the Ecological Society of America* 65:211.

Gaudreau, D. C., 1986, *Late-Quaternary Vegetational History of the Northeast: Paleoecological Implications of Topographic Patterns in Pollen Distributions,* Ph.D. Dissertation, Yale University, New Haven, Connecticut, unpublished.

Gaudreau, D. C., and Webb, T., III, 1985, Late-Quaternary Pollen Stratigraphy and Isochrone Maps for the Northeastern United States, in: *Pollen Records of Late-Quaternary North American Sediments* (V. M. Bryant, Jr., and R. G. Holloway, eds.), American Association of Stratigraphic Palynologists Foundation, Dallas, Texas, pp. 247–280.

Hughes, T., Borns, H. W., Jr., Fastook, J. L., Hyland, M. R., Kite, J. S., and Lowell, T. V., 1985, Models of Glacial Reconstruction and Deglaciation Applied to Maritime Canada and New England, in: *Late Pleistocene History of Northeastern New England and Adjacent Quebec* (H. W. Borns, Jr., P. LaSalle, and W. B. Thompson, eds.), Geological Society of America Special Paper, No. 197, pp. 139–150.

Jacobson, G. L., Jr., 1979, The Palaeoecology of White Pine (*Pinus strobus*) in Minnesota, *Journal of Ecology* 67:697–726.

Jacobson, G. L., Jr., and Bradshaw, R. H. W., 1981, The Selection of Sites for Paleoenvironmental Studies, *Quaternary Research* 16:80–96.

King, J. E., 1985, Palynological Applications to Archaeology: An Overview, in: *Archaeological Geology* (G. Rapp, Jr., and J. A. Gifford, eds.), Yale University Press, New Haven, Connecticut, pp. 135–154.

Küchler, A. W., 1964, *Potential Natural Vegetation of the Conterminous United States (Map and Manual)*, American Geographical Society Special Publication, No. 36.

Little, E. L., Jr., 1971, *Atlas of United States Trees*. Vol. 1. *Conifers and Important Hardwoods*, U.S.D.A. Forest Service Miscellaneous Publication, No. 1146.

Lull, H. W., 1968, *A Forest Atlas of the Northeast*, U.S.D.A. Forest Service, Northeastern Forest Experiment Station, Upper Darby, Pennsylvania.

Nicholas, G. P., 1980, Late Pleistocene–Early Holocene Occupation in New Hampshire Glacial Lakes: Paleoindian Settlement, Subsistence, and Environment, *New Hampshire Archeologist* 21:55–65.

Nicholas, G. P., 1983, A Model for the Early Postglacial Settlement of the Central Merrimack River Basin, New Hampshire, *Man in the Northeast* 25:43–63.

Nichols, G. E., 1935, The Hemlock–White Pine–Northern Hardwood Region of Eastern North America, *Ecology* 16:403–422.

Overpeck, J. T., and Fleri, E. C., 1982, The Development of Age Models for Holocene Sediment Cores: Northeast North American Examples, *AMQUA Abstracts* (American Quaternary Association) 1982:152.

Prentice, I. C., 1983, Postglacial Climatic Change: Vegetational Dynamics and the Pollen Record, *Progress in Physical Geography* 7:273–286.

Prentice, I. C., 1985, Pollen Representation, Source Area, and Basin Size: Toward a Unified Theory of Pollen Analysis, *Quaternary Research* 23:76–86.

Prentice, I. C., 1986, Vegetation Responses to Past Climatic Variation, *Vegetatio* 67:131–141.

Quigley, R. M., 1983, Glaciolacustrine and Glaciomarine Clay Deposition: A North American Perspective, in: *Glacial Geology* (N. Eyles, ed.), Pergamon Press, New York, pp. 140–167.

Sears, P. B., 1942, Postglacial Migration of Five Forest Genera, *American Journal of Botany* 29:684–691.

Solomon, A. M., and Silkworth, A. B., 1986, Spatial Patterns of Atmospheric Pollen Transport in a Montane Region, *Quaternary Research* 25:150–162.

Szafer, W. L., 1935, The Significance of Isopollen Lines for the Investigation of the Geographical Distribution of Trees in the Post-Glacial Period, *Bulletin International de l'Academie Polonaise des Sciences et des Lettres. Série B. Sciences Naturelles* I:235–239.

Watts, W. A., 1983, Vegetational History of the Eastern United States 25,000 to 10,000 Years Ago, in: *Late-Quaternary Environments of the United States*. Vol. 1. *The Late Pleistocene* (S. C. Porter, ed.), University of Minnesota Press, Minneapolis, Minnesota, pp. 294–310.

Webb, T., III, 1982, Temporal Resolution in Holocene Pollen Data, in: *Third North American Paleontological Convention, Proceedings*, Vol. 3, pp. 569–572.

Webb, T., III, 1986, Is Vegetation in Equilibrium with Climate? How to Interpret Late-Quaternary Pollen Data, *Vegetatio* 67:75–91.

Webb, T., III, and McAndrews, J. H., 1976, Corresponding Patterns of Contemporary Pollen and Vegetation in Central North America, *Geological Society of America Memoir* 145:267–299.

Webb, T., III, Laseski, R. A., and Bernabo, J. C., 1978, Sensing Vegetational Patterns with Pollen Data: Choosing the Data, *Ecology* 59:1151–1163.

Webb, T., III, Howe, S. E., Bradshaw, R. H. W., and Heide, K. M., 1981, Estimating Plant Abundances from Pollen Percentages: The Use of Regression Analysis, *Review of Palaeobotany and Palynology* 34:269–300.

Webb, T., III, Cushing, E. J., and Wright, H. E., Jr., 1983a, Holocene Changes in the Vegetation of the Midwest, in: *Late Quaternary Environments of the United States*. Vol. 2. *The Holocene* (H. E. Wright, Jr., ed.), University of Minnesota Press, Minneapolis, Minnesota, pp. 142–165.

Webb, T., III, Richard, P. J. H., and Mott, R. J., 1983b, A Mapped History of Holocene Vegetation in Southern Quebec, *Syllogeus* 49:273–336.

Whitehead, D. R., 1979, Late-Glacial and Postglacial Vegetational History of the Berkshires, Western Massachusetts, *Quaternary Research* 12:333–357.

Chapter 10

Ecological Leveling

The Archaeology and Environmental Dynamics of Early Postglacial Land Use

GEORGE P. NICHOLAS

1. INTRODUCTION

Environmental productivity has long been considered a principal factor of human land use. The historic association of modern nonagricultural groups with marginal environments may be one of the primary reasons why prehistoric hunter–gatherer groups have most frequently been described as small, mobile, and egalitarian (Lee and Devore 1968). Within highly productive landscapes, however, how would groups of gatherer–hunters[1] organize themselves spatially, socially, and economically? There is evidence to suggest that certain parts of the Northeastern interior, namely former glacial lake basins (Nicholas 1982, 1983), did at times possess an environment richer than at present, particularly during the early postglacial period. This possibility raises important questions about what the archaeological record of such basins actually represents. A number of

[1]As a response to what is being seen as greater economic flexibility in past human behavior, many researchers have replaced the term *hunter–gatherer* with *gatherer–hunter* (e.g., Bender 1978, 1985). This is not a minor point; as Marquardt correctly notes, "the way we write conditions (if not betrays) the way we think" (Marquardt 1985:84). He goes so far as to use the term "fisher–gatherer–hunter"; while in full agreement with his usage, I find *gatherer–hunter* a more convenient term at present, however, for exploring diversity in land-use behaviors at places like Robbins Swamp. Where the term *hunter–gatherer* is used in the text, it refers to the more traditional definition in the literature.

GEORGE P. NICHOLAS • Department of Anthropology, University of Massachusetts, Amherst, Massachusetts 01003, and Research Department, American Indian Archaeological Institute, Washington, Connecticut 06793.

studies have recently challenged the so-called anomalous nature of some groups of nonagricultural, complex societies, such as found on the Northwest Coast of North America, a region traditionally represented in the literature (e.g., Suttles 1968) as a locality where intensive maritime exploitation effectively replaced horticulture as the means for increased sedentism. The productivity of this environment was apparently a prime factor for social transformation and the establishment of ranked society (Ames 1981; Carlson 1983; Schalk 1981). Some researchers have gone so far as to suggest that the richness of Northwest Coast cultures may be a reasonable approximation of prehistoric hunter–gatherers in general (King 1978; see also Phillips and Brown 1983; Price and Brown 1985). The degree to which the Northeastern archaeological record of early hunter–gatherers supports propositions of environmentally rich landscapes and of greater diversity in the behaviors and social relations of land use is examined in this paper.

The archaeology of the early postglacial period in the Northeast (12,000–7000 B.P.) has much to contribute to the understanding of the dynamic relationship between hunter–gatherers and their environments over time. Variation in the regional archaeological record represents, to a large degree, shifts in the social, settlement, and subsistence systems as prehistoric populations responded to both short- and long-term environmental change. There is clearly no consensus, however, in explaining this variation; the early archaeological record of the Northeast has been examined within the context of both low productivity (e.g., Funk 1978; Funk and Wellman 1984) and high productivity (e.g., Dincauze and Mulholland 1977; Nicholas 1987) landscapes, each of which suggests entirely different sets of land-use behaviors.

Many archaeologists have interpreted diversity in the early prehistoric record by relying heavily on the use of modern environmental and ethnographic analogs. This type of interpretation frequently operates along evolutionary trajectories of low to high resource productivity or of simple to complex cultural systems. The most commonly used analog for the early postglacial period in the Northeast has been the ecologically restrictive landscapes of the northern tundra and boreal forests (Fitting 1968; Funk 1978), which are replaced only in southern New England by emerging deciduous forests during the early Holocene. Likewise, ethnographic analogs for Paleoindian and Early Archaic social, settlement, and subsistence systems have been provided by the Caribou Eskimo and Naskapi, among others (Funk 1978:18; MacDonald 1968; also Fitting 1977).

A strong argument has been made in recent years for the use of detailed paleoenvironmental models (Dincauze 1981) that address rigorously problems of definition, scale, and representative data sets in the study of cultural and environmental change. The development and application of such models has demonstrated that: (1) the early postglacial environment of northeastern North America was generally more diverse ecologically, had a broader resource base, and was climatically more seasonally distinct than had previously been assumed (Bryson

1985; Gaudreau, Chapter 9; Nicholas 1987); and (2) both the early archaeological record and the range of behaviors it represents have been underestimated significantly due to substantial interpretive and site-preservational biases (Dincauze and Mulholland 1977; Nicholas 1987). In fact, certain parts of the early Holocene landscape (namely glacial lake basins) may have been as productive as the coastal plain, long considered an important focus of settlement in the Northeast. If highly productive areas were indeed present on the early postglacial landscape, then the archaeological record of early hunter–gatherer land use should represent a set of behaviors and social relations similar to those of other productive environments. This hypothesis is examined critically using two models—the glacial lake basin mosaic model and ecological leveling—to explain changes in early postglacial environments and land use at the local (versus regional) level.

The glacial lake basin mosaic model (Nicholas 1982, 1983; see also Curran and Dincauze 1977; Handsman 1983) posits a series of landscape developments for specific parts of the early postglacial landscape against which the distribution of archaeological sites can be plotted, and diversity in the range of site types and behaviors in the use of that space mapped. This model is based on the postulate that wetland-dominated mosaics extremely rich in both resource diversity and biomass existed within many of the former glacial lake basins that are present across much of the Northeast. These basins thus provided focal points for settlement (i.e., *places* on the landscape) during the late Pleistocene and early Holocene periods, a time when the otherwise high productivity of the coastal plain was affected by transgressing sea levels. The behavioral patterns associated with such mosaics should be represented archaeologically by greater diversity over time in site location, and by increased local *site frequency,* number of sites per millennium (or other time unit), number of components per site, and length and season of occupation.

Ecological leveling is a concept used to examine changes in the composition, productivity, and environmental dynamics of former lake basins and in the associated cultural patterns of land use during the early postglacial period. The term itself refers to a postulated equalizing or reduction (i.e., "leveling") of differences in the resource base between basin and nonbasin areas through time as the result of developments in nonbasin biotic communities during the early Holocene. The increasing attractiveness of river valleys, the coastal plain, upland areas, and other landscape components by the end of the early Holocene would have resulted in shifts in land use primarily to maintain maximum foraging returns (Charnov 1976; Curran 1987:64–66). This may have been encouraged further by a decrease in basin resource productivity through long-term, geographically focused exploitation of the intrabasin resource base. The behaviors of land use associated with leveling should reflect a change in the way that these basins were used; i.e., parts of the landscape that had been *core areas* of intensive settlement became *peripheral areas* of less intensive or different land-use activities.

The way that these two models may explain long-term patterns in the relationship between early postglacial land use and dynamic landscape developments is discussed in relation to Robbins Swamp, an extensive glacial lake-based landscape of wetlands and rivers located in southwestern New England (Figure 1), where over 10,000 years of prehistoric occupation are represented. The following research questions frame a more extensive discussion:

1. What is ecological leveling and why does it differ conceptually from other approaches to the early Holocene period in the Northeast?

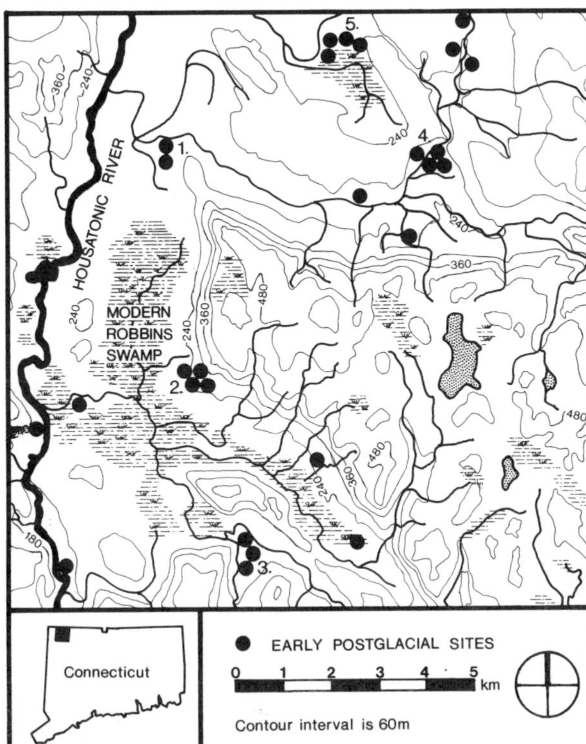

Figure 1. Location map of early postglacial archaeological sites and modern landscape features around Robbins Swamp, northwestern Connecticut. This figure illustrates only the southern portion of the project area which extends northward into Massachusetts. Redundancy in the use of the late Pleistocene/early Holocene landscape is evidenced by the high frequency of archaeological sites present and by a number of site "clusters": (1) Dead Exxon, (2) Foley, (3) Dean, (4) SWB/NWB, (5) Segalla.

2. What is the relationship between it and ideas associated with space, "place," and long-term patterns of land use?
3. How does the idea of leveling extend earlier work on postglacial landscape mosaics?
4. What do the mosaic model and leveling concept suggest about how to undertake research on core and periphery areas, on Robbins Swamp as a "site," and on the application of more traditional settlement/subsistence models?

Site-specific and land-use pattern data from the Robbins Swamp study area (Handsman 1983; Nicholas 1985, 1987; Nicholas and Handsman 1984a,b) are discussed in terms of their correspondence to other glacial lake basin and nonbasin data sets within the region (e.g., Mulholland 1984). Approaches to exploring the middle range between human behaviors and the archaeological record are also examined in terms of both theoretical considerations and field data. Several basic assumptions are used implicitly throughout this study: (1) sites are not located randomly in relation to critical resources; (2) groups organize themselves to take maximum advantage of highly productive and reliable habitats; (3) areas of resource concentration are temporally and spatially transgressive; and (4) shifts in the ecological composition of landscapes over time and space affect prehistoric land use. These and related premises are well grounded in ecological and anthropological studies (e.g., Charnov 1976; Colwell 1974; Curran 1987; Ellen 1982; Jochim 1981; Levin 1976; MacArthur and Pianka 1966; Moran 1982; Winterhalder and Smith 1981). Finally, the general orientation of this paper is directed explicitly to the general behaviors of land use; aspects of the social relations of land use, which are dealt with only superficially below except where the two types of behavior overlap (e.g., group size), are not addressed.

Clearly, the study of how landscapes and the behaviors of land use changed is important to an understanding of the dynamic relationship between early postglacial populations and their landscapes and in determining how best to do the archaeology of early gatherer–hunters. Concepts like ecological leveling are particularly appropriate because they provide a way to examine historical processes within the distant prehistoric past.

2. THE ARCHAEOLOGY OF EARLY GATHERER–HUNTER LAND USE

The early postglacial period in the Northeast was a time of relatively rapid and dramatic cultural and landscape developments. In fact, the dynamic character of this period has frequently been proposed as the major determinant of cultural, social, technological, and demographic development (e.g., Fitting 1968; Funk 1978). Aspects of the complex relationship between early gatherer–hunters and this dynamic environment should be evidenced in Paleoindian and Early Archaic land use, *although not necessarily at individual sites*. Land-use studies, versus single-site analyses, are important because they represent a class of data appropriate to the scale at which populations maintained the *status quo* over time and space, and provide the type of data necessary to develop robust models of Paleoindian and Early Archaic social, economic, and subsistence systems, whether in phase or out of phase with changing environmental conditions.

How variability in the archaeological record of land use is interpreted, however, remains subject to much debate (see Dincauze 1985 and Butzer 1971, 1982, respectively, for discussions on sociocultural and environmental determinants of variability). Some traditional models of early hunter–gatherers used variables sensitive to exogenous sources of change (e.g., migration, diffusion), coupled with the constraints of an essentially homogeneous environment, to explain a limited, regional archaeological record (Funk and Wellman 1984; Snow 1980). Other models sensitive to primarily endogenous social, technological, and economic opportunities, provided by a more heterogeneous environment, are used to explain variability in what may be a substantial local archaeological record (Dincauze and Mulholland 1977; Nicholas 1987). In either case, simply increasing the number of early sites will do little to end such debate. At locations where there are sufficient archaeological data, the appropriateness of any of these models to particular times and places in Northeastern prehistory can be addressed by evaluating: (1) the types of behaviors and social relations of land use that produced the early archaeological record; and (2) the processes by which land-use patterns changed in regard to shifts in resource availability or productivity or in other critical environmental factors. Adjusting the scale of interpretation from the number of sites to what those sites represents effectively also helps to break the circular reasoning trap of *limited site inventory ⇒ low site-density models ⇒ research designs that expect few sites ⇒ few sites*. The way that each of two explanatory models is addressed within the present study is discussed below.

2.1. The Prehistory of Early Gatherer–Hunters

For 20 years, the definition of hunter–gatherers loosely consolidated by the *Man the Hunter* symposium (Lee and DeVore 1968) has shaped significantly the ways that nonagricultural populations have been characterized. A variety of hunter–gatherers became the hallmarks of the prehistoric past. Prehistoric populations in northeastern North America, from the Paleoindian through Archaic periods, have since been characterized by small, egalitarian, highly mobile groups possessing a limited repertoire of social, technological, and organizational strategies. As a result, what is essentially an evolutionary scheme (Fitting 1977;

Service 1962) has been gradually superimposed upon cultural development during the Holocene. Paleoindians have been viewed, in effect, as less socially, technologically, and economically organized than subsequent Archaic populations, who were, in turn, correspondingly less well-organized than the semipermanent, horticulture-based Woodland populations, who possessed elaborate ceremonialism, class ranking, and ceramics. A growing number of studies have challenged the notion that hunter–gatherers are always mobile, band-sized, and noncomplex (e.g., Bender 1985; King 1978; Price and Brown 1985; Rick 1980; Rowley-Conwy 1983; also Schrire 1984). Recognizing the degree to which diversity is represented in the prehistoric records of the American Midwest (Phillips and Brown 1983), the Northwest and Northeast Coasts of North America and Asia, respectively (Ames 1985; Hayden *et al.* 1985; Koyama and Thomas 1981), or northwestern Europe (Clarke 1976) is especially important to an understanding of the basic range of behavioral responses available to past human populations within a variety of environmental conditions.

Some traditional approaches to the archaeology of hunter–gatherers may therefore be inappropriate for portions of the prehistoric record in the Northeast, particularly that of the early postglacial period (Nicholas 1987). Much of this archaeology has been done on the basis of assumptions of resource-poor landscapes, mobile and egalitarian populations, and limited site types that have characterized the literature on the early postglacial. Since the mid-1970s, some New England archaeologists have begun to reexamine the analytical and interpretive models that have been constructed about this period. These critical efforts have been stimulated, in part, by important discoveries elsewhere (e.g., Gardner 1983; Judge 1973; Schiffer 1975) that demonstrated that early archaeological remains in North America were not always inconsequential. Other studies have documented that early artifact assemblages and site distribution patterns are not always what was expected (e.g., Broyles 1971; Chapman 1975, 1985; Chapman and Adovasio 1977; Dincauze and Mulholland 1977).

At the same time, advances in other environmentally oriented disciplines have caused archaeologists to become more sensitized to the various scales of interaction that exist between the early populations and environments in the Northeast at a time of rapid and perhaps dramatic adjustment to newly emerged landscapes (see Bettinger 1980, Butzer 1982, and Dincauze 1981 for related discussions). As many of the papers in the present volume suggest, *obvious correlations* between changes in the paleoenvironmental record and in the archaeological record *are not always meaningful correlations*. Historical or processual approaches separately may be inadequate to isolating the "true" points of correlation between the two. An interesting parallel can be drawn between how paleoecologists are rethinking the dynamics of forest succession (see below) and how prehistorians are reviewing the dynamics of cultural and environmental change, which is particularly evident in the recognition of how individual organisms through to biome-level communities respond to a variety of types and scales of change (Dincauze 1981).

2.2. The Importance of Prehistoric Land-Use Studies

Studies of long-term land use are particularly well-suited for examining the relationship between cultural and environmental change. By mapping the behaviors and social relations of land use exhibited by a band of hunter–gatherers over the course of a generation, a detailed picture should emerge of the daily, seasonal, or annual dynamic structure of their lifeway[2] (e.g., Lee and Devore 1968). However, even if an ethnography of a Paleoindian or Early Archaic group were available, its usefulness would be restricted to describing only a limited set of short-term responses to change over several generations. If one is to make sense of these processes within a historical framework, then land-use patterns representing tens or hundreds of generations need to be observed.

Single sites are also normally inappropriate as the primary points of reference because populations respond relatively slowly to change; a single site usually represents only days or weeks of occupation, while multiple-component sites generally reflect repeated short-term use. Land-use patterns, however, evidence trends in site distribution, duration of use, intensity of resource exploitation, redundancy in the use of space, and geographic specificity that are the result of *long-term* human behavior. How a particular part of the landscape is used by populations over hundreds and thousands of years corresponds more closely to developments in local and regional landscape, biotic communities, and climate. When local archaeological and paleoenvironmental data are examined at *corresponding scales of interaction,* greater resolution becomes possible in explaining how human populations are affected by both internal and external-based change. Land-use data, unlike many other data sets, can be adjusted to correlate with specific environmental data by increasing their time depth.

Archaeologically defined patterns of land use reflect the repetition of certain types of prehistoric behavior (Binford 1982, 1983; White 1987). The use of parts of landscape for settlement and for social or subsistence-related activities can be defined as a *system pattern,* that is, the behaviors of land use. By focusing on the middle range between archaeologically and system-defined patterns of land use, the processes of prehistoric lifeways can be studied, and determinations on how best to do the archaeology of early societies be made (which will not be done in

[2]Clifford's (1974) account of Charlie York, a Maine Coast fisherman, provides an interesting twentieth century example of the degree to which historic "fisher–gatherer–hunters" are sensitive to relatively subtle environmental fluctuations. In this case, York responded to both short- and long-term shifts in resource productivity and location through a wide (i.e., generalist) array of logistical and exploitive strategies.

the same way as that of later societies). This is especially important in the long-term study of land use, where the behaviors and sites represented in a study area are the product of hundreds if not thousands of years of responses to cultural and environmental change (e.g., Schiffer 1983). While the interpretation of the early prehistoric record is complicated further by a host of environmental factors as well as by time, these same factors may provide new ways of linking the archaeological record to the dynamic behaviors it represents because they necessitate changing the scale of analysis (for examples of this diversity see Binford 1982, Clarke 1976, Hodder 1978, Tilley 1979, and Weissner 1982; for examples of the extensive related geographic literature see Chisholm 1962 and Hagget 1965).

2.3. Dynamic Models of Environmental Change

Coeval with theoretical developments in archaeology has been the advancement by biological scientists of new models stressing the dynamic nature of landscape transformations over time and space (e.g., Colinvaux 1987; Davis 1981a; Delcourt *et al.* 1983; Gaudreau 1984; Gutierrez and Fey 1980; Wright 1976). Significant advances in the modeling of past culturally and environmentally induced systems ecology have produced new ways of viewing local and regional landscape developmental processes, climatically defined changes in seasonality, interaction between climate and biotic communities, and biome succession (e.g., Bernabo and Webb 1977; Chapman *et al.* 1982; Davis 1981b; Patterson and Sassaman, Chapter 5; Webb 1981). Other examples of this type of intradisciplinary development are cited by both Gaudreau and Joyce (Chapters 8 and 9).

The development of new concepts of succession by ecologists provides an example pertinent to the present discussion on the dynamics of environmental change in the Northeast. *Succession* is the "process undergone by natural ecosystems such as lakes, grasslands, and forests as they proceed from early stages of development to maturity, when the community of plants and animals achieves a condition of equilibrium with its physical environment" (Gutierrez and Fey 1980:xiii). Knowledge of the changes that occurred in biotic composition and of the processes by which those changes occurred are equally important to the understanding of the effects of dynamic environmental parameters upon human populations. The classic models of succession (e.g., Clements 1916; Odum 1971) were generally historical in nature, and focused on a general (and usually direct and in-phase) interaction between regional vegetation, climate, geography, and related factors through time, with interpretive emphasis placed on the community. What has occurred in recent years is an increased emphasis on the processes of biotic community development based on the *individual* elements of a community, which interestingly parallels the anthropological observation that adaptation is a local phenomenon (Binford 1968:323)—a parallel that is perhaps not unrelated. This emphasis on individualistic responses has led some ecologists to view ecosystem development not as a linear and directed successive process (e.g., Odum 1969), but rather as "a continuum of overlapping sets of species, each responding to subtly different environmental cues" (Mitsch and Gosselink 1986:147; also Gleason 1917; McIntosh 1980; Whittaker 1967); wetlands, for example, may exhibit properties of both mature and immature ecosystems. It is also important to note that changes in vegetational composition, location, or fitness may not necessarily be in equilibrium with climatic or other large-scale shifts in environment (P. A. Delcourt and Delcourt 1983), and that the primary in-phase or out-of-phase response of vegetation to these changes is at the level of individual taxa (Gaudreau, Chapter 9; Webb 1986).

3. THE GEOGRAPHY OF EARLY POSTGLACIAL LANDSCAPES

Landscapes are seldom homogeneous, especially at the scale at which people interact with them. Local and regional heterogeneity has only infrequently been emphasized in the context of prehistoric studies, however (e.g., Curran 1979, 1987), despite a long-lived correlation between particular cultural traditions and regional environmental zones (e.g., Paleoindian/spruce-tundra, Pollen Zone A). How the landscape is viewed is important not only for providing a backdrop against which to view prehistoric populations, but for identifying those factors (i.e., geological, hydrological, environmental, climatic) that may figure importantly in understanding the processes of prehistoric settlement, subsistence, and social transformation. Neither these environmental data, however, nor the paradigms from which they originate are permanent fixtures within paleoecology, and they should not become so within archaeological applications. For example, using the closed boreal forest type of south-central Canada as an analog, Fitting (1968) explained the paucity of early postglacial archaeological sites in the Northeast as being a function of a resource-poor environment. At the time, this model was sensitive to the types of paleoenvironmental data available, and to the scales of interaction then recognized, and could be considered relatively progressive. Substantial changes have since occurred in the regional archaeological and paleoenvironmental data base and in how environmental change is being modeled by paleoecologists, so that the utility of Fitting's model has become increasingly limited. Changes of this kind are often slow in coming, nonetheless, and frequently there is substantial lag time in interdiscipline communication.

3.1. The Glacial Lake Basin Mosaic Model

The traditional idea of inhospitable, homogeneous early postglacial environments (Fitting 1968; Funk 1978; Snow 1980) is currently being questioned by both archaeologists and paleoecologists (Brown and Cleland 1968; Curran and Dincauze 1977; Dincauze and Mulholland 1977; Gaudreau, Chapter 9; Nicholas 1983, 1987). Their work suggests that the limitations of early Holocene environments may have been overestimated, at least in some ecological settings. It can be argued that in some areas in New England, glacial lake basins, most active between 14,000 and 11,000 B.P., would have provided hydrological settings for the subsequent development of extensive lake/pond/wetland/river basin mosaics between 10,000 and 7000 B.P. Such systems have been reconstructed for basins in Connecticut (Nicholas and Handsman 1984b, 1986), Massachusetts (Curran and Dincauze 1977), northern Maine (Nicholas *et al.* 1981), and New Hampshire (Nicholas 1983), and were evidently important foci for early prehistoric settlement.

The primary postulate of the glacial lake mosaic model (Nicholas 1982, 1983) is that when compared to other parts of the contemporaneous landscape, the basin mosaics would have had the highest levels of resource diversity and productivity at the time, even higher than that of the interbasin riverine systems (Curran and Dincauze 1977). Within these basins, which appear to have been widely distributed across glaciated New England, the presence of additional water in shallow ponds and wetlands as well as in flowing streams created complex mosaics of vegetation and faunal resources that would have increased the diversity of certain early postglacial environments. Such mosaics would have been capable of supporting particular land-use patterns not normally associated with highly mobile foragers. Thus, it is possible that in particular settings at particular times, a mosaic of early Holocene environments created *landscapes of opportunity*. The archaeological record of early postglacial land use in the Northeast should show a strong correlation between glacial lake basins and Paleoindian and Early Archaic sites, while site-specific data should indicate more than a pattern of rare, amorphous scatters of tools and flakes as a result of longer term and/or more intensive occupation and exploitation of the area. The development of these systems and their implications for early settlement and subsistence developments are further discussed below (also Nicholas 1987).

What is critical to an understanding of the Paleoindian and Early Archaic occupation of places like Robbins Swamp, however, is not the archaeological patterns of land use per se, but the behaviors of land use they represent. Given the resource base present in the type of wetland mosaics outlined above, these places on the landscape would be expected to have been used differently from other types of places. This should be evidenced not only in the distribution of sites, located to take advantage of the diversity of resources present, but also in the way such landscapes were used. For example, if there was a greater degree of redundancy in the use of places like Robbins Swamp as the focus for repeated behaviors of resource exploitation or other uses, then this should be represented at individual sites by the presence of multiple early components, or throughout the basin area by a high early site frequency. As this pattern of land use intensifies over time, Robbins Swamp would represent, in effect, a different scale of archaeological "site." Another way to view a redundant and intensive use of these basin mosaics would be as a *core area* of geographically focused settlement and subsistence (e.g., base locale) from which less intensively used *peripheral areas* were exploited by special-activity groups; archaeological indicators of such a division of space would be reflected in site size and function, length and number of occupations, and overall site frequency.

3.1.1. Wetland Mosaic Productivity

The landscape sequence suggested by this model that occurred within the larger lake basins following postglacial stabilization, and which may have continued for substantial periods during the Holocene, can be characterized by the presence of lowered or fluctuating lakes, ponds, extensive wetlands, and an emerging riverine system. Empirical support for this model is provided by modern primary production estimates made on local, regional, and worldwide studies of the modern landscape (e.g., Bernard and Gorham 1978; Bradbury and Grace 1983; Good *et al.* 1978; Hofsetter 1983; Mitsch and Gosselink 1986; Moss 1980; Speight and Blackith 1983; Tallis 1983; Westlake 1963; Whittaker 1975), by surficial geology (Kite *et al.* 1982; Wright and Glasser 1983), and other sources (e.g., Kratz *et al.* 1979). Although compositional data for early postglacial landscapes are available through pollen spectra, productivity data are not, so these figures must be used with caution. What these and other studies indicate is that the net primary productivity and available biomass of freshwater wetlands alone is second only to that of tropical rain forests and may even exceed that of some forms of intensive agriculture (Table 1); in fact, much of the animal biomass in tropical forests (and, to a lesser degree, in temporal forests) occupies the canopy and is therefore not easily accessible. Although no figures are available for the productivity of freshwater wetland-based *mosaics* (i.e., interrelated biomes) of the glacial lake mosaic model, it is predicted to be very high, and should rank well above that of *individual* temperate deciduous wetlands, forest, and riverine zones of the early through late Holocene landscapes (Bradbury and Grace 1983:305), and be equal to or frequently above that of estuarine zones (Whigham *et al.* 1978:18; Thorbahn and Cox, Chapter 7, Table 3).

The landscape features or vegetation zones within and around the lake basins would each have had an associated biotic assemblage (Brown and Cleland 1968; Curran and Dincauze 1977: Table 3, Figure 5; Gosselink and Turner 1978:66). The diversity represented within these floral and faunal communities would have been further increased by the edge effect (Odum 1971) of contiguous biotic assemblages in close geographic proximity. Although animal production and biomass are considerably lower within wetlands than in many other biomes

Table 1. Estimated Net Primary Production and Plant and Animal Biomass for Selected Biomes[a]

Ecosystem	Net primary productivity per unit area[b] (g/m^2 year)		Biomass per unit area[c]		Animal production[c] (10^6 tons/year)	Animal biomass[c] (10^6 tons)
	Normal range	Mean	Normal range	Mean		
Tropical rain forest	1000–3500	2200	6–80	45	260	330
Swamp and marsh	800–3500	2000	3–50	15	32	20
Estuaries	200–3500	1500	0.01–6	1	48	21
Temperate evergreen forest	600–2500	1300	6–200	35	26	50
Temperate deciduous forest	600–2500	1200	6–60	30	42	110
Boreal forest	400–1200	800	6–40	20	38	57
Woodland and shrubland	250–1200	700	2–20	6	30	40
Temperate grassland	200–1500	600	0.2–5	1.6	80	60
Tundra and alpine	10–400	140	0.1–3	0.6	3	3.5
Lake and stream	100–1500	250	0–0.01	0.02	10	10
Cultivated land	100–3500	650	0.4–12	1	9	6

[a]Adapted from Whittaker 1975:83. Biomass measured in dry grams, kilograms, or metric tons.
[b]Grams per square meter × 10 = kg/ha.
[c]Planetary biomass in metric tons (10^6).

(Table 1), this should not be a critical factor, due to the presence of forested areas, grasslands, and riverine zones elsewhere in the mosaic; wetlands would provide a major feeding station for non-wetland-based fauna, however. In addition, many of these basins are bordered by or contain high ground, so that a wide variety of both lowland and upland resources would be available within a geographically constrained area.

The reliability of the resource base provided by these mosaics may also have been generally higher than that of other parts of the larger interior upland zone. Although subject to the same types of seasonal or longer term shifts as nonbasin areas, intrabasin mosaics are expected to react differently to environmental fluctuations. At the level of individual taxa, leaf water stress, for example, in wetland plants is less acute than that observed in other habitats (Bradbury and Grace 1983:298–306; Scholander *et al.* 1965). At the community level, wetland basins may have a higher reserve against the effects of short- or long-term climatic stress, such as drought, as a result of both a considerable below-ground (roots, shoots, etc.) biomass (Bradbury and Grace 1983: Table 8.4) and their capacity as a reservoir for upland drainage in-flow. The holding capacity of some glacial lake-based wetlands may, in fact, be increased by the presence of lake clays lining the basin floor. Upland deciduous forest, on the other hand, would be affected in drier periods not only by reduced moisture availability, but by increased evapotranspiration. Thus, during periods of environmental stress, the intrabasin mosaic habitat would have provided a more reliable habitat for many of the floral and faunal resources upon which prehistoric populations were dependent than many other areas.

An example of the type of long-term stress that would have affected regional landscapes occurred between about 9000 and 5000 B.P., during the hypsithermal episode. This period has long been thought to have been warmer and drier than at present, based primarily on regional/continental pollen studies (e.g., Wright 1976). Additional support is provided by a series of recent studies of changes in the earth's orbital parameters over the last 18,000 years, in terms of received versus reflected solar radiation (Kutzbach 1981, 1984; Kutzbach and Guetter 1984; Kutzbach and Street-Perrott 1985; also Bryson 1985). What this means is that prior to about 9000 B.P., the seasonal character of northern hemisphere climates was comparable or rising to modern values; between about 9000 and 6000 B.P., however, seasonality was enhanced, producing hotter summers, warmer falls, and colder winters than at present. Changes in contrastive seasonality have important short- and long-term implications for type, rate, and direction of local and regional biome development (Webb 1984; Webb *et al.* 1983:322), would affect the availability, growing season, and habitat range of various resources important to human subsistence, and would modify precipitation patterns through changes in storm track positions. The diversity of resources present within glacial lake basins, and the resiliency of at least some of those species, would have supported the continued, relatively high productivity of these locations, if not encouraged its continued exploitation by humans at the expense of other areas less resilient to seasonal or climatic stress.

Environmental data sets that support the interpretation of landscape mosaic should be evidenced by paired pollen cores between highland/lowland and ecotonal/nonecotonal habitats (Gaudreau 1986; and Chapter 9), and by comparisons of pollen spectra influx and absolute values from local and regional cores (e.g., J. White 1985). Basin size, type, and location criteria for sampling (Jacobson and Bradshaw 1981) must be chosen carefully; since cores are taken from bogs, lakes, and wetlands, the relationship of these features to adjacent glacial lake basin mosaics and to other topographic features relating to various stages of Holocene landscape development must be determined so that the necessary ecotonal/nonecotonal contrasts are maintained. A series of large-diameter cores taken from various locations in Robbins Swamp, for example, will provide the basis for a detailed study of intrabasin pollen when analyzed and compared to other local pollen diagrams (e.g., Black Spruce Bog/Mohawk Pond [Gaudreau and Suter 1984]).

3.2. Early Postglacial Land-Use Patterns in Robbins Swamp

If the type of landscape mosaic postulated for glacial lake basins in the Northeast is an accurate approximation of the regional resource base during the early postglacial period, a different archaeological record from that predicted by low-carrying-capacity/low-population models should be expected, because a different set of land-use behaviors would be expected—namely high site frequency, assemblage diversity, and a variety of site/landform associations. To evaluate the predictive potential of this alternative model, several sources of bias must first be accounted for. Of these, site location and site preservation biases are thought to figure significantly in the composition of the regional archaeological record (Nicholas 1987); e.g., despite the importance of Mulholland's (1984) compilation of site data for southern New England, his results are affected by the integrity of the site survey forms upon which it is based.

During four seasons of early site-directed fieldwork in Robbins Swamp, approximately 500 sites have been identified that represent over 10,000 years of occupation within the study area (which covers over 125 km^2, although actual survey coverage has focused on less than 40 km^2). Of these, Paleoindian and/or Early Archaic components have been positively or tentatively identified at about 40 sites, based on the recovery of fluted, bifurcated, Kirk, Palmer, Hardaway, and other points, and gravers, scrapers, John's Bridge-type choppers (Thomas and Robinson 1980), and *pièces esquillées*. Early site/landform associations range from lake shoreline and upper river terraces to wetland margins and upland springs, and include large, multiple, early component sites, and small, single-component, special-activity sites. In all, a substantial range of artifacts, site types, and locations is represented for the period between about 11,000 and 7000 B.P. within the study area (Nicholas 1987).

These data indicate that the early postglacial land use associated with Robbins Swamp was both extensive and long term. This is based both on the temporal range of sites identified, which is indicated by a wide range early artifact types, and the diversity represented in site location, size, number of components, and other land-use factors. At one location within the study area, the contiguous SWB and NWB site complexes, a number of Paleoindian and Early Archaic components are in close proximity or overlap (Table 2), suggesting a redundant use of this location, either for base camp or special function site purposes (in southern Ontario, some Paleoindian site locations appear redundant [Jackson 1983: Figure 3]). Redundancy in site location, however, does not mean redundancy in site function, and there is no reason to suppose that this locus was used in the same way throughout the early postglacial period or at other times during the Holocene. Interpreting evidence of function, length and season of occupation, or even size at any of these sites is complicated by the subsequent occupation of

Table 2. Summary of Early Postglacial Site Data, Robbins Swamp

Cluster[a]	Site	Component[b]	Artifact types[c]
N/A	21-001	Early Archaic	Untyped bifurcated points, miscellaneous tools
Dead Exxon	21-005	?	Flakes under Vosburg component
	21-008	?	*Pièce esquillée*
N/A	21-009	Early Archaic	Kanawha-type bifurcated point
N/A	21-010	Paleo/Early Archaic	Hardaway-like point
N/A	21-019	?	Untyped bifaces; miscellaneous tools
Foley	21-024	Early Archaic	Corner-notched point; miscellaneous tools
	21-027-N	?	Multiple graver, untyped point; miscellaneous tools
	21-027-S	?	Miscellaneous tools
	21-028	Early Archaic	Kirk-type point; miscellaneous tools
	21-087	?	Corner-notched point
N/A	21-034	?	Scrapers; *pièce esquillée*
Dean	21-038-B	?	Possible graver; serrated flake tool
	21-039	?	Backed flake blade
	21-040	?	Untyped biface
N/A	21-CLP5	?	Biface/chopper
N/A	21-060	?	Lanceolate point base; John's Bridge-type "chopper"
N/A	100-023	?	Chopper
N/A	100-032	Early Archaic	Corner-notched point; miscellaneous tools
N/A	100-038	?	Miscellaneous tools
N/A	100-093	?	Corner-notched point
N/A	100-096	Early Archaic	Untyped bifurcated point
SWB/NWB	100-083	Early Archaic	MacCorkle/St. Albans type point
	100-100	?	*Pièce esquillée*
	100-101	Paleoindian	Fluted points
		Early Archaic	Kanawha-type point
		?	Miscellaneous tools
	100-102	Paleoindian	Fluted point, miscellaneous tools
		Early Archaic	Untyped bifurcated point
		?	*Pièce esquillée*; possible graver
N/A	100-097	?	Retouched chert flake/graver
Segalla	100-128-Locus A	Paleo/Early Archaic	Hardaway side-notched point

Table 2. (*cont.*)

Cluster[a]	Site	Component[b]	Artifact types[c]
	100-128-Locus B	Early Archaic	Kirk-type point, miscellaneous tools
	100-128-Locus AA	Early Archaic	Kirk-type point, miscellaneous tools
	100-127	?	Biface/chopper
N/A	100-129	?	Spurred scraper
N/A	Unnamed-84-UN	Early Archaic	Untyped bifurcated points
??	Unnamed-84-CBa	Paleo/Early Archaic	Hardaway side-notched points
	Unnamed-84-CBb	Early Archaic	Untyped bifurcated points
??	Unnamed-84-BCa	Paleoindian	Fluted points
	Unnamed-84-BCb	Early Archaic	Untyped bifurcated point
N/A	Unnamed-84-H	Early Archaic	Untyped bifurcated points
N/A	128-018	Early Archaic	MacCorkle/St. Albans-type point

[a]Clusters denoted sites in close geographic proximity to each other, usually <100 m. Sites were defined as groupings of artifacts that were spatially discrete (i.e., 50 m of sterile space between areas). This definition is clearly arbitrary, but provides the means to separate archaeological patterns into their smallest recognizable unit within a study area that is essentially one very large site. In the event that individual "sites" are found to represent contemporary activity loci, they can be easily combined. This degree of resolution in mapping sites is vital where questions of land-use redundancy are posed. N/A = Not applicable (i.e., no discernible clustering).
[b]"?" denoted unspecified Paleoindian or Early Archaic component.
[c]Early artifact designations, or possibility thereof, based on comparison to documented Paleoindian and Early Archaic artifact assemblages within the region and elsewhere (e.g., Chapman 1975, 1978; Coe 1964; Funk and Wellman 1985; McNett 1985; Thomas and Robinson 1980).

this location throughout much of the Holocene, as evidenced by hundreds of Archaic and Woodland components (Nicholas 1985:2).

Determination of the range of site types represented in the early archaeological record at Robbins Swamp is clearly dependent upon careful analysis of field data. In fact, the actual number of early sites remains subject to debate in the absence of dated components. There is also little doubt that certain "diagnostic" artifact types have a wider range in time than is convenient for research purposes (e.g., Rogers' 1985 discussion on gravers). Although a number of the "early" point types recovered during fieldwork or examined in local collections are not (yet) diagnostic, their tentative classification is supported by a similarity to a large number of less well-known Early Archaic point types (i.e., nonbifurcated), as well as by the range of morphological diversity represented at documented early sites both within the region (Funk and Wellman 1985) and within the extensive southeastern literature (e.g., Chapman 1975). Other support for considering some of these sites as early Holocene in association is provided by a series of presumed early point types extracted from collections from Connecticut and New York by D. F. Dincauze (personal communication, 1985), as well as at the Lewis-Walpole site by Starbuck (1980), who also summarizes the development of the Connecticut Archaic sequence.

3.2.1. Mosaic-Related Land-Use Behaviors

The number and frequency of prehistoric sites in the Robbins Swamp area raise several questions about the relationship between environmental change and archaeological land-use patterns, and even more questions about the behaviors and social relations those land-use patterns represent. One that is particularly relevant here is: What are the general behavioral implications of the types of environmental change associated with the glacial lake basin mosaic model at places like Robbins Swamp? In other words, to what degree can certain patterns in land use be explained by reference to such ecological factors as resource diversity, predictability, and productivity or increasing or decreasing seasonality?

A number of generalizations are proposed for the early postglacial land use associated with glacial lake basin mosaics and are listed below with a summary of supportive archaeological data from Robbins Swamp. These generalizations are based to varying degrees on the implicit assumption that human populations will organize themselves advantageously within attractive environmental niches (e.g., Curran 1987; Winterhalder 1981; also Hayden 1981). The archaeological data cited are preliminary, as analysis of field data recovered in the 1984–1986 seasons is still underway.

During periods when intrabasin resource productivity, predictability, and diversity were greater than those of nonbasin areas, as postulated for the early Holocene, prehistoric utilization of these basins could be expected to be:

1. *More focused than it would be elsewhere at the same time due to differences between lake basins and other parts of the landscape* (e.g., higher site frequency): The number of Paleoindian and Early Archaic sites identified within the Robbins Swamp study area exceeds general expectations for early site distribution based on the regional baseline (Mulholland 1984). A second series of clustered Paleoindian and Early Archaic sites appears about 50 km downstream from Robbins Swamp in the New Milford and Danbury areas, also associated with a glacial lake basin located within the Housatonic drainage. At various *inter*-lake basin locations in northwestern Connecticut, early sites are more widely spaced in association with a variety of landforms (e.g., Lake Waramaug; Bantam, Naugatuck, and Shepaug drainages).

2. *More regular through time as a result of both the seasonal and long-term character of the intrabasin resources* (e.g., evidence of more frequent and longer occupations; decreased mobility; more redundant use of both sites and parts of the basin landscape): Several areas within the study area show evidence of repeated use during the early Holocene. The first is situated in the area of a now inactive confluence of two rivers (SWB/NWB site complexes); the second is located on an old lake shoreline above the modern wetland (Foley complex); and the third is in an inter-tributary/basin area (Segalla complex). In the absence of detailed, on-site data, little information is available on degree of mobility represented or length of occupation, although site frequency and the redundant use of space are suggestive of a general decrease in mobility.

3. *More intensive due to the range of resources and landscape components present* (e.g., more parts of the landscape used; more aspects of the economic and social infrastructure represented): As noted above, sites are associated with a wide variety of landforms, including lake and wetland shorelines, upper river terraces, and upland springs. More than one of these or other landscapes features are located within any square kilometer in the study area. All of the sites noted are located within or immediately adjacent to the main lake basin area; sites situated in upland areas are still associated with the basin mosaic. If site location is a function of access to particular resources at particular times (i.e., special-activity sites), then the diversity represented in site location may indicate some degree of economic diversity, particularly when associated with a generalist economy. When site location is determined more by socially oriented behaviors (e.g., base camp location), then less diversity in site location may be expected. Analysis of artifacts and debitage recovered from these sites should reveal some information on site function (e.g., use as a hunting station, tool workshop area, habitation area, etc.). The availability of those data at sites such as those included in the SWB/NWB complexes may provide the means to distinguish between special-activity and base camp areas within a series of contiguous or overlapping surfically mapped sites.

4. *More specific due to the spatial relationship between basin mosaics and other parts of the landscape* (e.g., particular parts of the basin landscape comprise core areas of settlement, subsistence, and social activities; others are used as peripheral areas for specialized activities): Several areas within the basin appear to have been more intensively used during the early postglacial period than were other places. For example, there are at least three and possibly more locations containing "concentrations" of Paleoindian or Early Archaic sites in the eastern, northern, and northeastern parts of the study area (SWB/NWB, Segalla, and Foley complexes). Other sites are relatively isolated within the area of the basin they occupy (e.g., sites 21-112, 122-18/19), but may actually be part of a less discrete cluster where a number of sites are associated with a branch of a tributary or part of the former lake basin.

What is most evident in the application of field data to the predicted archaeological record (beyond the absence of detailed site data on size, depth of deposit, range of materials, length and season of occupation, etc.) is the degree of redundancy evidenced in the distribution of early archaeological sites, which is most evident at those sites included in Table 2. This is discussed in more detail below.

3.3. Comparison to Regional Land-Use Patterns

Support for the type of land-use patterns indicated by the glacial lake mosaic model is also provided by archaeological data from across a broad area of New England. A comprehensive study of the distribution of Paleoindian and Early Archaic sites in New Hampshire (Nicholas 1983) revealed a very strong correlation between glacial lake basins and early sites. Similar associations may be noted for sites in the Connecticut drainage (although for different reasons [Curran and Dincauze 1977]), in the Taunton, Merrimack, and other drainages in eastern Massachusetts (Johnson 1986; D. Ritchie 1986; Simon 1986), and in northern Maine (Bonnichsen *et al.* 1980; Kite *et al.* 1982). Data evaluated by Mulholland (1984) reveal a relatively wide distribution of early sites across southern New England, with several areas of concentration (Mulholland 1984: Figures 6–11). A weak correlation between these sites and lake or wetland basin elements identified by Mulholland (1984: Table 12; although see Table 104) is expected to be more the result of how these sites were originally recorded—in the context of middle and late Holocene (i.e., essentially modern) landscape features—than a lack of actual association. This problem can be avoided by reevaluating the paleoenvironmental associations of previously recorded sites or by using field methods developed specifically for the location of early sites on late Pleistocene and early Holocene-aged landforms to locate systematically early postglacial sites (e.g., Bonnichsen *et al.* 1980; Chapman 1978; Nicholas *et al.* 1981). In the case of the Robbins Swamp Project, systematic surveys have been made separately of late Pleistocene/early Holocene landforms (Nicholas 1987) and of middle/late Holocene landforms (Handsman 1983). This has been augmented by additional site information provided by both additional surveys outside of the basin area and by local collectors; the resultant data base is considered representative of the prehistoric period.

4. ECOLOGICAL LEVELING AND THE DYNAMIC NATURE OF EARLY HOLOCENE LANDSCAPE DEVELOPMENT

The glacial lake mosaic model is a particularly useful tool in early postglacial research. It proposes a geography of highly productive biotic mosaics within what may otherwise have been a resource-poor regional landscape. By isolating spatial variability in the archaeological record of one of these mosaics, the behaviors and social relations of land use associated with an environmentally unique place on the landscape can be examined. During the early postglacial period, and particularly during the early Holocene, the contrasts between basin and nonbasin areas in terms of resource productivity, reliability, and diversity would have been significant.

The composition of these basins would also have varied substantially through time and space; some basins may have had unusually high levels of resource productivity over an extended period, while others may have operated at a much lower level of productivity for a shorter period. Parts of the local landscape, in fact, could have been intentionally manipulated to maintain the attractiveness of the area (e.g., periodic burning to encourage new growth to attract game or to maintain resource diversity). In general, however, it is expected that during the early Holocene, average basin composition was substantially more productive than adjacent nonbasin areas (even when these were part of a larger heterogeneous landscape), based on modern and modeled biome productivity estimates, and on palynological, climatological, hydrological, and geomorphological sources (cited above). Nonbasin riverine areas not necessarily associated with lake systems would probably rank somewhere between the two. Resource reliability (i.e., predictability/constancy/contingency [Colwell 1974]) would also be expected to be highest in lake basins, providing a relatively stable and potentially long-term resource base that would, in conjunction with other factors, allow such basins to function as buffer zones against local and regional environmental fluctuations. Glacial lake basins may thus have represented focal areas of resource distribution, or *prime locations,* spaced more or less regularly across the early landscape.

4.1. Ecological Leveling

The concept of ecological leveling provides a way to explain long-term changes in the patterns of land use and social systems by focusing on the dynamics of landscape development around and within the glacial lake/wetland system now defined as Robbins Swamp (Figure 1). Based upon the mosaic model, it examines a postulated high degree of resource productivity, reliability, and diversity associated with the post-Glacial Lake Falls Village basin (Nicholas and Handsman 1984b). The model suggests that at some point following the peak in intrabasin resource productivity, a change occurred between basin and nonbasin areas in which the productivity of nonbasin areas "leveled" with that of basin areas. Identifying this change provides a critical point of reference for developing a history of land-use behaviors represented in the early postglacial archaeological record.

Ecological leveling is a way to evaluate developments between coeval, contrastive environmental settings over time in terms of their productivity for relatively low technology-equipped human populations. In the present study, the model is bounded by specific geographic and ecological constraints that are defined by changes in the degree of contrast in resource productivity occurring between glacial lake basin mosaics and the surrounding nonmosaic areas during the early Holocene. The use of particular types of ecological factors (e.g., hydrology; net primary biotic productivity; vegetative succession) provides a framework within which to explore early postglacial land-use behaviors and how those behaviors changed in response to landscape changes over time.

The type of early landscape mosaic presented here is not considered static. At some points in time, the "uniqueness" of glacial lake basins (as measured by degree of biotic heterogeneity, resource productivity, diversity, and reliability) may have diminished *in relation to landscape composition changes beyond the basin.* Such changes may have occurred at various points during the early Holocene as the result of vegetative succession (e.g., an increase in oak), for example, or through climatic events such as the early/middle Holocene xerothermic episode. In the latter case, changes in precipitation, temperature, and evaporation rate may have resulted in wetland shrinkage and vegetative shifts (decrease in wetland species; increase in upslope species) away from lowland areas between about 9000 and 5000 B.P. (Davis *et al.* 1980; Gaudreau and Webb 1985).

4.2. Processes of Holocene Landscape Change

Understanding the relationship between the lake basin and nonbasin environments is critical to defining shifts in land-use patterns. Despite its status as a long-term landscape feature, the type of lake basin mosaic modeled here is considered internally dynamic in character; continuing developments in the individual components of the mosaic and in the relationship between components would have helped to maintain a high level of resource productivity:

> Climax (mature) ecosystems typically exhibit low productivity and high stability, while seral (immature) ecosystems are typically very productive, but also very unstable. Therefore a mature ecosystem such as a climax forest maintains itself and protects itself from environmental perturbations, but does not produce much excess biomass that can be harvested frequently by man. (Gutierrez and Fey 1980:4)

Time histories of primary productivity, biomass, and diversity for successional changes within relatively stable oak–pine forest (Gutierrez and Fey 1980: Figure 7:10, Whittaker 1975: Figure 4.3) indicate that *these three characteristics tend to increase during many successions,* although during early and not late stages. Landscape developmental processes within former lake basins would also be affected by geological and hydrological changes as the late Pleistocene–early Holocene lake and river systems stabilized; in addition, "while there are some common patterns in wetlands vegetation change, there are many variations: some of which may have been the result of regional climatic differences" (Tallis 1983:341; Weller 1981). Changing edaphic conditions as the result of soil development, topography, and related factors would have maintained dynamic microenvironments distinct to each basin (Mitsch and Gosselink 1986; also Birks and Birks 1980: Chapter 12). The overall result is that for much of the early Holocene, and in some cases the middle and even late Holocene periods, these basins would have continued to contrast with adjacent nonbasin areas that reached maturity more quickly.

4.3. Critical Paleoenvironmental Data Sets

The primary source of data for documenting changes in the degree of ecological contrast between basin and nonbasin areas is from paired pollen cores (Gaudreau 1984, 1986; also Birks and Birks 1980). Paired or multiple cores should also record changes in the degree of species heterogeneity within the former lake basins at various times during the Holocene. As noted above, paired pollen cores are currently not available for Robbins Swamp, but are expected.

What is apparent in examining single core data elsewhere within the area (e.g., Black Spruce Bog, Bingham Pond, Mt. Tom Pond [see White 1985]) is that there is a rapid decline of spruce, a more gradual reduction in pine, herbs, and shrubs, and a significant increase in oak pollen percentages in Zone B (9000–7000 B.P.). The increase in oak pollen is particularly relevant to the leveling concept, as it is seen as a major component of the nonbasin woodlands. The utility of these unpaired core data to more detailed discussion here beyond general trends, however, is limited by the range of elevational gradients, ecotonal associations, and topographic settings they represent (see Gaudreau, Chapter 9).

5. THE DYNAMICS OF THE PAST AND THE ARCHAEOLOGY OF THE PRESENT

Archaeological sites represent a static and discontinuous record of behaviors that were dynamic and continuous through time and space. In recent years, prehistorians have begun to address rigorously the black hole that lies between the past and the present, and to identify certain of the behaviors responsible for particular archaeological assemblages (Binford 1977; Dincauze and Curran 1983; Issac 1981; also see Raab and Goodyear 1984). This is clearly a problem of substantial proportions, and one that archaeologists face constantly. Binford, among others, has had reasonable success in identifying several of the linking mechanisms between the physical record produced by modern populations of hunter–gatherers and the behaviors responsible for it (Binford 1983; cf. Weissner 1982), and has documented substantial differences in Nunamiut land use during just one person's lifetime over a band territory of 13,600 km^2 (Binford 1983). In fact, much of the ethnoarchaeological literature that focuses on this middle range has developed on a contemporary baseline where the artifacts, their spatial distribution, and the activities involved could be observed by archaeologists, and frequently even commented on by the participants themselves (e.g., Binford 1976, 1978a, 1978b, 1980, 1981, 1982; Gould 1980; Hayden 1979; Hayden and Spurling 1984).

"Opening" the middle range for the more distant past, however, is complicated by a number of factors, some of which archaeologists can have no control over. One that is especially significant is time. With increasing time, the relationship between behaviors and their physical remains becomes more fragmentary, indiscrete, and obscured by the cumulative effects of subsequent cultural events and environmental processes. This obfuscation may encourage the interpretation of this archaeological record using a baseline of ethnographic and modern analogs for past behaviors and social relations, so that what is actually being accomplished is an archaeology of the present. There is, in addition, no steady rate of change in the prehistoric record, especially in the early Holocene period, but relatively inactive periods of cultural and environmental interaction, punctuated by periods of extensive "disequilibrium" (see Joyce, Chapter 8). Archaeologists do have control over the recognition of these interpretive and preservational biases in the interpretation of the past, however. For example, ethnographic records of historic or modern hunter–gatherers, which have long served as analogs to prehistoric populations, reflect significant biases inherent in a way of life that may be more the product of historic influences than of "natural" development (Schrire 1984; Wobst 1978). Ethnoarchaeological, experimental, and replicative studies, on the other hand, carry a different, but equally persuasive, set of biases. What all of this means is that, while the behaviors of the past can be accessed using methods that are theoretically explicit and logically constructed (i.e., the middle range), it does becomes increasingly difficult to "operationalize" some of Binford's ideas, for example, about prehistoric land use the further back in time one goes.

5.1. Behavioral Implications and Archaeological Indicators

A series of general developmental stages can be outlined for Robbins Swamp, beginning with deglaciation by 14,000 B.P. and ending with the development of the essentially modern landscape by about 4000 B.P. These stages and their major features are outlined below; the dates are extrapolated from a variety of local and regional geological and palynological sources:

1. *An initial late Pleistocene period of deglaciation* (?–13K): initial biotic resettlement of landscape; short-lived meltwater lakes; maximum extent of stages of glacial lakes in area; upland areas generally ice-free before lowlands.
2. *The late Pleistocene–early Holocene stabilization of the local watershed* (13–10K): development of post-lake maxima drainage systems; regional landscape colonized by tundra, pioneer arboreal and nonarboreal plant communities, followed by open spruce woodlands; strength of seasonality before 9K comparable or rising to modern values.
3. *The early Holocene development of mosaic landscapes within former glacial lake basins* (10–8K): expected maximum contrast in resource diversity and productivity between lake basin and nonbasin parts of the landscape; establishment of mixed deciduous (high oak percentage) and coniferous forests throughout the uplands and river valleys; upslope migration of earlier vegetation patterns; strength of seasonality enhanced compared to modern values.
4. *The early/middle Holocene decline of ecological uniqueness of glacial lake basin mosaics* (8–7K?): "leveling" of differences between the productivity of basin/nonbasin areas due to dominance of full deciduous forests, landscape stabilization, vegetative succession, and climatic factors.
5. *Continued Holocene development of landscape* (7–0K): climate, hydrology, and landscape composition trending toward present.

The concept of leveling is directed here only to landscape changes occurring during the early Holocene and early/middle Holocene periods (Stages 3 and 4 above). Some of the behavioral implications of these environmental developments are outlined in Tables 3 and 4. Regardless of their interpretation, the land-use behaviors associated with these developmental stages must be evaluated in terms of a period of maximum resource diversity of basin mosaics, a period when the uniqueness of the mosaics declined in relation to the "resource productivity" of nonbasin areas, and a period of increased seasonality (hotter summers/colder winters).

One result of this leveling may have been a redivision of early space; that is, the way parts of the landscape were used, whether defined by the location of base camps and special activity areas or of core and peripheral areas. During the early Holocene, Robbins Swamp may have become a *place* (i.e., a culturally or environmentally fixed position in space) on the landscape as the result of intensive, long-term land use (Table 3). The development and expansion of the wetland-based mosaic produced a unique environmental setting that became the focus of human

Table 3. Early Holocene Patterns of Land-Use Behaviors within Glacial Lake Basin Mosaics: "Robbins Swamp As a Place on the Landscape"[a]

Landscape composition

1. Lowering of lake levels; stabilization of local landscape; primary reduction in lake configuration; initial establishment of intrabasin landscape based on hydrological, microtopographic, edaphic, and vegetative elements
2. Development and expansion of wetland-based mosaic
3. Robbins Swamp and similar locations become ecologically unique within the larger, early Holocene landscape of the developing river valleys and uplands and the coastal plain

Expected land-use patterns (see text)

1. Development of general foraging subsistence pattern not limited to a small number of seasonally specific resources
2. Increasing redundancy in the use of parts or all of the mosaic landscape
3. Robbins Swamp becomes a place on the landscape as a result of its increasing use as a "core" area of settlement, subsistence, economic, and social activities
4. The landscape system stabilizes and, through the intrabasin mosaic, becomes physically distinct and culturally separable from other areas
5. More regular use is made of local resources and geographic features
6. Robbins Swamp becomes a "place" of repeated and regular use
7. Shift toward more formal/structured patterns of use; definition of core and peripheral areas

Archaeological visibility

1. Increased diversity in artifact assemblages or a more generalized tool kit as a response to the breadth of the resource base
2. Presence of both base camps and special-function sites within watershed due to a significant portion of the annual social and economic system being focused within or adjacent to the basin
3. Presence of burials related to a long-term pattern of land use or similar commitment (e.g., territoriality) to this location on the landscape
4. Presence of storage pits and caches, and other reused features reflecting an expected return to the sites within a relatively short period
5. Changes in degree of site "openness" and other indicators of site structure reflecting shifts in site function, types of social activities, group size, and intra/intergroup relations
6. Diversity in seasonal representation and length of individual site occupation representing a longer and perhaps nonseasonally specific use of the basin
7. Evidence of site reuse/multiple early components and of increased site density as use of the basin intensifies over time
8. Increase in midden deposits representing longer and more intense use of sites

[a]This table does not include variables associated with social relations of land use, e.g., group size and composition, class structure and grave goods, social transformation, aggregation, social relations of production and resource access, etc.

occupation. If settlement and subsistence behaviors were directed to highly productive areas of the landscape, then land-use patterns should be characterized by more regular, intensive, and repeated use of Robbins Swamp as a core area. If generalized subsistence patterns were oriented to the diversity of resources available in Robbins Swamp throughout much of the year, site frequency would be expected to be high, evidenced both by base camps (some reflecting repeated and longer use) and by a variety of specialized activity sites. At this point, Robbins Swamp could be considered a site of intensive and focused land use.

In response to changes in landscape composition resulting from the "inherent changeability of the wetlands habitat" (Tallis 1983:341), the uniqueness of Robbins Swamp is expected to have diminished toward the end of the early Holocene as a result primarily of increases in the resource diversity, productivity, and reliability of nonbasin areas (Table 4). In other words, it is not so much that the productivity of the basin mosaics declined, as that that of upland areas

Table 4. Early/Middle Holocene Patterns of Land-Use Behaviors within Glacial Lake Basin Mosiacs: "Ecological Leveling Period/Decline of Lake Basin Uniqueness"[a]

Landscape composition
1. Reduction of ecological contrasts between Robbins Swamp and adjacent, nonbasin areas
2. Increase in resource productivity, diversity, and reliability of areas surrounding former lake basin mosaic

Expected land-use patterns (see text)
1. Emerging seasonality and decreased intensity of basin use
2. Portions of basin become parts of different catchment areas
3. Increase in nonredundancy of site or basin use
4. Redefinition of the basin mosaic as divisible space
5. Emergence of a dominating pattern of more specific/perceived use of landscapes adjacent to basin
6. Decrease in use/identification of Robbins Swamp as a "core" area; increase in use as "peripheral" areas of other "cores"
7. The swamp becomes "another" settlement pattern component of a different land-use pattern

Archaeological visibility
1. Decrease in diversity in artifact assemblages or increase in tool kit specialization as a more specific resource base is exploited
2. Reduction in the maximum size of sites
3. Decrease in the number of early components as use of basin becomes less spatially intensive
4. Formalization of intrasite space and changes in degree of site "openness" and other indicators of site structure as basin sites become more specialized
5. Increase in seasonally specific/limited use of basin mosiac
6. Decrease in number of base camps; increase in number of special-activity sites as a greater portion of social and economic activities are conducted outside of the basin
7. Decrease in number and variety of storage pits and caches and other reused features, including cemeteries, as both sites and the general areas are used less frequently and with less regularity

[a]This table does not include variables associated with social relations of land use, e.g., group size and composition, class structure and grave goods, social transformation, aggregation, social relations of production and resource access, etc.

increased. If settlement and subsistence were oriented to certain levels of habitat productivity, such a leveling would be expected to have resulted in a less intensive or at least different pattern of land use. What was once a specific *place* (or core area) on the landscape (i.e., Robbins Swamp) may thus have been redefined to become *space* (or peripheral areas) incorporated within a number of new *places* on that same landscape (e.g., Blackberry/Whiting Rivers, Housatonic River/Twin Lakes, or Wangum Lake areas; Figure 1). The use of this space would have been redefined, in part, by the emergence of a more obvious or specialized subsistence–settlement pattern that was more seasonally specific.

5.2. Exploring the Middle Range for Land-Use Behaviors

In evaluating the relationship between the system patterns of land use, archaeological patterns of land use, and the transgressive nature of places like Robbins Swamp, those behavioral variables responsive to local environmental change must be identified. If there were areas of high productivity on the early postglacial landscape, as suggested by the glacial lake mosaic model, then the archaeology of these basins, at both site-specific and land-use pattern levels, provides a critical test of various assumptions of early gatherer–hunter spatial and economic organization, social complexity, and degree of mobility. Models such as ecological leveling, which examine diversity (or lack thereof) in these behaviors and social relations within dynamic landscapes over time, provide a type of history of early gatherer–hunters. The way that populations may have responded to certain types of environmental change in places like Robbins Swamp is illustrated by discussing one trajectory of land-use change.

5.3. Redundancy in the Archaeological Record

At some point between the early Holocene and the middle Holocene in Robbins Swamp, changes in the behaviors of land use are expected to have occurred in degree of sedentism/mobility, site reuse, specificity and seasonality of core and peripheral site use, and other variables as the relative ecological uniqueness of the basin declined. The archaeological implications of these shifts in land-use behaviors should include new modes of site function, site size, site location, artifact assemblage composition, site frequency within the area, length of occupation, and frequency of reoccupation, among other things. Coeval social or technological transformations would have occurred in conjunction with, or as a result of, shifts in land use, population size, and resource availability; these may have resulted in changes in group membership and social alliances, exchange networks, craft specialization, or the rise of temporary elite to control certain resources. Although the social relations of land use represent a critical factor in the overall development of land-use patterns, they are generally less accessible in the archaeological record, and thus have not been discussed here. [Identifying and explaining the social relations of land use within the prehistoric record is, however, an important component of the Robbins Swamp Project (Handsman 1985; Nicholas and Handsman 1984a).]

Changes in the way that prehistoric space was used in Robbins Swamp during the early and early/middle Holocene periods can be examined in terms of four variables of land-use behavior (Table 5): focus, redundancy, intensity, and specificity. The degree to which each variable relates to specific behaviors of land use is evidenced by specific types and scales of archaeological data. For example, *focus of land use* can be measured by site frequency; *redundancy of land use* by evidence of reuse of parts of the landscape; *intensity of land use* by artifact assemblages and indicators of seasonality; and *specificity of land use* by internal site patterning. A listing of some of the individual elements of one variable, land-use redundancy (Table 6), illustrates how the middle range between certain prehistoric behaviors and their archaeological record can be examined.

Redundancy in land use is a good indicator of how human populations organized themselves in relation to specific parts of the landscape. The degree to which sites or parts of the landscape are reused over time is especially important to an understanding of the dynamic relationship between land-use behaviors and local environmental change. If Robbins Swamp was a *core area* of focused, long-

Table 5. General Behavioral Variables of Land Use

Focus
- Site frequency = relative frequency of sites on given landform type ÷ relative frequency of occurrence of landform type (Dincauze 1978)

Redundancy of use (see Table 6)
1. Reuse of specific parts of the landscape
2. Reuse of sites
3. Reuse of activity areas and features
4. Cemetery reuse
5. Use of cache pits and storage technology

Intensity
1. Substantial portion of annual life cycle represented in basin mosaic area
2. Diverse social, economic, and technological activities
3. Use of basin areas not limited to one season

Specificity of use
1. Consistency in base camp organization
2. Specialization between collecting localities
3. Seasonally specific use of particular locations
4. Periodic burning of portions of landscape for future resource management

Table 6. Indicators of Redundancy in Land Use

Duration of occupation
1. Midden-type deposits
2. Storage facilities/caches
3. Technological recycling of tools and materials
4. Resource depletion (e.g., firewood)

Frequency of use/reuse
1. High site frequency, multiple-component sites, features, cemeteries
2. Diversity in site types represented
3. Increase in size of sites

Nucleation (definition of core/peripheral areas)
1. Base camps and clustered sites located in specific parts of basin
2. Large, heavily used sites or areas reflecting areas of resource concentration
3. Locally linear distribution of small, limited-use, extractive, or special-activity sites or landscape areas around basin

Economic and technological organization
1. Evidence of a generalized pattern of foraging/collecting
2. Emergence of specialized tool kits
3. Similar patterns of internal site structure between clustered base sites

term settlement, then indicators of redundant land use identifiable within the archaeological record should reflect duration of occupation of sites or site areas (i.e., parts or all of Robbins Swamp), frequency of site/area use and reuse, nucleation in site distribution, and type of economic/technological organization. The repeated use of all or parts of Robbins Swamp at various times in the past should be represented archaeologically by at least some of these indicators (Figure 1), although not all evidence of redundant land use would be visible (e.g., firewood depletion).

Land-use patterns associated with glacial lake mosaics during the early Holocene (Table 3) would be characterized by a redundant use of places, if populations organized themselves to take advantage of the broad, stable resource base available there at that time. As a *core area,* Robbins Swamp would be the focus of a wide variety of economic, social, and subsistence activities, which would be evidenced both by base camps and special-activity sites. A generalized subsistence orientation would probably be most ecologically sensible within this type of mosaic, where general biotic diversity and productivity were not restricted to particular seasons (although individual species might; e.g., migrating fowl). The archaeological indicators of such basin mosaic-focused behaviors include increased diversity in artifact assemblages, multiple or spatially contiguous early components (Figure 1), presence of storage pits, caches, and other reused features, presence of cemeteries (either a specific site or area), and diversity in seasonal evidence of site occupation.

As the ecological uniqueness of Robbins Swamp declined at the end of the early Holocene, the behaviors of land use are expected to have shifted accordingly (Table 4). The landscape would have been redefined both by changes in the local environment and in the use of parts of the new environment. Subsistence-oriented activities would be more seasonal as more and different resources became available in the nonbasin areas. The previous advantages of focusing so much time and energy into Robbins Swamp for social, subsistence, or economic reasons would have been lost, or at least modified, by the new opportunities provided by other areas. The archaeological indicators of this transition in land-use behaviors would include decreases in artifact assemblage diversity and in the number and variety or reuseable features, the presence of isolated burials, and increased specificity in seasonal use of sites. The general absence of ceramics at Woodland Period sites in the study area illustrates this type of land-use change, suggesting that those activities associated with ceramics (e.g., base camps) were located elsewhere, perhaps near the broad alluvial terraces of the Housatonic to the north or south.

Redundancy in early postglacial land use will be falsified if it can be demonstrated that (1) many of the proposed early sites are Archaic or Woodland in association; (2) early site clusters (e.g., the Segalla Complex) represent contemporary activity loci and not separate sites or components; and (3) the number of early sites identified within the project area represents so many nonrelated groups, in which case there is no redundancy in land-use behaviors, only in the spatial patterning of archaeological sites—an important distinction. The first two falsification criteria are being carefully examined; the third criterion is unlikely to be met, however, given the apparent continuity in the local prehistoric record. Other criteria are also being sought. Continued redundant use of Robbins Swamp during the later Holocene may not be used to falsify the land-use hypotheses, because the location would still have been intensively used, albeit for different reasons.

6. CONCLUSIONS AND SUMMARY

The relationship between early gatherer–hunter land-use behaviors and the changing postglacial landscapes of which they were a part is both complex and dynamic (Hayden 1981). This means that making sense of the archaeological record of this period is very much a challenge to archaeologists; the reward, however, is a unique set of data vital to understanding the range and complexity of human responses to types of environmental change having no modern analog. This paper has discussed several of the ways that the archaeology of the early postglacial period is currently being approached at one location in the Northeast, Robbins Swamp. The development and application of models such the *glacial lake basin mosaic* and *ecological leveling* are critical to this type of research because they provide a historical sense to the processes of landscape development. The utility of these models, particularly ecological leveling, is enhanced by the fact that they are not restricted to specific research problems, but have wide applicability throughout the prehistoric period in the Northeast, wherever and whenever contrasts in environmental zones are present (e.g., Narragensett Bay area [Thorbahn and Cox, Chapter 7] or Great Montezuma Swamp [W. A. Ritchie 1944]). Similarly, it would be useful to examine the high frequency and spatial redundancy of early postglacial sites in the lakes region of southern Ontario (Jackson 1983) within the context of the glacial lake mosaic model.

The early postglacial component of the Robbins Swamp Project is now entering its sixth year. Analyses are presently underway on site survey, testing, and excavation data recovered on 40 positively or tentatively identified Paleoindian and Early Archaic sites, as well as upon those relating to the hundreds of later Holocene-aged sites identified. The total inventory of about 500 late Pleistocene through late Holocene sites represents a valuable set of archaeological data pertaining to *local* adaptations to changing environmental and cultural conditions, balancing the *regional* baseline provided by Mulholland (1984) and others. Although the results of the formal analysis of early postglacial land-use data represented at Robbins Swamp will be presented elsewhere, some preliminary observations have been noted above. These indicate that there are significant differences between the early archaeological record of Robbins Swamp and those of other areas within the region; this is particularly evident in terms of site frequency, redundancy of land use, and diversity in artifact assemblages represented. Such patterns can be explained, in part, by reference to detailed models of local landscapes that have been described both in this paper and elsewhere. Furthermore, the middle and late Holocene archaeological records at Robbins Swamp indicate that general land-use patterns shifted as the local landscapes trended toward their present state.

If glacial lake basin mosaics represented a highly productive, long-term, and reliable resource base on the early postglacial landscape of the Northeast, then it may be useful to examine the archaeological record of this period within a context of *affluent* gatherer–hunters, such as found on the Northwest Coast. Not only would land use tend to focus on such places on the landscape during periods of maximum basin uniqueness, but resource availability would also facilitate certain types and levels of social and economic structures normally not associated with small, mobile, and egalitarian groups. As Schalk (1981:53) argues for the highly productive maritime zone of the Northwest Coast, home range size, logistic mobility from base camps, and group size generally increase in direct proportion to resource clumping, which also encourages increasingly complex resource/consumer relationships, although "organizational complexity bears no simple relationship to population density or food abundance." Schalk's study is appropriate here because in the absence of a highly productive, early Holocene marine zone in the Northeast, glacial lake basins may have represented the most attractive landscape zone *at that time*: "Most of the principal marine resource species available to foragers of the temperate and higher latitudes are highly clumped in their spatial and temporal distribution. If there is a single characteristic of marine resources that accounts for the organizational complexity of foragers who exploit them, *it is this property of their distributional structure in the environment*, rather than their abundance" (Schalk 1981:70 [emphasis mine]). It is this type of resource distribution across the Northeast that is seen as the single most important factor affecting early postglacial land-use behaviors.

There is still much work to be done in determining the relationship between the environmental history, behaviors of land use, and archaeological records at places like Robbins Swamp and in testing rigorously any proposed correlations between them. This paper has focused on spatial and temporal variability in the archaeological record at the local level within what had been a geographically and ecologically distinct part of the early postglacial landscape. There are undoubtedly other parts of the Northeastern landscape that may have a similar archaeological and environmental history as Robbins Swamp. Identifying and studying these will provide not only another test of some of the ideas discussed here, but also the means to review and revise the histories of early gatherer–hunters in the Northeast.

Acknowledgments

This paper is the product of a long and fruitful collaboration with Russell Handsman over the last 5 years; many of the ideas presented above are the result of discussions concerned with the Robbins Swamp Project. The development of several of the themes expressed in this paper was facilitated by discussions with Catherine Carlson, Mary Lou Curran, Dena Dincauze, William Fawcett, Robert Funk, Denise Gaudreau, Ann McMullen, Mitchell Mulholland, Dolores Root, Brooke Thomas, Peter Thomas, and Martin Wobst. I would also like to thank the staff of the American Indian Archaeological Institute, James Chiarelli (Center for Field Studies), David Poirier (Connecticut Historical Commission), and our many friends in the project area for their contributions, especially John Feathers and the Quenville family. I am grateful to the Robbins Swamp fieldcrews for their efforts: K. Ames, B. Baker, L. Boccuzi, C. Carlson, S. Coe, J. Disviscour, P. Fiacco, B. Guralnik, R. Hampton, E. Larkins, L. Markowtiz, J. Maymon, T. Prindle, B. Pavlovski, P. Quinn, H. Starwalker, and G. Whitbeck. Research was funded by the Center for Field Research (Earthwatch) (1985); the Connecticut Historical Commission (1982 and 1984); the Diebold Foundation (1985), and AIAI's Friends of Research.

7. REFERENCES

Ames, K. M., 1981, The Evolution of Social Ranking on the Northwest Coast of North America, *American Antiquity* 46:789–805.

Ames, K. M., 1985, Hierarchies, Stress, and Logistical Struggles among Hunter–Gatherers in Northwestern North America, in: *Prehistoric Hunter–Gatherers: The Emergence of Cultural Complexity* (T. D. Price and J. A. Brown, eds.), Academic Press, New York, pp. 155–180.

Bender, B., 1978, Gatherer–Hunter to Farmer: A Social Perspective, *World Archaeology* 10:361–392.

Bender, B., 1985, Emergent Tribal Formations in the American Midcontinent, *American Antiquity* 50:52–62.

Bernabo, J. C., and Webb, T., III, 1977, Changing Patterns in the Holocene Pollen Record from Northeastern North America: A Mapped Summary, *Quaternary Research* 8:64–96.

Bernard, J. M., and Gorham, E., 1978, Life History Aspects of Primary Production in Sedge Wetlands, in: *Freshwater Wetlands* (R. E. Good, D. F. Whigman, and R. L. Simpson, eds.), Academic Press, New York, pp. 39–52.

Bettinger, R. L., 1980, Explanatory/Predictive Models of Hunter–Gatherer Adaptation, in: *Advances in Archaeological Method and Theory,* Vol. 3 (M. B. Schiffer, ed.), Academic Press, New York, pp. 189–255.

Binford, L. R., 1968, Post-Pleistocene Adaptations, in: *New Perspectives in Archaeology* (L. R. Binford and S. Binford, eds.), Aldine, Chicago, Illinois, pp. 313–341.

Binford, L. R., 1976, Forty-Seven Trips: A Case Study in the Character of Some Formation Processes of the Archaeological Record, in: *Contributions to Anthropology: The Interior Peoples of Northern Alaska* (E. Hall, ed.), National Museum of Man, Archaeological Survey of Canada, Mercury Series, No. 49, Ottawa, Canada, pp. 299–351.

Binford, L. R., 1977, Introduction, in: *For Theory Building in Archaeology* (L. R. Binford, ed.), Academic Press, New York, pp. 1–10.

Binford, L. R., 1978a, *Nunamiut Ethnoarchaeology,* Academic Press, New York.

Binford, L. R., 1978b, Dimensional Analysis of Behavior and Site Structure: Learning from an Eskimo Hunting Stand, *American Antiquity* 43:330–361.

Binford, L. R., 1980, Willow Smoke and Dogs' Tails: Hunter–Gatherer Settlement Systems and Archaeological Site Formation, *American Antiquity* 45:4–20.

Binford, L. R., 1981, Behavioral Archaeology and the "Pompeii Premise," *Journal of Anthropological Research* 37:195–208.

Binford, L. R., 1982, The Archaeology of Place, *Journal of Anthropological Archaeology* 1(1):5–31.

Binford, L. R., 1983, Long Term Land Use Patterns: Some Implications for Archaeology, in: *Lulu Linear Punctated: Essays in Honor of George Irving Quimby* (R. C. Dunnell and D. K. Grayson, eds.), Anthropological Papers, No. 72, University of Michigan, Ann Arbor, Michigan, pp. 27–54.

Birks, H. J. B., and Birks, H. H., 1980, *Quaternary Paleoecology,* Edward Arnold, London.

Bonnichsen, R., Konrad, V., Clay, V., Gibson, T., and Schnurrenberger, T., 1980, Archaeological Research at Munsungun Lake: 1980 Preliminary Technical Report of Activities, Institute for Quaternary Studies, University of Maine, Orono, Maine.

Bradbury, I. K., and Grace, J., 1983, Primary Production in Wetlands, in: *Mires: Swamp, Bog, Fen and Moor—Ecosystems of the World,* Vol. 4A: *General Studies* (A. J. P. Gore, ed.), Elsevier, Amsterdam, pp. 285–310.

Brown, J., and Cleland, C., 1968, The Late Glacial and Early Post-Glacial Faunal Resources in Midwestern Biomes Newly Opened to Human Adaptation, in: *The Quaternary of Illinois* (R. Bergstrom, ed.), University of Illinois College of Agriculture Publication Series, No. 14, pp. 114–122.

Broyles, B., 1971, Second Preliminary Report: The St. Albans Site, Kanawha County, West Virginia, West Virginia Geological and Economic Survey, Report of Archaeological Investigations, No. 3.

Bryson, R. A., 1985, On Climatic Analogs in Paleoclimatic Reconstruction, *Quaternary Research* 23:275–286.

Butzer, K., 1971, *Environment and Archaeology: An Ecological Approach to Prehistory,* Aldine, Chicago, Illinois.

Butzer, K., 1982, *Archaeology As Human Ecology,* Cambridge University Press, New York.

Carlson, R. L., 1983, Prehistory of the Northwest Coast, in: *Indian Art Traditions of the Northwest Coast* (R. L. Carlson, ed.), Archaeology Press, Simon Fraser University, Burnaby, British Columbia, Canada, pp. 13–32.

Chapman, J., 1975, The Rose Island Site, Report of Investigations, No. 14, Department of Anthropology, University of Tennessee, Knoxville, Tennessee.

Chapman, J., 1978, The Bacon Farm Site and a Buried Site Reconnaissance, Report of Investigations, No. 23, Department of Anthropology, University of Tennessee, Knoxville, Tennessee.

Chapman, J., 1985, Archaeology and the Archaic Period in the Southern Ridge-and-Valley Province, in: *Structure and Process in Southeastern Archaeology* (R. S. Dickens, Jr. and H. T. Ward, eds.), University of Alabama Press, University, Alabama, pp. 137–153.

Chapman, J., and Adovasio, J., 1977, Textile and Basketry Impressions from Icehouse Bottom, Tennessee, *American Antiquity* 42:620–625.

Chapman, J., Delcourt, P., Cridlebaugh, P. A., Shea, A. B., and Delcourt, H. R., 1982, Man–Land Interaction: 10,000 Years of American Indian Impact on Native Ecosystems in the Lower Little Tennessee River Valley, Eastern Tennessee, *Southeastern Archaeology* 1(2):115–121.

Charnov, E. L., 1976, Optimal Foraging: The Marginal Value Theorem, *Theoretical Population Biology* 9:129–136.

Chisholm, M., 1962, *Rural Settlement and Land Use: An Essay in Location,* Wiley, New York.

Clarke, D. L., 1976, Mesolithic Europe: The Economic Basis, in: *Problems in Economic and Social Anthropology* (G. Sievking, I. Longworth, and K. Wilson, eds.), Duckworth, London, pp. 449–481.

Clarke, D. L. (ed.), 1978, *Spatial Archaeology,* Academic Press, London.

Clements, F. E., 1916, Plant Succession, Carnegie Institution of Washington, Publication 242.

Clifford, H. B., 1974, *Charlie York: Maine Coast Fisherman,* International Marine Publishing Company, Camden, Maine.

Coe, J. L., 1964, The Formative Cultures of the Carolina Piedmont, *Transactions of the American Philosophical Society* 54(5).

Colinvaux, P. A., 1987, The Changing Forests: Ephemeral Communities, Climate and Refugia, *Quarterly Review of Archaeology* 8(1).

Colwell, R. K., 1974, Predictability, Constancy, and Contingency of Periodic Phenomena, *Ecology* 55:1148–1153.

Curran, M. L., 1979, Studying Human Adaptation at a Paleo-Indian Site: A Preliminary Report, in: *Ecological Anthropology of the Middle Connecticut River Valley* (R. Paynter, ed.), Department of Anthropology Research Reports, No. 18, University of Massachusetts, Amherst, Massachusetts, pp. 14–26.

Curran, M. L., 1987, *The Spatial Organization of Paleoindian Populations in the Late Pleistocene of the Northeast,* Ph.D. Dissertation, University of Massachusetts, Amherst, Massachusetts, unpublished.

Curran, M. L., and Dincauze, D. F., 1977, Paleoindians and Paleo-Lakes: New Data from the Connecticut Drainage, *Annals of the New York Academy of Science* 288:333–348.

Davis, M. B., 1981a, Quaternary History and the Stability of Forest Communities, in: *Forest Succession* (D. C. West, H. H. Shugart, and D. B. Bodkin, eds.), Springer-Verlag, New York, pp. 132–177.

Davis, M. B., 1981b, Outbreaks of Forest Pathogens in Quaternary History, in: *Proceedings of the Fourth International Palynological Conference,* Vol. 3, pp. 216–227.

Davis, M. B., Spear, R. W., and Shane, L. C. K., 1980, Holocene Climate of New England, *Quaternary Research* 14:240–250.

Delcourt, H. R., Delcourt, P. A., and Webb, T., III, 1983, Dynamic Plant Ecology: The Spectrum of Vegetational Change in Space and Time, *Quaternary Science Review* 1:153–175.

Delcourt, P. A., and Delcourt, H. R., 1983, Late-Quaternary Vegetational Dynamics and Community Stability Reconsidered, *Quaternary Research* 19:265–271.

Dincauze, D. F., 1978, Predicting Site Encounter: Two Techniques for Landscape Analysis, *American Society of Conservation Archaeology Newsletter* 5(3):2–11.

Dincauze, D. F., 1981, Paleoenvironmental Reconstruction in the Northeast: The Art of Multidisciplinary Science, in: *Foundations of Northeast Archaeology* (D. R. Snow, ed.), Academic Press, New York, pp. 51–96.

Dincauze, D. F., 1985, Perspectives: Rejoinder, *Quarterly Review of Archaeology* 6(1):15–16.

Dincauze, D. F., and Curran, M. L., 1983, Paleoindians As Flexible Generalists: An Ecological Approach, Paper presented at 48th Annual Meeting, Society for American Archaeology, Pittsburgh, Pennsylvania.

Dincauze, D. F., and Mulholland, M. T., 1977, Early and Middle Archaic Site Distributions and Habitats in Southern New England, *Annals of the New York Academy of Science* 288:439–456.

Ellen, R., 1982, *Environment, Subsistence and System: The Ecology of Small-Scale Social Formations*, Cambridge University Press, New York.

Fitting, J. E., 1968, Environmental Potential and the Postglacial Readaptation in Eastern North America, *American Antiquity* 33:441–445.

Fitting, J. E., 1977, Social Adaptations of the Paleoindian Adaptation in the Northeast, *Annals of the New York Academy of Science* 288:369–374.

Funk, R., 1978, Post-Pleistocene Adaptations, in: *Handbook of North American Indians: Northeast:* Vol. 15 (B. G. Trigger, ed.), Smithsonian Institute, Washington, D.C., pp. 16–27.

Funk, R., and Wellman, B., 1984, Evidence of Early Holocene Occupations in the Upper Susquehanna Valley, New York State, *Archaeology of Eastern North America* 12:81–109.

Gardner, W. M., 1983, Stop Me If You've Heard This One Before: The Flint Run Paleoindian Complex Revisited, *Archaeology of Eastern North America* 11:49–64.

Gaudreau, D. C., 1984, Features of Holocene Plant Population Dynamics Inferred from Mapped Pollen Data from Northeastern North America, *Bulletin of the Ecological Society of America* 65:211.

Gaudreau, D. C., 1986, *Holocene Vegetational History of Central New England: The Use of Elevational Contrasts and Latitudinal Gradients in Pollen Data for Paleoecological Interpretation*, Ph.D. Dissertation, Yale University, New Haven, Connecticut, unpublished.

Gaudreau, D. C., and Suter, S. M., 1984, Holocene Vegetational Contrasts in Southern New England: Two Radiocarbon-Dated Pollen Records, *AMQUA Abstracts* (American Quaternary Association) 8:47.

Gaudreau, D. C., and Webb, T., III, 1985, Late Quaternary Pollen Stratigraphy and Isochrone Maps for the Northeastern United States, in: *Pollen Records of Late-Quaternary North American Sediments* (V. M. Bryant and R. G. Holloway, eds.), American Association of Stratigraphic Palynologists Foundation, Dallas, Texas, pp. 247–280.

Gleason, H. A., 1917, The Structure and Development of the Plant Association, *Bulletin of the Torrey Botanical Club* 44:463–481.

Good, R. E., Whigman, D. F., and Simpson, R. L. (eds.), 1978, *Freshwater Wetlands: Ecological Resources and Management Potential*, Academic Press, New York.

Gosselink, J. G., and Turner, R. E., 1978, The Role of Hydrology in Freshwater Wetland Ecosystems, in: *Freshwater Wetlands: Ecological Resources and Management Potential* (R. E. Good, D. F. Whigman, and R. L. Simpson, eds.), Academic Press, New York, pp. 63–78.

Gould, R. A., 1980, *Living Archaeology*, Cambridge University Press, Cambridge.

Gutierrez, L. T., and Fey, W. R., 1980, *Ecosystem Succession: A General Hypothesis and a Test Model of a Grassland*, MIT Press, Cambridge, Massachusetts.

Hagget, P., 1965, *Locational Analysis in Human Geography*, Macmillan of Canada, Toronto.

Handsman, R. G., 1983, Towards Archaeological Histories in Robbins Swamp, *Artifacts* (American Indian Archaeological Institute, Washington, Connecticut) XI(3):1–20.

Handsman, R. G., 1985, History and Communal Class Struggles among Early Gatherer-Hunters, Paper presented at the 84th Annual Meeting of the American Anthropological Association, Washington, D.C.

Hayden, B., 1979, *Paleolithic Reflections: Lithic Technology and Ethnographic Excavations among the Australian Aborigines,* Canberra, Australia.

Hayden, B., 1981, Subsistence and Ecological Adaptations of Modern Hunter/Gatherers, in: *Omnivorous Primates: Gathering and Hunting in Human Evolution* (R. S. O. Harding and G. Teleki, eds.), Columbia University Press, New York, pp. 344–421.

Hayden, B., and Spurling, B., 1984, Ethnoarchaeology and Intrasite Spatial Analysis: A Case Study from the Australian Western Desert, in: *Intrasite Spatial Analysis in Archaeology* (H. Hietala, ed.), Cambridge University Press, Cambridge, pp. 224–241.

Hayden, B., Eldridge, M., Eldridge, A., and Cannon, A., 1985, Complex Hunter-Gatherers in Interior British Columbia, in: *Prehistoric Hunter–Gatherers: The Emergence of Cultural Complexity* (T. D. Price and J. A. Brown, eds.), Academic Press, New York, pp. 181–200.

Hodder, I. R. (ed.), 1978, *The Spatial Organization of Culture,* Duckworth, London.

Hofsetter, R. H., 1983, Wetlands in the United States, in: *Mires: Swamp, Bog, Fen and Moor—Ecosystems of the World,* Vol. 4B: *Regional Studies* (A. J. P. Gore, ed.), Elsevier, Amsterdam, pp. 201–244.

Issac, G. Ll., 1981, Stone Age Visiting Cards: Approaches to the Study of Early Hominid Land Use Patterns, in: *Patterns of the Past* (I. Hodder, ed.), Cambridge University Press, Cambridge, pp. 131–155.

Jackson, L. J., 1983, Geochronology and Settlement Deposition in the Early Palaeo-Indian Occupation of Southern Ontario, Canada, *Quaternary Research* 19:388–399.

Jacobson, G. L., and Bradshaw, R. H. W., 1981, The Selection of Sites for Paleovegetational Studies, *Quaternary Research* 16:80–96.

Jochim, M. A., 1981, *Strategies for Survival: Cultural Behavior in an Ecological Context,* Academic Press, New York.

Johnson, E., 1986, Bifurcate Base Projectile Points in Massachusetts: Distribution and Raw Materials, Paper presented at the 24th Annual Meeting of the Northeastern Anthropological Association, Hartford, Connecticut.

Judge, W. J., 1973, *Paleoindian Occupation of the Central Rio Grande Valley in New Mexico,* University of New Mexico Press, Albuquerque, New Mexico.

King, T., 1978, Don't That Beat the Band: Nonegalitarian Political Organization in Prehistoric Central California, in: *Social Archaeology, Beyond Subsistence and Dating* (C. Redman, M. Berman, E. Curtin, W. Langhorne, N. Versaggi, and J. Wasner, eds.), Academic Press, New York, pp. 225–248.

Kite, S. J., Lowell, T. V., and Nicholas, G. P., 1982, *Quaternary Studies in the Upper St. John River Basin, Maine and New Brunswick,* New Brunswick Quaternary Association, Fredericton, New Brunswick, Canada.

Koyama, S., and Thomas, D. H. (eds.), 1981, *Affluent Foragers: Pacific Coasts East and West,* Senri Ethnological Series, No. 9, National Museum of Ethnology, Osaka, Japan.

Kratz, T. K., Friedman, R. M., and DeWitt, C. B., 1979, A Spatial Simulation Model of Lake-Edge Wetland Formation, IES Report 107, Institute for Environmental Studies, University of Wisconsin, Madison, Wisconsin.

Kutzbach, J. E., 1981, Monsoon Climate of Early Holocene: Climate Experiment Using the Earth's Orbital Parameters for 9000 Years Ago, *Science* 214:59–61.

Kutzbach, J. E., 1984, The Seasonal Nature of Climatic Forcing and Responses on Quaternary Time Scales, *AMQUA Abstracts* (American Quaternary Association) 8:70–71.

Kutzbach, J. E., and Guetter, P. J., 1984, Sensitivity of Late-Glacial and Holocene Climates to the Combined Effects of Oribital Parameter Changes and Lower Boundary Condition Changes: "Snapshot" Simulation with a General Circulation Model for 18-, 9-, and 6000 B.P., *Annals of Glaciology* 5:85–87.

Kutzbach, J. E., and Street-Perrott, F. A., 1985, Milankovitch Forcing of Fluctuations in the Level of Tropical Lakes from 18 BP to 0 K Yr BP, *Nature* 317(12):130–134.

Lee, R., and Devore, I. (eds.), 1968, *Man the Hunter,* Aldine, Chicago, Illinois.

Levin, S. A., 1976, Population Dynamic Models in Heterogeneous Environments, *American Scientist* 54:421–431.

MacArthur, R. H., and Pianka, E. R., 1966, On Optimal Use of a Patchy Environment, *American Naturalist* 100(916):603–609.

MacDonald, G. F., 1968, Debert: A Paleo-Indian Site in Central Nova Scotia, National Museum of Canada, Anthropological Papers, No. 16, Ottawa, Canada.

Marquardt, W. H., 1985, Complexity and Scale in the Study of Fisher–Gatherer–Hunters: An Example from the Eastern United States, in: *Prehistoric Hunter–Gatherers: The Emergence of Cultural Complexity* (T. D. Price and J. A. Brown, eds.), Academic Press, New York, pp. 59–98.

McIntosh, R. P., 1980, The Background and Some Current Problems of Theoretical Ecology, *Synthese* 43:195–255.

McNett, C. W., Jr. (ed.), 1985, *Shawnee-Minisink: A Stratified Paleo-Archaic Site in the Upper Delaware Valley of Pennsylvania,* Academic Press, New York.

Mitsch, W. J., and Gosselink, J. G., 1986, *Wetlands,* Van Nostrand Reinhold, New York.

Moran, E., 1982, *Human Adaptability: An Introduction to Ecological Anthropology,* Westview Press, Boulder, Colorado.

Moss, B., 1980, *Ecology of Fresh Waters,* Wiley, New York.

Mulholland, M. T., 1984, *Patterns of Change in Prehistoric Southern New England: A Regional Approach,* Ph.D. Dissertation, Department of Anthropology, University of Massachusetts, Amherst, Massachusetts, unpublished.

Nicholas, G. P., 1982, Former Glacial Lake Basins and Early Postglacial Settlement, *AMQUA Abstracts* (American Quaternary Association) 7:148.

Nicholas, G. P., 1983, A Model for the Early Postglacial Settlement of the Central Merrimack River Basin, New Hampshire, *Man in the Northeast* 25:43–63.

Nicholas, G. P., 1985, Overview of the 1984 Field Season around Robbins Swamp, *Artifacts* (American Indian Archaeological Institute, Washington, Connecticut) XIII(1):1–4.

Nicholas, G. P., 1987, Rethinking the Early Archaic, *Archaeology of Eastern North America* 15:99–124.

Nicholas, G. P., and Handsman, R. G., 1984a, Working at the Archaeology of Early Societies in Robbins Swamp, *Artifacts* (American Indian Archaeological Institute, Washington, Connecticut) XII(4):11–13.

Nicholas, G. P., and Handsman, R. G., 1984b, Early Postglacial Landscape Development and Archaeology at Robbins Swamp, Northwestern Connecticut, *AMQUA Abstracts* (American Quaternary Association) 8:149.

Nicholas, G. P., and Handsman, R. G., 1986, Early Postglacial Land-Use Patterns at Robbins Swamp, Northwestern Connecticut, *AMQUA Abstracts* (American Quaternary Association) 9:158.

Nicholas, G. P., Kite, S. J., and Bonnichsen, R., 1981, Archaeological Survey and Testing of Late Pleistocene–Early Holocene Landforms in the Dickey-Lincoln School Reservoir, Northern Maine, Institute for Quaternary Studies, University of Maine, Orono, Maine, Final Report submitted to the U.S. Army Corps of Engineers, New England Division, Waltham, Massachusetts.

Odum, E. P., 1969, The Strategy of Ecosystem Development, *Science* 164:262–270.

Odum, E. P., 1971, *Fundamentals of Ecology,* 3rd ed., Saunders, Philadelphia, Pennsylvania.

Phillips, J. L., and Brown, J. A. (eds.), 1983, *Archaic Hunters and Gatherers in the American Midwest,* Academic Press, New York.

Price, T. D., and Brown, J. A. (eds.), 1985, *Prehistoric Hunter–Gatherers: The Emergence of Cultural Complexity,* Academic Press, New York.

Raab, L. M., and Goodyear, A. C., 1984, Middle-Range Theory in Archaeology: A Critical Review of Origins and Applications, *American Antiquity* 49:255–268.

Rick, J. W., 1980, *Prehistoric Hunters of the High Andes,* Academic Press, New York.

Ritchie, D., 1986, Musketaquid, 8000 Years B.P., Paper presented at the 24th Annual Meeting of the Northeastern Anthropological Association, Hartford, Connecticut.

Ritchie, W. A., 1944, The Pre-Iroquoian Occupation of New York State, Rochester Museum Memoir, No. 1, Rochester Museum of Arts and Sciences, New York.

Rogers, R. A., 1985, The Use of Gravers through Time: A Distributional Pattern on Stream Terraces, *Plains Anthropologist* 30(109):265–268.

Rowley-Conwy, P., 1983, Sedentary Hunter–Gatherers: The Ertebølle Example, in: *Hunter-Gatherer Economy: A European Perspective* (G. Bailey, ed.), Cambridge University Press, Cambridge, pp. 111–126.

Schalk, R. F., 1981, Land Use and Organizational Complexity among Foragers of Northwestern North America, in: *Affluent Foragers: Pacific Coasts East and West* (S. Koyama and D. H. Thomas, eds.), Senri Ethnological Studies, No. 9, National Museum of Ethnology, Osaka, Japan, pp. 53–76.

Schiffer, M. B., 1975, An Alternative to Morse's Dalton Settlement Pattern Hypothesis, *Plains Anthropologist* 20(70):253–266.

Schiffer, M. B., 1983, Toward the Identification of Formation Processes, *American Antiquity* 48:675–706.

Scholander, P. F., Hammel, H. T., Bradstreet, E. D., and Hemmingsen, E. A., 1965, Sap Pressure in Vascular Plants, *Science* 148:339–346.

Schrire, C. (ed.), 1984, *Past and Present in Hunter–Gatherer Studies,* Academic Press, New York.

Service, E. R., 1962, *Primitive Social Organization: An Evolutionary Perspective,* Random House, New York.

Simon, B., 1986, The Double P Site, Bridgewater, Massachusetts, Paper presented at the 24th Annual Meeting of the Northeastern Anthropological Association, Hartford, Connecticut.

Snow, D. R., 1980, *The Archaeology of New England,* Academic Press, New York.

Speight, M. C. D., and Blackith, R. E., 1983, The Animals, in: *Mires: Swamp, Bog, Fen and Moor—Ecosystems of the World,* Vol. 4A: *General Studies* (A. J. P. Gore, ed.), Elsevier, Amsterdam, pp. 349–366.

Starbuck, D. R., 1980, The Middle Archaic in Central Connecticut: The Excavation of the Lewis-Walpole Site (6-HT-15), in: *Early and Middle Archaic Cultures in the Northeast* (D. R. Starbuck and C. E. Bolian, eds.), Occasional Publications in Northeastern Anthropology, No. 7, pp. 5–38.

Suttles, W., 1968, Coping with Abundance: Subsistence on the Northwest Coast, in: *Man the Hunter* (R. B. Lee and I. Devore, eds.), Aldine, Chicago, Illinois, pp. 56–68.

Tallis, J. H., 1983, Changes in Wetland Communities, in: *Mires: Swamp, Bog, Fen and Moor—Ecosystems of the World,* Vol. 4A: *General Studies* (A. J. P. Gore, ed.), Elsevier, Amsterdam, pp. 311–348.

Thomas, P. A., and Robinson, B. S., 1980, The John's Bridge Site: VT-FR-69: An Early Archaic Period Site in Northeastern Vermont, Department of Anthropology Report, No. 23, University of Vermont, Burlington, Vermont.

Tilley, C. Y., 1979, Post-Glacial Communities in the Cambridge Region: Some Theoretical Approaches to Settlement and Subsistence, *BAR British Series* 66.

Webb, T., III, 1981, The Past 11,000 Years of Vegetational Change in Eastern North America, *Bioscience* 31:501–506.

Webb, T., III, 1984, Late Quaternary Changes in Climatic Seasonality in Eastern North America, *AMQUA Abstracts* (American Quaternary Association) 8:138–139.

Webb, T., III, 1986, Is the Vegetation in Equilibrium with Climate? How to Interpret Late-Quaternary Pollen Data, *Vegetatio* 67(2):75–92.

Webb, T., III, Richard, P. J., and Mott, R. J., 1983, A Mapped History of Holocene Vegetation in Southern Quèbec, in: *Climate Change in Canada,* Vol. 3 (C. R. Harrington, ed.), Syllogeus, No. 49, National Museum of Natural Sciences, National Museums of Canada, Ottawa, Canada, pp. 273–336.

Weissner, P., 1982, Beyond Willow Smoke and Dog's Tails: A Comment on Binford's Analysis of Hunter-Gatherer Settlement Systems, *American Antiquity* 47:171–178.

Weller, M. W., 1981, *Freshwater Marshes: Ecology and Wildlife Management,* University of Minnesota Press, Minneapolis, Minnesota.

Westlake, D. F., 1963, Comparisons of Plant Productivity, *Biological Review* 38:385–425.

Whigham, D. F., McCormick, J., Good, R. E., and Simpson, R. L., 1978, Biomass and Primary Production in Freshwater Tidal Wetlands of the Middle Atlantic Coast, in: *Freshwater Wetlands: Ecological Processes and Management Potential* (R. E. Good, D. G. Whigman, and R. L. Simpson, eds.), Academic Press, New York, pp. 3–20.

White, R., 1987, Glimpses of Long-Term Shifts in Late Paleolithic Land Use in the Périgord, in: *The Pleistocene Old World: Regional Perspectives* (O. Soffer, ed.), Plenum Press, New York, pp. 263–278.

White, J., 1985, Paleoenvironmental Data Sets and the Development of a Vegetational Mosaic in Southern New England during the Early Holocene, Research Manuscript Series, American Indian Archaeological Institute, Washington, Connecticut.

Whittaker, R. H., 1967, Gradient Analysis of Vegetation, *Biological Review* 42:640–643.

Whittaker, R. H., 1975, *Communities and Ecosystems,* Macmillan, New York.

Winterhalder, B., 1981, Optimal Foraging Strategies and Hunter-Gatherer Research in Anthropology: Theory and Methods, in: *Hunter–Gatherer Foraging Strategies: Ethnographic and Archeological Analyses* (B. Winterhalder and E. A. Smith, eds.), University of Chicago Press, Chicago, Illinois, pp. 13–35.

Winterhalder, B., and Smith, E. A. (eds.), 1981, *Hunter–Gatherer Foraging Strategies: Ethnographic and Archeological Analyses,* University of Chicago Press, Chicago, Illinois.

Wobst, H. M., 1978, The Archaeo-Ethnography of Hunter-Gatherers, or The Tyranny of the Ethnographic Record, *American Antiquity* 43:303–309.

Wright, H. E., Jr., 1976, The Dynamic Nature of Holocene Vegetation, *Quaternary Research* 6:581–596.

Wright, H. E., Jr., and Glasser, P. H., 1983, Postglacial Peatlands of the Lake Agassiz Plain, Northern Minnesota, in: *Glacial Lake Agassiz* (J. T. Teller and L. Clayton, eds.), Special Paper 26, Geological Association of Canada, pp. 375–390.

Afterword

George P. Nicholas

A Zen *koan* provides an unexpected but appropriate comment on how we observe the operation of human ecosystems. Two monks walking through the countryside with Enō, a Zen Master, stopped to observe a banner fluttering in the wind. The first student said, "Look, it is the banner that moves." The second said, "No, it is the wind that moves." They argued back and forth without ever reaching the core of the matter. Finally, Enō, as all Zen Masters will, hit each on the head with his stave to clear their minds, and said, "It is not the wind that moves. It is not the banner that moves. It is your mind that moves." What Enō was ultimately referring to, at least when translated into Western terms, is that things are shaped not only by natural processes, but by the very process of observation.

This is a lesson that we must learn as human ecologists—to look at the interaction between people and places in ways that are as dynamic as the ecosystem itself, and to be more aware of our own role in explaining change (see Figure 1 for one perspective). The ways we define and explain variability in the archaeological and environmental record are very much a product of the paradigm within which we work, whether it is archaeology, ethnology, climatology, or history; human ecology is, of necessity, an interdisciplinary subject. How can variability in human behavior be explained? It is not simply a matter of correlating shifts in pollen spectra with archaeological land-use patterns. We need to know how the entire system operates, or at least how the primary parts fit together (Jochim 1979). This requires rethinking (sometimes radically) our re-

GEORGE P. NICHOLAS • Department of Anthropology, University of Massachusetts, Amherst, Massachusetts 01003, and Research Department, American Indian Archaeological Institute, Washington, Connecticut 06793.

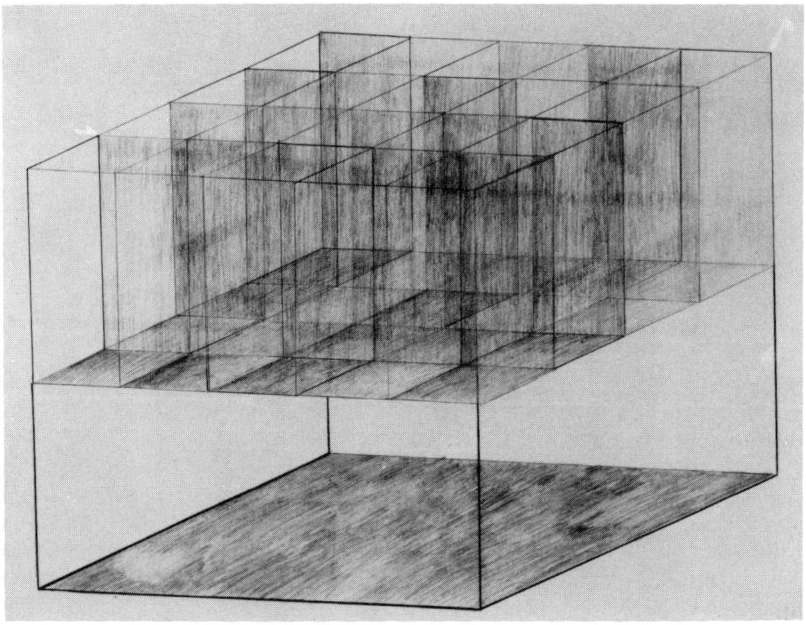

Figure 1. *Passageways* 1985 by Betsy Guralnik (18″ × 24″, prismacolor). Our understanding of the human environment is the product not only of the variables we use to describe the landscape, but also of the different "human ecologies" we impose upon it. (Photograph provided by the American Indian Archaeological Institute, Washington, Connecticut.)

search goals, in addition to revising analytical methods to include multiple lines of inquiry, more rigorous hypothesis testing, and increased sensitivity to the varying scales of human interaction with cultural and environmental variables. Our subject matter is not static, so there is no reason why our research methods should be.

The need for greater awareness of how ecological process is observed is made clear by recent developments in Quaternary paleoecology concerning vegetational continua (Colinvaux 1987; Davis 1986; Webb 1987). There is a growing consensus that forest associations are ephemeral: "There are no permanent associations in the complex vegetation of eastern North America, only shifting alliances as climate and circumstance allow different taxa to colonize habitats simultaneously. The world is Gleasonian, not Clementsian" (Colinvaux 1987:6). This new series of regional reconstructions of late Quaternary plant communities is based on a revised model of forest development that focuses on the collective life histories of individual members of a biotic community. This differs radically from those associated with more traditional "superorganismic" models. Among other things, these new reconstructions suggest that there is even greater diversity in the human habitat at the local level: "In the *patchwork dynamics* of modern ecology, plant communities suffer almost ceaseless local change as successions endlessly colonize gaps" (Colinvaux 1987:6, emphasis in original). The point here is that rethinking the concept of plant associations has resulted in alternative histories of late Quaternary forest ecology that have important implications for understanding how ecosystems developed over time.

The contributors to this volume have each focused on specific aspects of the human ecosystem. The research hypotheses, analytical methods, and data sets incorporated into these studies are derived from diverse anthropological, historical, and paleoecological perspectives and motives, yet the common goal has been to understand the workings of the human ecosystem. All of the authors have been concerned with how their subject matter fits within the larger system, as well as with such mundane matters as the representativeness of their data sets and the scale of their analysis. Scale has proved critical in human ecology because it allows us to identify the points of contact between the components of the system (Dincauze 1987). Another common point of discussion has been the variety of human behaviors and habitats present in the region. Diversity is indeed the stuff of human ecology, and the authors have offered explanations for the some of the behavioral variability implicit in—or evidenced by—certain habitats and archaeological patterns. Finally, most authors would probably agree that a regional perspective provides a useful frame of reference in investigating human adaptation (also see Soffer 1987). Brief comments on several other issues discussed in this collection follow.

1. CAUTIONARY TALES

A number of cautionary tales are presented on the pitfalls of environmental reconstruction and other problems that beset human ecologists in the Northeast. Baron's study, for example, indicates that we cannot make *a priori* assumptions about the impact of climatic events on human communities. The type of weather patterns tracked in this study apparently had little impact within what was already a highly variable environment (Braun and Plog 1982:506). In this case, a major climatic event was even not perceived in New England, so alternative explanations are required to explain shifts in, for example, farm economy and rural social organization. One explanation for the population decreases recorded for some parts of New England during this period is the development of the railroad and westward expansion. The impact of the Little Ice age in other more marginal areas was substantial, however (Ingram *et al.* 1981; Lamb 1977; McGovern 1981). Can we expect similar nonresponses to short-term and gradual environmental perturbations in other heterogeneous habitats? At other times during the Holocene, climate may certainly have had a more significant impact on prehistoric lifeways, as both Joyce and Nicholas point out, and a wide variety of lessening responses (e.g., Bowden *et al.* 1981) may be reflected in the archaeological record.

Joyce is also concerned with climate and with how some archaeologists have approached the Hypsithermal episode—a postglacial xerothermic climatic shift—in their reconstruction of human habitat. This warm/dry episode has been used to explain a variety of changes in the North American archaeological record during the middle Holocene. Recent climate modeling and environmental reconstructions (e.g., Davis *et al.* 1980) indicate, however, that the period of maximum warmth and dryness occurred earlier than previously supposed, so other explanations must be found for the rapid shifts noted in Late Archaic land-use behaviors and social organization. In addition, Joyce's critical review of the Victor Carbone school of paleoenvironmental reconstruction, which has dominated the greater Northeast region for the last decade, provides a useful example of how shifts in ecological theory ultimately affect the study of human ecology.

The cautionary tales provided by Barber and Kellogg stress the importance of interdisciplinary methods and the need for very detailed environmental data in the reconstruction of human habitat. Land snails are highly sensitive paleoenvironmental indicators because they are niche-specific. Barber's study thus emphasizes both the importance of life histories in ecological studies and the processes of hypothesis evaluation to demonstrate the linkages between snail habitat and various environmental factors. He also questions why gastropod data have generally been ignored in the Northeast. Kellogg effectively argues that an accurate understanding of postglacial sea-level rise is critical to environmental reconstruction in coastal areas. Use of incorrect sea-level data also obfuscates the study of human coastal adaptation through false relative dates on the age of a site or inaccurate information on the presence of clam flats and other resources.

2. LANDSCAPE HISTORIES

Landscapes are an integral component of the human ecosystem. They provide not only a place where people live, but also the basis of their survival and frequently the motivation for social transitions. As landscapes change, aspects of human lifeways will either change or remain the same in response to new or lost opportunities. To understand the relationship between shifts in environment and human land-use behavior at particular points in time, the ecology of the landscape must be known in detail and the relationship between habitat and site formation behaviors must be made explicit. Once again the scale of observation is important; the Nunamiut cycle of land use (Binford 1983), for example, takes generations to complete. In order to relate archaeological patterns of land use to ecological factors, landscape histories must also be constructed that track that relationship over hundreds or thousands of years (e.g., White 1987).

Kellogg has demonstrated that detailed and accurate reconstructions of past environmental conditions are the prerequisite to explaining human behavior. Gaudreau, Joyce, and Nicholas focus more on the processes of landscape development at local and regional levels, and attempt to map the relationship between climate, vegetation, and topography at various times. All three studies indicate that environmental heterogeneity was an important characteristic of the Holocene environment. Because patchy or mosaic-like environments are generally productive resource locations, they should figure importantly in subsistence strategies and settlement location—an hypothesis examined by Nicholas and by Thorbahn and Cox. As resource availability or productivity at these locations changed in response to processes of forest succession, subsistence/settlement patterns would be expected to shift to other productive locations or to other types of resources. Such shifts should be apparent in the prehistoric record if archaeological indicators of those behaviors can be defined.

The fire-management strategies discussed by Patterson and Sassaman may have provided one means to maintain critical resources during periods of declining productivity, and may even have been used to produce artificial patchy environments. Controlled burning was also used to increase the amount of arable land as horticulture became a widespread and intensive subsistence strategy, as noted by Mulholland (see Chapman *et al.* 1982 for another view of the impact of this type of land use). The clearing of forests by burning could also have provided at least a temporary alternative to intergroup competition for increasingly scarce arable land.

3. THE ECOLOGY OF SOCIAL AND SUBSISTENCE SYSTEMS

Environmental factors have had a significant influence on the ecology of small-scale social formations in the Northeast and elsewhere (for overviews see Ellen 1982 and Stoltman and Baerreis 1983). Papers in this collection have examined that ecology in two ways: first, by defining the relationship between social organization variables and specific environmental factors; and second, by using environmental factors to explain some of the variations noted in the lifeways of hunter–gatherers, emergent horticulturalists, and historic period fishermen and farmers.

Carlson has provided a more precise understanding of the role that anadromous fisheries played in aboriginal social and economic organization in the Northeast. The resident salmon population here was apparently dissimilar to that of the Northwest Coast, where the highly productive anadromous fishery is thought to have contributed to increased sedentism, social differentiation, technological specialization, and other behaviors normally associated with "complex" societies (e.g., Ames 1981; however, see Carlson 1982). The conclusions of this study question the need for seasonal transhumancy given the abundant year-round resources of the coast, as well as *a priori* assumptions about site function (e.g., sites located at falls are not always fishing stations). Carlson does suggest that some marine fish, particularly cod, may be an important element in the development of aboriginal social systems in the Northeast.

Horticulture also played an important role in the transformation of human lifeways during the late Holocene. It was clearly one cause of the shifts in demography, social relations, territoriality, and even health of prehistoric populations evidenced in the prehistoric record. The negative effects of intensive horticulture included degradation of the local landscape and increased competition for arable land; because of the limited distribution of lands suitable for cultivation, intensive horticulturalists were more limited in their response to those pressures than were either forager–horticulturalists or foragers. Weather aberrations undoubtedly had an effect on the success of the crops, as did shifts in temperature gradients, especially near the geographic tolerance limits of the crops. Variability in interannual weather patterns would also have had an impact upon agricultural productivity, labor practices, and even folk knowledge about planting and harvesting, and one would expect that there were mechanisms in place to alleviate problems resulting from a poor harvest, e.g., trade, raids on other villages, alternative resources and extractive technologies, group fission (Engelbrecht 1987; also see Jochim 1981).

In some respects, fire management (Patterson and Sassaman) is a subsistence strategy similar to horticulture. Resources are selectively controlled and then collected or harvested. There is a delayed return on the investment, so land-use range and annual cycles must be adjusted to periods of maximum productivity if this is a primary resource procurement strategy. Alternatively, burning off land peripheral to the primary subsistence base area may also be an effective fallback strategy for times when resource productivity has declined. The fire-managed land could also be viewed as a resource itself, and access to it may have been open to the entire group or controlled by a single family—like fishing locations on rivers in the Northwest Coast—and may even have been defended territory.

These and other adaptive strategies suggest that there is substantial diversity in the prehistoric lifeways of the Northeast. There appear to be not only hunter–gatherers in the traditional sense, but also hunter–gatherer–resource managers (Patterson and Sassaman), fisher–hunter–gatherers (Carlson), foraging horticulturalists (Mulholland), and gatherer–hunters (Nicholas). In some cases, specific resources such as cod or maize may be a key to understanding cultural developments during the Holocene; in other instances, it may have been the way places on the landscape were used or managed (e.g., fire-managed areas or the redundant use of productive places like Robbins Swamp) that resulted in more complex social relations.

4. CONCLUSIONS

If we are ever to reach the core of the matter, whether as scientists or humanists, we must understand that there are no simple causative mechanisms or noncomplex linkages between cultural and environmental systems. They are each dynamic and complex and only infrequently operate at corresponding scales of interaction. At the same time, what can be known about either past or present human lifeways is a function of the questions asked and of the scales of observation used. The way that we "define" relationships within the human ecosystem thus has a great deal to do with what the "reality" of that ecosystem becomes. The way we think about human ecology therefore needs to be as dynamic and clear-headed as Enō's observation.

ACKNOWLEDGMENTS

I am very grateful to Betsy Guralnik for showing me new ways of visualizing people on a distant landscape, and for allowing me to reproduce *Passageways* here.

5. REFERENCES

Ames, K. M., 1981, The Evolution of Social Ranking on the Northwest Coast of North America, *American Antiquity* 46:789–805.

Binford, L. R., 1983, Long-Term Land-Use Patterning: Some Implications for Archaeology, in: L. R. Binford, *Working at Archaeology,* Academic Press, New York, pp. 379–386.

Bowden, M. J., Kates, R. W., Kay, P. A., Riebsame, W. E., Warrick, R. A., Johnson, D. L., Gould, H. A., and Weiner, D., 1981, The Effect of Climate Fluctuations on Human Populations, in: *Climate and History* (T. M. L. Wigley, M. J. Ingram, and G. Farmer, eds.), Cambridge University Press, Cambridge, pp. 479–513.

Braun, D. P., and Plog, S., 1982, The Evolution of "Tribal" Social Networks: Theory and Prehistoric North American Evidence, *American Antiquity* 47:504–525.

Carlson, C. C., 1982, A Reevaluation of the Role of Salmon to the Evolution of Ranked Society on the Northwest Coast, Paper presented at the 15th Annual Meeting of the Canadian Archaeological Society, Hamilton, Ontario.

Chapman, J., Delcourt, P. A., Cridlebaugh, P. A., Shea, A. B., and Delcourt, H. R., 1982, Man–Land Interaction: 10,000 Years of American Indian Impact on Native Ecosystems in the Lower Little Tennessee River Valley, Eastern Tennessee, *Southeastern Archaeology* 1(2):115–121.

Colinvaux, P. A., 1987, The Changing Forests: Ephermeral Communities, Climate and Refugia, *Quarterly Review of Archaeology* 8(1).

Davis, M. B. (ed.), 1986, Vegetation–Climate Equilibrium, *Vegetatio* 67:65–141 (Special issue).

Davis, M. B., Spear, R. W., and Shane, L. C. K., 1980, Holocene Climate of New England, *Quaternary Research* 14:240–250.

Dincauze, D. F., 1987, Strategies for Paleoenvironmental Reconstruction in Archaeology, in: *Advances* in *Archaeological Method and Theory,* Vol. 11 (M. B. Schiffer, ed.), Academic Press, New York, pp. 255–336.

Ellen, R., 1982, *Environment, Subsistence and System: The Ecology of Small-Scale Social Formations,* Cambridge University Press, New York.

Engelbrecht, W., 1987, Factors Maintaining Low Population Density among the Prehistoric Iroquois, *American Antiquity* 52:13–27.

Ingram, M. J., Farmer, G., and Wigley, T. M. L., 1981, Past Climates and Their Impact on Man: A Review, in: *Climate and History* (T. M. L. Wigley, M. J. Ingram, and G. Farmer, eds.), Cambridge University Press, Cambridge, pp. 3–49.

Jochim, M. A., 1981, *Strategies for Survival: Cultural Behavior in an Ecological Context,* Academic Press, New York.

Lamb, H. H., 1977, *Climate: Past and Present.* Vol. 2: *Climate History and the Future,* Methuen, London.

McGovern, T., 1981, The Economics of Extinction in Norse Greenland, in: *Climate and History* (T. M. L. Wigley, M. J. Ingram, and G. Farmer, eds.), Cambridge University Press, Cambridge, pp. 404–433.

Soffer, O. (ed.), 1987, *The Pleistocene Old World: Regional Perspectives,* Plenum Press, New York.

Stoltman, J. B., and Baerreis, D. A., 1983, The Evolution of Human Ecosystem in the Eastern United States, in: *Late-Quaternary Environments of the United States* (H. E. Wright, Jr., ed.), Univeristy of Minnesota Press, Minneapolis, Minnesota, pp. 252–268.

Webb, T., III, 1987, The Appearance and Disappearance of Major Vegetational Assemblages: Long-Term Vegetational Dynamics in Eastern North America, *Vegetatio* 67:75–92.

White, R., 1987, Glimpses of Long-Term Shifts in Late Paleolithic Land Use in the Périgord, in: *The Pleistocene Old World: Regional Perspectives* (O. Soffer, ed.), Plenum Press, New York, pp. 263–278.

Index